Chemical Kinetics of Small Organic Radicals

Volume III
Reactions of Special Radicals

Editor

Zeev B. Alfassi, Ph.D.
Professor
Department of Nuclear Engineering
Ben Gurion University of the Negev
Beer Sheva, Israel

CRC Press, Inc.
Boca Raton, Florida

Library of Congress Cataloging-in-Publication Data

Chemical kinetics of small organic radicals.

Includes bibliographies and indexes.
1. Free radical reactions. 2. Free radicals
(Chemistry) 3. Chemical reaction, Rate of
I. Alfassi, Zeev B.
QD471.C49 1988   547.1'394   87-21812

Direct all inquiries to CRC Press, Inc., 2000 Corporate Blvd., N.W., Boca Raton, Florida, 33431.

© 1988 by CRC Press, Inc.

International Standard Book Number 0-8493-4362-3 (v. 1)
International Standard Book Number 0-8493-4363-1 (v. 2)
International Standard Book Number 0-8493-4364-x (v. 3)
International Standard Book Number 0-8493-4365-8 (v. 4)

Library of Congress Card Number 87-21812
Printed in the United States

INTRODUCTION

The reactions of organic radicals is involved in many processes both in the liquid and the gaseous phase. These reactions are responsible for processes such as coal liquefaction and gasifications, air and water pollution, stratosphere chemistry, combustion, gaseous products formation in several synthetic processes, and consequently detonation hazards, CCl_4 toxication, and so on. The knowledge of the rate constants of these reactions and their temperature dependence is important both from theoretical points of view to check various theories on chemical reactivities and from practical points of view in order to predict the behavior of unknown new systems.

These volumes will discuss the subject of reactions of free radicals both from its general point of view as general methods of formation and detection, thermochemisry, and structure and from the point of view of specific radicals giving detailed measurement of rate constants of reactions of the more common radicals. The important reactions of each radical will be given correlation between rate constants, and other chemical data will be drawn in order to enable estimation of unmeasured rate constants for other reactions. The possibility to predict the behavior of complex chemical systems by numerical simulation of all the possible reactions using known reaction constants will be given in general. A special treatment will be given to the modeling of air pollution and combustion.

THE EDITOR

Prof. Zeev B. Alfassi, Ph.D. was born in Tel Aviv, Israel. He received his undergraduate and M.Sc. degrees from the Hebrew University in 1964 and 1965, repectively. He moved to the Soreq Nuclear Research Center in collaboration with the Weizman Institute and received his Ph.D. degree in 1970. Prof. Alfassi stayed on with the Soreq Nuclear Research Center as a research associate and later as a senior scientist, while lecturing for the Department of Nuclear Sciences at Ben Gurion University in Beer Sheva, Israel. He became a senior lecturer in 1974 and was named an Associate Professor in the Department of Nuclear Engineering in 1980. Later, Prof. Alfassi was named Professor and Chairman of the department,

Prof. Alfassi has been a visiting professor and scientist at many universities in the U.S. and Germany and has written many journal articles and technical reports.

CONTRIBUTORS

Zeev B. Alfassi, Ph.D.
Professor
Chairman of Department of Nuclear
  Engineering
Ben Gurion University of the Negev
Beer Sheva, Israel

L. Batt, Ph.D.
Professor
Department of Chemistry
Aberdeen University
Aberdeen, Scotland

Takaaki Dohmaru, Dr. Eng.
Senior Researcher
Department of Chemistry
Radiation Center of Osaka Prefecture
Sakai, Osaka, Japan

Osamu Ito, Ph.D.
Associate Professor
Chemical Research Institute of
  Nonaqueous Solutions
Tohoku University
Sendai, Katahira, Japan

M. C. Lin, Ph.D.
Senior Scientist for Chemical Kinetics
Chemistry Division
Naval Research Laboratory
Washington, D.C.

Minoru Matsuda, Dr.Sc.
Professor
Chemical Research Institute of
  Nonaqueous Solutions
Tohoku University
Sendai, Miyagi Pref., Japan

Shlomo Mosseri, Ph.D.
Chemist
Chemical Kinetics Division
National Bureau of Standards
Gathersburg, Maryland

William A. Sanders, Ph.D.
Associate Professor
Department of Chemistry
The Catholic University of America
Washington, D.C.

J. C. Scaiano, Ph.D.
Senior Research Officer
Department of Chemistry
National Research Council
Ottawa, Ontario, Canada

TABLE OF CONTENTS

Volume I

Volume II

Volume III

Volume IV

Chapter 11

REACTIONS OF METHYL RADICALS

L. Batt

TABLE OF CONTENTS

I. INTRODUCTION

The reaction of methyl radicals with other species must be one of the most well-studied processes in the field of chemical kinetics. For this reason, in order to provide a review of a reasonable size, only gas-phase reactions are considered. In addition, the literature survey from 1970 to the present time was confined to a number of key journals together with the A.C.S. Chemical Abstracts. Previous reviews have been made by Kerr and Trotman-Dickenson[1] — reactions of alkyl radicals; Gray, Herod, and Jones[2] — H-atom abstraction reactions; and Kerr and Parsonage[3] — addition reactions.

Studies of methyl radical reactions have important implications for RRKM theory, atmospheric chemistry, combustion, and the concomitant atmospheric pollution. The first of these topics is illustrated by Waage and Rabinovitch's article on the pressure dependence of the combination of methyl radicals.[4]

$$Me + Me + M \rightarrow C_2H_6 + M \qquad (1)$$

The second and fourth topics are illustrated for the troposphere at least by the article by Cox et al. on the photo-oxidation of methane in the presence of nitric oxide and nitrogen dioxide.[5] The third topic is illustrated by the article from Baldwin et al. on the combustion of methane/hydrogen mixtures.[6] Although this article is concerned with methane, the reactions of the methyl radical that take place are also relevant to any species that produce methyl radicals upon pyrolysis or reaction in the presence of oxygen.

Although hydrogen atom abstraction processes play a dominating role in the reactions of methyl radicals, combination cum addition processes are also important. Although a number of papers have appeared on theoretical aspects of all methyl radical reactions this review will deal mainly with the interpretation of experimental results. However, in respect of hydrogen atom abstraction reactions, tunneling has to be considered. S.I. units are adapted for thermodynamic quantities but for biomolecular rate constants the units of liter moles^{-1} seconds^{-1} are used. These are given by symbols M^{-1} sec^{-1}. In terms of S.I. units this would be dm^3 mol^{-1} sec^{-1}. The rate constant k is expressed in the form k/M^{-1} sec^{-1} = A(exp − B/T). Although it is clearly possible to obtain a rate constant for a particular methyl radical reaction from that for the reverse process together with the thermodynamic quantities involved, this review mainly avoids this temptation. However, in connection with this, since Benson has emphasized the importance of thermochemical aspects of chemical reactions,[7] some space is devoted first to the heat of formation and the entropy of the methyl radical.

II. THERMOCHEMISTRY

The obvious starting point for a consideration of the thermochemistry of the methyl radical are the JANAF Thermochemical Tables[8] (see also reference 104). The value quoted here for ΔH°_f (Me) depends upon photoionization spectra. Previous to these studies ΔH°_f (Me) was based upon Benson's iodination studies as reviewed by Golden and Benson.[9] This depends essentially on the equilibrium

$$CH_4 + I_2 \rightleftarrows MeI + HI \qquad (2)$$

and yields a value of 144.4 ± 5.9 kJ mol^{-1}. The photoionization studies lead to 145.7 ± 0.8 kJ mol^{-1}. Two further similar studies gave 149.4 ± 0.5[10] and 143.9 ± 0.5[11] kJ mol^{-1}, respectively. In order to increase the precision of his equilibrium measurements, Benson used a very low pressure reactor (VLPR).[12] This is essentially a modification of his very low pressure pyrolysis apparatus (VLPP)[13] which is confined to high temperatures. The equilibrium chosen for this study was

$$Cl + CH_4 \rightleftarrows HCl + Me \qquad (3)$$

Lower temperatures were achieved by using a microwave discharge to generate chlorine atoms from molecular chlorine. One of the problems associated with the VLPP apparatus is surface catalyzed processes. Benson sought to avoid this problem by using a fluorocarbon coating in the VLPR. This study leads to a value for ΔH°_f (Me) = 146.9 ± 0.6 kJ mol^{-1}. Pacey and Wimalasena[4] have extracted a value from a study of the initial stages in the pyrolysis of ethane together with the appropriate thermochemistry which gives ΔH°_f (Me) = 147 ± 2 kJ mol^{-1}. Finally, Whittle[15] has considered a number of equilibria which involve the methyl and a number of polyfluorinated (R') radicals and their iodides

$$Me + R'I \rightleftarrows MeI + R' \qquad (4)$$

These data are consistent with ΔH°_f (Me) = 145.0 ± 0.5 kJ mol^{-1}. It is this value which is chosen as the recommended value. Other recent values are clearly very close to this value.

The value for S$^\circ$ (Me) = 194.1 ± 1.3 J mol^{-1} K^{-1} and C$^\circ_p$ (Me) = 38.7 J mol^{-1} K^{-1} are taken from Reference 8. The arguments used there are fully endorsed.

III. REACTION WITH ATOMS

Two types of reactions may be expected to take place between a methyl radical and an atom (X). The first is a biomolecular process (5) and the second is a combination process which may be pressure dependent.[6]

$$Me + X \rightarrow CH_2 + HX \qquad (5)$$

$$Me + X(+M) \rightarrow MeX(+M) \qquad (6)$$

Over the time period allocated for the literature survey no data was found for the interaction of methyl radicals with halogen atoms. Benson[7] assumes that all methyl radical + atom combination processes have rate constants that are one quarter of the value calculated from simple collision theory, i.e., $10^{11.5}$ M^{-1} sec^{-1} when X = halogen, this value is in reasonable accord with both experimental and theoretical work (Table 1). Patrick, Pilling and Rogers[17] measured k_6 for X = H at 300 K over the pressure range 50 to 1000 torr using either argon or sulfur hexafluoride. The latest value for k_6 (∞) is $10^{11.6}$ M^{-1} sec^{-1}.[17] Some concern was raised about the implications for the rate constant for the reverse process k_7. In particular the calculated value

$$CH_4 \rightarrow Me + H \qquad (7)$$

for k_7 was greater than literature values. In fact there are a number of conflicting features about the decomposition of methane in respect of k_7. A value has been obtained for k_6 at low pressures.[20] This is given by k_6/M^{-2} sec^{-1} = 12.59 ± 0.01 + (133 ± 4/T). No information is available for Reaction 5 in respect of either halogen or hydrogen atoms. In terms of the thermochemistry for Reaction 5 (Table 2) reaction with a fluorine atom is exothermic whereas that with all other halogen atoms is endothermic. For reaction with a hydrogen atom, two theoretical studies[21,22] for the reverse process (Reaction 8).

$$CH_2(3B_1) + H_2 \rightarrow Me + H \qquad (8)$$

Table 1
RATE CONSTANTS FOR THE
COMBINATION OF Me AND H

$k/10^{11}\ M^{-1}\ sec^{-1}$	T/K	Remarks	Ref.
2.0 ± 0.9	308		16
0.9 ± 0.4^a	298		17
3.9^a	298		17
1.2	298		18
0.25 ± 0.07		Stiff Morse P. E. surface	
	300		
3.0 ± 0.5		Standard Morse P. E. surface	

^{a} k_∞.

Table 2
THERMOCHEMISTRY FOR Me
+ X → HX + CH$_2$ (REACTION 5)

X	$\Delta H°/kJ\ mol^{-1}$	$E_5/kJ\ mol^{-1}$
F	-110.0	—
Cl	27.2	—
Br	92.0	—
I	159.8	—
H	22.2	27.8

derive a value of 50 kJ mol^{-1} for the barrier height which makes E_5 for X = H 27.8 kJ mol^{-1}. Over the temperature range 1800 to 2500 K, which corresponds to shock tube measurements, Just[23] quotes a value for k_5 (H) = $10^{11.9}\ M^{-1}$ sec^{-1}. In view of the above calculated activation energy, this value for k_5 (H) looks too high.

The reaction of methyl radicals with oxygen atoms is an important step in the high temperature oxidation of methyl radicals and in the fate of both of these species in the higher atmosphere. The reaction is written as a two-step process but in fact must be classed as a type (6) reaction. However the methoxy radical adduct contains 380 kJ mol^{-1} of excess energy. Since there are so few vibrational degrees of freedom the hot adduct must fly apart within a few vibrations to form the products formaldehyde and a hydrogen atom. No pressure dependence for this reaction is ever considered

$$Me + O \rightarrow (MeO^\bullet) \rightarrow CH_2O + H \tag{9}$$

Niki and Morris[24] used a discharge flow apparatus coupled with a time of flight mass spectrometer to study Reaction 9. They were able to show that formaldehyde was a major product. The rate constant k_9 was determined by an indirect method. Gutman et al.[25] used a fast-flow system coupled with a photoionization mass spectrometer. They generated methyl radicals by reacting oxygen atoms with ethylene. In this way they were able to measure k_9 directly. Washida and Bayes[26] used the same system as Gutman et al. but, whereas the latter confined their measurements to 300 K, Washida and Bayes covered the temperature range 259 to 341 K. This work was later extended by Washida.[27] Just[23] considers this reaction at shock tube temperatures (1900 K) and appreciates that MeO may be an intermediate. One other intermediate is the isomer of MeO, i.e., CH$_2$OH[28].

$$MeO \rightarrow CH_2OH \tag{10}$$

Table 3
RATE CONSTANTS FOR THE REACTION
Me + O → CH$_2$O + H (REACTION 9)

T/K	k/10^{11} M^{-1} sec^{-1}	Ref.
300	> 0.2	24
300	1.1 ± 0.2	25
259 — 341	0.6 ± 0.1	26
298 ± 3	0.83 ± 0.3	27
1900	0.85 ± 0.1	23

Washida and Boyes[26] found that there was no discernible temperature variation for k_9. Inspection of Table 3 shows that there is considerable uncertainty over the value of k_9. This important reaction warrants further study. One other channel has been suggested for the reaction of methyl radicals and oxygen atoms. This involves the production of water and the CH radical, a remarkable process by modern standards!

$$Me + O \rightarrow H_2O + CH \tag{11}$$

Becker and Klay[29] reject this route as a source of CH radicals and Just[23] considers Reaction 11 as of very minor importance in the high temperature oxidation of methyl radicals.

IV. REACTION WITH SMALL MOLECULES

A. Me + O$_2$

The reaction of methyl radicals with oxygen is both the simplest and most important process that can take place in the oxidation and combustion of organic species. Apart from its importance in the oxidation of methane, many other compounds break down either photochemically or thermally to produce methyl radicals. At room temperature the exclusive process is simple addition which is expected to take place with little or no activation energy since it involves the combination of two radical species. The process is important up to temperatures of at least 1500 K. However at very high temperatures redissociation of the methyl peroxy radical is so fast that its concentration becomes very low. For this reason other reactions between the methyl radical and oxygen have been put forward which will be discussed later. Because of the reversible nature of the addition process both the Me –O$_2$ bond dissociation energy and the equilibrium constant should be considered. Four systems are considered in respect of D(Me –O$_2$). The first depends upon the heat of formation of the hydroperoxy radical which has been reviewed recently by Benson and Shum.[30] They recommend a value of 14.6 ± 3 kJ mol^{-1}. By making use of the values $\Delta H°_f$ (H$_2$O$_2$) = − 136.1 kJ mol^{-1} and $\Delta H°_f$ (H) = 217.6 kJ mol^{-1} (8), D(HO$_2$ –H) = 368.3 ± 5 kJ mol^{-1}. Benson and Shaw assumed that D(HO$_2$ − H) = D(RO$_2$ − H) where R is an alkyl group. Benson and Shaw also made use of the principle of group additivity[7] to determine $\Delta H°_f$ (MeO$_2$H) since there are no reliable experimental values available. Thus $\Delta H°_{12} = 0$. With $\Delta H°_f$ (MeO)$_2$ = − 125.5 kJ mol^{-1},[32]

$$(MeO)_2 + H_2O_2 \rightleftarrows 2MeO_2H \tag{12}$$

$\Delta H°_f$ (MeO$_2$H) = − 130.8 kJ mol^{-1} and hence $\Delta H°_f$ (MeO$_2$) = 19.9 ± 7.5 kJ mol^{-1}. Using $\Delta H°_f$(Me) = 145.0 kJ mol^{-1} as already discussed makes −$\Delta H°_{13}$ = D(Me −O$_2$)

$$Me + O_2 \rightleftarrows MeO_2 \tag{13}$$

= 125.1 ± 7.5 kJ mol^{-1}. Kondo and Benson[33] studied the equilibrium:

Table 4
VALUES FOR THE RATIO [MeO₂]/
[Me] IN THE EQUILIBRIUM
Me + O₂ ⇄ MeO₂ (REACTION 13)

$\Delta H° = -129.3$ kJ mol^{-1}			$\Delta S° = -126.1$ J mol^{-1}K^{-1}
$\Delta C°_p$/J mol^{-1} K^{-1}	K_c	T/K	K_p or [MeO₂]/[Me]
−6.0	2.09×10^{17}	300	8.50×10^{15}
19.0	3.71×10^8	500	9.05×10^6
48.1	59.1	1000	0.72
62.3	15.8	1500	0.13
—	0.94	2000	5.73×10^{-3}
—	5.01×10^{-3}	2500	2.44×10^{-5}

$$Br + MeO_2H \rightleftarrows HBr + MeO_2 \tag{14}$$

This leads to $\Delta H°_f$ (MeO₂) = 23.0 ± 4.2 kJ mol^{-1} and D(Me −O₂) = 122 ± 4.2 kJ mol^{-1}. Both methods suffer from the uncertainty in $\Delta H°_f$ (MeO₂H). Khachatryan et al.[34] studied the equilibrium (Reaction 13) via the oxidation of methane over the temperature range 706 to 786 K. They obtained the result $\Delta H°_{13}$ = −119.2 ± 10.2 kJ mol^{-1}. It turns out that $\Delta C°_p$ (13) = 14.7 J mol^{-1} K^{-1} [35] and hence $-\Delta H°_{13}$ (300) = D(Me −O₂) = 125.5 ± 10.2 kJ mol^{-1}. Very recently Gutman and Slagle[36] measured $\Delta H°_{13}$ directly using photoionization mass spectroscopy over the temperature range 694 to 811 K and find that $\Delta H°_{13}$ = −129.3 ± 10.5 kJ mol^{-1}. All four values are in essential agreement within experimental error although there is some uncertainty in the value for D(Me −O₂). The value ranges from 122 to 129 kJ mol^{-1}. Since the last value arises from a direct determination this value is recommended. It is instructive to consider the ratio of [MeO₂]/[Me] in the equilibrium (Reaction 13) because this has an important bearing on the mechanism for the oxidation of methyl radicals in for instance the CH₄/O₂ system.[37] These values are given in Table 4 over the temperature range 300 to 2500 K. $\Delta C°_p$ values up to 1500 K were calculated from the data in Reference 35. Other values were calculated by extrapolating those results to higher temperatures. One other thing that has to be stipulated is the concentration of oxygen which is taken to be 1 atm and, therefore, clearly varies over the chosen temperature range. Inspection of Table 4 shows that up to 500 K virtually no methyl radicals exist whereas virtually no methylperoxy radicals exist at 2500 K. The ratio [Me₂]/[Me] becomes unity just below 1000 K.

Studies of the addition of methyl radicals to oxygen show that the process is pressure dependent under normal experimental conditions. Basco, James, and James[38] studied Reaction 15

$$Me + O_2 + M \rightarrow MeO_2 + M \tag{15}$$

at 295 ± 2 K over the pressure range 25 to 380 torr using nitrogen and pentane as inert gases. Methyl radicals were produced from the flash photolysis of azomethane and its concentration monitored in absorption. Values for k_{15} (0) and k_{15} (∞) are given in Tables 5 and 6. Parkes covered essentially the same pressure range (20 to 760 torr) using a photolytic molecular modulation apparatus.[39] Hochanadel et al.[40] measured the rate of formation and

Table 5
VALUES FOR k_{15} (0), Me + O_2
+ M → MeO_2 + M AT 295 K

$k_o/10^{11} M^{-2}$ sec^{-1}	M	Ref.
0.94 ± 0.03	N_2	38
3.6 ± 0.3	Pentane	38
1.12 ± 0.3	N_2	39
5.44 ± 2.9	Neopentane	39
0.6	He	42
1.23 ± 0.4	He	44
0.65	Ar	20

Table 6
VALUES FOR k_{15} (∞) Me + O_2 →
MeO_2 (REACTION 15)

$k_\infty/10^8 M^{-1}$ sec^{-1}	T/K	M	Ref.
3.1 ± 0.3	295	N_2, pentane	38
7.2 ± 3.6		N_2, neopentane	39
13 ± 2			40
11			41
12.1			46
10.2	298	N_2, Ar, He	59

decay of the methyl peroxy radical. They concluded that process (Reaction 15) was close to its high pressure limit at 1 atm. Their value for k_{15} (∞) was a factor of 3 higher than Basco, James, and James[38] but in agreement with an earlier study by Callear and van den Bergh.[40] Another study gave a much lower value for k_{15} (∞).[41] This was probably due to data being too far removed from the high pressure limit and the use of a Lindemann plot. An extension of this work[42] led to a value for k_{15} (0) in poor agreement with other work (Table 5).

Both sets of work[41,42] are complicated by a spurious alternative path for the reaction between methyl radicals and oxygen.[43] Ryan and Plumb[44] find a value of $(1.23 \pm 0.4) \times 10^{11} M^{-2}$ sec^{-1}. Pratt and Wood[20] used a discharge flow apparatus to generate methyl radicals via Reactions 16 and 17 over the temperature range 230 to 568 K. They used a competitive

$$H + C_2H_5 \rightarrow C_2H_6^* \tag{16}$$

$$C_2H_6^* \rightarrow 2Me \tag{17}$$

technique — combination of methyl radicals — to determine a value for k_{15}. Extrapolation to zero pressure was made using the Oref-Rabinovitch technique.[45] Over the temperature range examined the rate constant k_{15} (0) was expressed as k_{15} (0) = $10^{10.46 \pm 0.06}$ exp(243 ± 20/T M^{-2} sec^{-1}. This corresponds to a value of $6.5 \times 10^{10} M^{-2}$ sec^{-1} at 298 K emphasizing the uncertainty yet again. This system clearly merits further study.

At much higher temperatures (Table 4) no net reaction takes place and two other reactions have been proposed in place of Reaction 15. The first involves a complex rearrangement that produces formaldehyde and a hydroxyl radical.

$$Me + O_2 \rightarrow CH_2O + OH \tag{18}$$

Using shock waves over a temperature range 1100 to 1400 K and a time-of-flight mass spectrometer for analysis, Clark et al.[47], for example, find a value of $2 \times 10^7\ M^{-1}\ sec^{-1}$ for k_{18}. Two specific experimental tests[48,49] revealed no evidence for Reaction 18 and it has been largely discarded.[50] The most convincing argument against Reaction 18 is that it has to involve the methyl peroxy radical adduct in the rearrangement with a four-center

$$MeO_2 \rightarrow \left[\begin{array}{c} H_2C\cdots O \\ \vdots \quad \vdots \\ H-O \end{array} \right]^{\ddagger} \rightarrow CH_2O + OH \tag{19}$$

transition state (Reaction 19). At temperatures that it may be expected to take place >1500 K,[51,37] the concentration of the methyl peroxy radical is so low that the rate of reaction becomes miniscule. The second reaction involves an oxygen atom abstraction process by the methyl radical.

$$Me + O_2 \rightarrow MeO + O \tag{20}$$

This process would be followed by the almost instantaneous dissociation of the methoxy radical at these very high temperatures (Reaction 21) and therefore the production of hydrogen atoms. Simultaneous measurement of oxygen and

$$MeO + M \rightarrow CH_2O + H + M \tag{21}$$

hydrogen atoms would therefore be desirable. Just et al.[51] were able to study Reaction 20 by using a shock tube together with atomic resonance absorption spectrometry (ARAS) over the temperature range 1700 to 2300 K. The estimated rate constant for Reaction 20 would be $A_{20} = 10^{9.5 \pm 0.5}\ M^{-1}\ sec^{-1}$ and $E_{20} > \Delta H°_{20} = 121\ kJ\ mol^{-1}$. In fact the experimental value is given by $k_{20} = 10^{9.85} \exp(-12910/T)\ M^{-1}\ sec^{-1}$ such that $E_{20} = 107\ kJ\ mol^{-1}$ somewhat less than $\Delta H°_{20}$. At 300 K and 1500 K $\Delta C°_p = -10$ and $+5\ J\ mol^{-1}\ K^{-1}$, respectively, but it is difficult to obtain an accurate value for $\Delta C°_p$ up to 2000 K. However, the measured value for E_{20} looks too low. No other direct values for Reaction 20 have been reported.

B. Me + NO, NO$_2$

The combination of methyl radicals with nitric oxide is both of current interest in terms of fuel combustion in air[49] and of historic importance because of its use in classical studies of the inhibition of gas-phase, free-radical reactions.[52,53] However the use of nitric oxide in inhibition experiments totally confused the issues in the pyrolyses of paraffins.[54] One reason for this is that the nitrosomethane adduct is extremely unstable at the high temperatures involved and also isomerises to fomaldoxime.[55] The latter may not always lead to stable products. The rate of redissociation

$$Me + NO + M \rightleftarrows MeNO + M \rightarrow CH_2NOH + M \tag{22}$$

of nitrosomethane depends upon D(Me −NO). The most reliable value is based upon $\Delta H°_f$ (MeNO)[56] which leads to D(Me −NO) = 165 ± 3 kJ mol^{-1}.

Four recent studies on the combination of methyl radicals with nitric oxide (Reaction 23)[57-60] all used flash photolysis techniques.

Table 7
VALUES FOR $k_{23}(\infty)$ Me + NO + M $\rightarrow$ MeNO +
M (REACTION 23)

$k/10^9 M^{-1} sec^{-1}$	Pressure/torr	T/K	Ref.
2.4 ± 0.2	50 — 400 (N_2)	298	57
1.1	20 — 1009 (C_3H_8)	295	58
	45 — 440 (N_2)		
19	50 — 700 (He,Ar,N_2)	298	59
7.2 ± 0.6	50 — 1000 (Ar,SF_6)	298	60

$$Me + NO + M \rightarrow MeNO + M \qquad (23)$$

The process is pressure dependent under the conditions studied 10 to 1000 torr. The results for k_{23} (∞) are summarized in Table 7. Both Basco et al.[57] and Callear and Van den Bergh[58] monitored the reaction via the decay methyl radicals measured in absorption. Because of problems associated with this method Laufer and Bass[59] and Pilling et al.[60] preferred to use a competitive method and end-product analysis. Basco et al. used the unsatisfactory Lindemann plot to determine k_∞ whereas Callear and Van den Bergh used an RRKM extrapolation. However, they used an erroneously high value for D(Me − NO). The other two sets of workers also used an RRKM extrapolation with D(Me − NO) = 165 kJ mol^{-1}. Laufer and Bass used a Gorin model[61,62] for their RRKM calculations whereas Pilling et al. used a Morse potential. This resulted in a very tight complex and gave a value for k_∞ which was less than their experimental extrapolation. These considerations lead to a recommended value for k_{23} (∞) of $10^{10.0}$ M^{-1} sec^{-1}. Pratt and Veltman[63] used a fast flow discharge system to determine k_{23} at a total pressure of 8 torr (He) over the temperature range 325 to 521 K. Under their conditions the process was essentially third order. They found that the rate constant was given by $k_{23} = 10^{12.5 \pm 0.5} \exp(-23.5 \pm 0.08/T)$ M^{-2} sec^{-1}.

By contrast with nitrosomethane, D(Me − NO_2) is much stronger. On the basis that ΔH°_f (MeNO$_2$) = −74.7 ± 0.6 kJ mol^{-1} (64), ΔHo_f (Me) = 145.0 ± 0.5 kJ mol^{-1} (15) and ΔHo_f (NO_2) = 33.1 ± 0.4 kJ mol^{-1} (8), D(Me − NO_2) = 252.8 ± 0.9 kJ mol^{-1}. There are two possible channels for the reaction of methyl radicals with nitrogen dioxide depending upon whether combination takes place to form either a

$$Me + NO_2 + M \rightarrow MeNO_2 + M \qquad (24)$$

$$Me + NO_2 \rightarrow MeO + NO \qquad (25)$$

C–N or C–O bond. Both processes are important in the pyrolysis of nitromethane.[55] No measurements have been made for Reaction 24 but using shock tube data for the decomposition process (Reaction 26), Glänzer and Troe[65] find

$$MeNO_2 \rightarrow Me + NO_2 \qquad (26)$$

k_{24} (∞) = $10^{10(T/1000)-0.6}$ M^{-1} sec^{-1} (300 to 1400 K). By monitoring the build-up and decay of nitrogen dioxide at 400 nm in absorption, they find that $k_{25} = 1.3 \times 10^{10}$ M^{-1} sec^{-1}. Phillips and Shaw used a competitive technique to measure k_{25} relative to combination of methyl radicals with nitric oxide — Reaction 23 at 323 and 363 K.[66] Using $k_{23} = 10^{10}$ M^{-1} sec^{-1} make $k_{25} = 3 \times 10^{10}$ M^{-1} sec^{-1} independent of temperature. Gutman et al. used a flow system at 295 K to determine k_{25} using photoionization mass spectrometry.[67] Their value is in good agreement with Glänzer and Troe.[66] The data is summarized in Table 8.

Table 8
RATE CONSTANTS FOR THE REACTION Me +
$NO_2 \rightarrow MeO + NO$ (REACTION 25)

$k/10^{10}\ M^{-1}$ sec^{-1}	T/K	Ref.
3	323 and 363	66
1.3	900 — 1500	65
1.5 ± 0.3	295	67

The question arises whether the process in Reaction 25 is an atom transfer process as assumed or whether a hot methyl nitrite species is formed as an intermediate. Gutman et al. find that as the methyl radical concentration falls the nitric oxide concentration increases. They consider that this is good evidence for the direct atom transfer process. A similar atom transfer process is proposed for the same reaction with trifluoromethyl radicals.[68]

$$CF_3 + NO_2 \rightarrow CF_3O + NO \tag{27}$$

However, for higher alkyl radicals (Et, Pr, Bu) there is clear evidence for alkyl nitrite formation.[69] It would be a curious result that the reaction with methyl radicals involves a different transition state. The alternative explanation is that under the conditions involved the hot methyl nitrite formed always spontaneously dissociates to methoxy radicals and nitric oxide such that the channel (see equation) is never observed.

$$Me + NO_2 \rightarrow MeONO^* \nearrow^M MeONO + M$$
$$\searrow MeO + NO$$

C. Me + CO, CO$_2$

The reaction of methyl radicals with carbon monoxide is of interest because the product, the acetyl radical, plays an important role in hydrocarbon oxidation. Reaction with oxygen leads to peracetic acid which provides the basis for the modeling of cool flame behavior in the oxidation of acetaldehyde. Early work on the addition process (Reaction 28), which is summarized

$$Me + CO + M \rightarrow MeCO + M \tag{28}$$

by Wakins and Ward,[70] was based upon product analysis. Usually azomethane is photolyzed in the presence of carbon monoxide and k_{28} was determined by measuring the yields of acetone and ethane. More recently, using molecular modulation spectroscopy, Parkes[71] was able to follow the process directly by measuring both the decay of methyl radicals and the formation of the acetyl radical. Anastasi and Maw[72] extended this study and combined molecular modulation spectroscopy with product analysis. At high pressures of carbon monoxide the mechanism is necessarily complex because four combination processes are involved.

$$Me + CO + M \rightarrow MeCO + M \tag{29}$$

$$Me + Me \rightarrow C_2H_6 \tag{30}$$

$$MeCO + Me \rightarrow MeCOMe \tag{31}$$

Table 9
REACTION OF METHYL RADICALS WITH CARBON MONOXIDE
Me + CO + M → MeCO + M (REACTION 28)

log [$A_{28}(\infty)/M^{-1}$ sec^{-1}]	E_{28} (∞)/kJ mol^{-1}	T/K	k(298)/M^{-1} sec^{-1}	Ref.
8.2	25	260 — 413	6.6 × 10³	70
7.9	276	263 — 343	1.2 × 10³	72

$$2MeCO \rightarrow (MeCO)_2 \tag{32}$$

All studies show that the additon process (Reaction 28) is pressure dependent. Thus, k_{28} varies from 2.2×10^3 M^{-1} sec^{-1} at 100 torr to 3.6×10^3 M^{-1} sec^{-1} at 750 torr (298 K).[71] The most extensive pressure studies were carried out by Watkins and Ward[70] who covered a pressure range of 700 to 2100 torr. Nevertheless, a long extrapolation was required to determine k_{28} (∞). The latter two sets of work[70,71] (Table 9) were in reasonable agreement* but there was considerable disagreement with the most recent work on the reverse decomposition process Reaction 33.[73] This was in terms of the measured activation E_{33} which together with ΔH°_{33} was

$$MeCO + M \rightarrow Me + CO + M \tag{33}$$

some 19 kJ mol^{-1} higher than predicted by E_{28}. The reason for this is not clear. No information is available for the reaction of methyl radicals with carbon dioxide. In solution acetate radicals, which are formed as a result of the decomposition of acetyl peroxide, rapidly decompose to give methyl radicals. In the gas phase methoxy radicals are converted to methyl radicals

$$MeCOO \rightarrow Me + CO_2 \tag{34}$$

via reaction with carbon monoxide.[7] The process appears to be direct atom transfer although intermediate MeOCO formation cannot be ruled out.

$$MeO + CO \rightarrow (MeOCO) \rightarrow Me + CO_2 \tag{35}$$

D. Me + SO$_2$

Since sulfate aerosol formation is known to be important in smog formation, one possible precursor may be the product from an interaction between methyl radicals and sulfur dioxide. In fact early work was hampered by just such aerosol formation.[75] James et al.[75] avoided the problem by going to very small extents of conversion and measuring the reaction directly. Methyl radicals were generated by the flash photolysis of azomethane and the decay of methyl radicals were measured in absorption at 216.4 nm. MeSO$_2$ radical formation was assumed. The rate of reaction was independent of pressure

$$Me + SO_2 + M \rightarrow MeSO_2 + M \tag{36}$$

(Ar and N$_2$ — 50 to 200 torr). They found that the rate constant was given by $k_{36} = 1.75 \pm 0.25 \times 10^8 \cdot M^{-1}$ sec^{-1} (295 ± 2 K). The fate of the MeSO$_2$ radical is unknown, but presumably sulfate is formed via a complex oxidation process summed up by (Reaction 37)

---

* At 298 K there is a factor of ∼5 between the rate constants (Table 9).

$$MeSO_2 + O_2 \rightarrow MeSO_4 \tag{37}$$

E. Me + O_3

The reaction of methyl radicals with ozone may be of stratospheric significance. The most likely event is simple addition. The adduct MeO_3 carried excess vibrational energy and two possibilities occur for the

$$Me + O_3 \rightarrow MeO_3 \tag{38}$$

decomposition of an adduct of this type. The first is simple dissociation of form a methoxy radical and molecular oxygen. The second possibility is

$$MeO_3^* \rightarrow MeO + O_2 \tag{39}$$

cyclization resulting in formaldehyde and HO_2 formation.

$$MeO_3^* \rightarrow \begin{bmatrix} & O & \\ & \diagup \diagdown & \\ O & & O \\ | & & \vdots \\ CH_2 \cdot & \cdots \cdots & \cdot H \end{bmatrix}^{\ddagger} \rightarrow CH_2O + HO_2 \tag{40}$$

The second process involves a 5-center transition state and therefore has an A factor $\sim 10^{12}$ sec^{-1} and the internal hydrogen atom abstraction requires a suitable activation energy despite the overall exothermicity of the process (385 kJ mol^{-1}). This makes Reaction 39 the favored step. Since the overall process is exothermic (272 kJ mol^{-1}) the methoxy radical carries enough energy to dissociate spontaneously.

$$MeO^* + M \rightarrow CH_2O + H + M \tag{41}$$

Alternatively reaction with oxygen could take place,

$$MeO + O_2 \rightarrow CH_2O + HO_2 \tag{42}$$

such that Reaction 38 is the rate-determining step. Simonaitis and Heicklen studied the photolysis of a mixture of $CH_4/O_2/O_3$ at 254 nm. Although this was necessarily a complex system, they were able to extract a value for $k_{38} = 10^{9.5} \exp(-1050/RT)$ M^{-1} sec^{-1}. At 298 K, $k_{38} = 5.6 \times 10^8$ M^{-1} sec^{-1}. The high A factor is in agreement with analysis of the mechanism. Washida et al.[77] generated methyl radicals from oxygen atoms and ethylene which may not be a clean source. They used a flow system and analyzed directly for methyl radicals using a photoionization mass spectrometer. This detector also allowed them to demonstrate that formaldehyde was a major product. They found that $k_{38} = (4.2 \pm 1.6)$ $\times 10^8$ M^{-1} sec^{-1} (298 K) in good agreement with Simonaitus and Heicklen's work. Also, over the pressure range used — 2 to 6 torr He, admittedly small — k_{38} was independent of pressure in agreement with the proposed mechanism. The rate constant is comparable to that of k_{15} (∞) for reaction of methyl radicals with oxygen. However because of the pressure dependence of Reaction 15, despite the unfavorable ratio of $[O_3]/[O_2]$, Reaction 38 may have a role at stratospheric levels.

F. Me + N_2O

No direct studies have been made of the reaction between methyl radicals source of oxygen

Table 10
VALUES OF k_∞ FOR 2 Me $\rightarrow$ C$_2$H$_6$ (REACTION 44)

$k_\infty/10^{10}$ M^{-1} sec^{-1}	T/K	Method	Ref.
2.4	400	Sector	86
2.7	293	FP of AM and DMM, KS	87
5.9	298	FP of AM, DMM and ketene/H$_2$, KS	82
3.4	293	FP of AM, KS (photoelectric)	88
2.4	313	FP of MeI, MS	89
2.4	250 — 450	MMS of AM	90
3.2	298	FP of AM, KS	91
2.81 — 2.12	296 — 577	FP of AM, KS	83
0.4	1120 — 1400	ST of AM and DMM, MS	85
1.0	1200 — 1500	ST of AM, KS	80

Note: FP, flash photolysis; ST, shock tube; MMS, molecular modulation spectroscopy; MS, mass spectroscopy; KS, kinetic spectroscopy; AM, azomethane; DMM, dimethyl mercury.

atoms in a study of the oxidation of methane. At lower temperatures — 900 to 2000 K — it was realized that there may be a direct reaction between methyl radicals and nitrous oxide. This reaction is

$$Me + N_2O \rightarrow MeO + N_2 \qquad (43)$$

therefore a high temperature source of methoxy radicals. Thus, Falconer and co-workers[78] were able to determine k_{43} relative to the combination of methyl radicals, albeit from a complex scheme. At 873 K they found that $k_{43} = 1.4 \pm 0.3 \times 10^4$ M^{-1} sec^{-1}. Borisov et al.[79] studied the shock-induced reaction of methane/nitrous oxide/argon mixtures over the temperature range 1000 to 2000 K. They were able to obtain values for k_{43} as a result of modeling the system. At 873 K their results extrapolated to a value for $k_{43} < 10^{4.9}$ M^{-1} sec^{-1} in reasonable agreement with the previous result. These values must be regarded as tentative.

V. Me + Me

A study of the combination of methyl radicals is important for a number of reasons. It is used as a reference reaction for the study of both methyl radical addition and hydrogen atom abstraction reactions. It is a central issue in both hydrocarbon pyrolysis and combustion in particular methane.[6] It is used as a test reference point for the theories of unimolecular reactions. Although RRK theory — in terms of the number of effective oscillators — is still considered in the chosen timescale of this review, most emphasis is laid on RRKM theory. Any such consideration of the rate constant for the combination of methyl radicals is inevitably bound up with the reverse process and it would be wrong not to consider this here.

Some values for k_{44} (∞) are given in Table 10. This is not a complete

$$Me + Me + M \rightarrow C_2H_6 + M \qquad (44)$$

compilation but earlier values are given by Waage and Rabinovitch[4] and more extensive tables can be found in References 80 and 81. At room temperature values in Table 10 lead to a mean value k_{44} (∞) = $2.8 \pm 0.4 \times 10^{10}$ M^{-1} sec^{-1} which is the recommended value.

One value was not used in calculating this mean value. This refers to the work of Bass and Laufer[82] which leads to a considerably higher value for k_{44} (∞). The reason for this may be the use of the wavelength 150.4 nm to monitor the decay of methyl radicals in absorption. Most other workers use 216.4 nm. The precision of k_{44} will also be affected by the uncertainty in the absorption cross-section (5). However, extrapolation procedures can also affect the value for k_{∞}. Although many workers have found that the rate constant has a value which is independent of temperature, the very careful work of Pilling et al.[83] has shown that the rate constant slowly falls with temperature. Equally careful work by Trenwith[84] on the reverse decomposition process (Reaction 45) came to the same

$$C_2H_6 + M \rightarrow 2Me + M \qquad (45)$$

conclusion. This is supported by shock tube data at much higher temperatures.[80,85] On the assumption that simple collision theory applies, i.e., a $T^{1/2}$ term, the rate constant has a negative activation energy of 3.8 kJ mol^{-1}.[84] Pilling et al.[83] prefer to represent the variation of k_{44} (∞) by the term $(1.67 \pm 0.11) \times 10^{10}$ exp(154 $\pm$ 22/T) M^{-1} sec^{-1}. Notwithstanding this conclusion about the variation of k_{44} (∞), the value of $2.8 \pm 0.4 \times 10^{10} M^{-1}$ sec^{-1} is still recommended as a reference rate constant independent of temperature, since other errors will be greater in those types of studies.

Another way to determine k_{44} (∞) is via the rate of decomposition of ethane. One may write a fairly straightforward Rice-Herzfeld mechanism in view of the simplicity of the products

$$Me + C_2H_6 \rightarrow CH_4 + C_2H_5 \qquad (46)$$

$$C_2H_5 + M \rightarrow C_2H_4 + H + M \qquad (47)$$

$$H + C_2H_6 \rightarrow H_2 + C_2H_5 \qquad (48)$$

It is clear that the rate of formation of methane gives a direct measure of k_{45}. Values of k_{45} determined this way are listed in Table 11. Baulch and Duxbury[92] have reviewed both methods of determining k_{45} (∞) and recommend the value $(2.4 \pm 0.4) \times 10^{10} M^{-1}$ sec^{-1}. It is relevant here to point out that Waage and Rabinovitch noted that the experimental A factors A_{45} were less than those calculated from thermodynamic values together with k_{exp} for the combination processes. Most of these discrepancies disappear when ΔH°_o (45) = 367.2 kJ mol^{-1} derived from Chupka's data,[104] is used.[80,94] This value for ΔH°_o (45) is in agreement with the value for ΔH°_f (Me) recommended earlier. In terms of any theory of unimolecular reactions, an isomerization process in both the forward and reverse directions usually consists of stepwise collisional activation/deactivation processes. The ethane, methyl radical system differs from this in that although one can still consider the decomposition process in the same way, the combination of methyl radicals is a chemical activation process by virtue of the energy released upon forming the C–C bond. This aspect of this system is discussed by Waage and Rabinovitch[4] but is not discussed further here. However, it is relevant to point out that most of the nascent ethane molecules are simply at the threshold of decomposition. A different situation is realized as a result of the combination of methyl radical with a trifluoromethyl radical when there is a second decomposition path of lower energy which eliminates hydrogen fluoride.

$$Me + CF_3 \rightleftarrows MeCF_3^* \rightarrow CH_2{=}CF_2 + HF \qquad (49)$$

There is also a second path available for the decomposition of the ethane molecule which

Table 11
VALUES FOR k_∞ FOR 2Me $\rightarrow$ C_2H_6 (REACTION
44) DETERMINED FROM THE REVERSE STEP
$C_2H_6 \rightarrow$ 2Me (REACTION 45)

$k_\infty/10^{10}$ M^{-1} sec^{-1}	T/K	Method	Ref.
		Experimental studies	
2.0	1400	S-T, single pulse	92
1.6	900	Static system	93
2.1	838	Static system	94
1.5	900	Static system	84
2.8	298	S-T, single pulse extrap. to 298 k	95
		Theoretical studies	
1.3	300		
6.48	900	RRKM, ΔH°_0 = 361.1 kJ mol^{-1}	4
12.7	1400		
0.04	300		
0.5	900	RRKM, ΔH°_0 = 357.7 kJ mol^{-1}	92
0.9	1400		
0.8	300		
10	900	RRKM, ΔH°_0 = 367.2 kJ mol^{-1}	96
25	1400		
3.1	300	Statistical adiabatic channel	
3.0	900	ΔH°_0 = 367.2 kJ mol^{-1}	96
2.9	1400		

involves the breaking of the C–H bond but with $A_{50} \sim 10^{14}$ sec^{-1} and E_{50} = 417 kJ mol^{-1}, this process never competes with the decomposition

$$C_2H_6 \rightarrow C_2H_5 + H \tag{50}$$

process (Reaction 45). The rate constant for Reaction 45 is given by k_{45} = $10^{16.72 \pm 0.17}$ exp(-88850 ± 680/RT) sec^{-1}.[84] However, at high temperatures an equilibrium situation can be reached between the ethane molecule and methyl radicals such that some of the ethane molecules slowly "leak" away to form an ethyl radical and a hydrogen atom. This provides an alternative mechanism to that provided by Roth and Just in their shock

$$Me + Me \rightleftarrows C_2H_6 \rightarrow C_2H_5 + H \tag{51}$$

tube study of the decomposition of ethane from 1650 to 2100 K.[97] Here they consider that the interaction of two methyl radicals result in the direct formation of the same products such that k_{52} = 8.0×10^{11} exp(-13.4×10^3/T) M^{-1} sec^{-1}.

$$Me + Me \rightarrow C_2H_5 + H \tag{52}$$

The reverse of this process (Reaction 50) of course will provide a chemically activated ethane molecule in terms of Reaction 45. Recently another system has

$$C_2H_5 + H \rightarrow C_2H_6^* \tag{53}$$

been used to produce chemically activated ethane.[98] This involves the production of methylene via the photolysis of diazomethane at 436 nm. Reaction with methane produces ethane with an energy of excitation of 481 ± 8 kJ mol^{-1}. The value of $k_{55} = (4.6 \pm 1.2) \times 10^9$ sec^{-1} obtained is in

$$CH_2 + CH_4 \rightarrow C_2H_6^* \qquad\qquad (54)$$

$$C_2H_6^* \rightarrow 2Me \qquad\qquad (55)$$

agreement with other chemically activated ethane data. However, no matter what model was taken, the system leads to a value for k_{44} at 298 K some 8 to 60 times lower than the accepted value. This disagreement of theory with experiment is reflected in the results shown in Table 11 which is only slightly modified by the value of ΔH°_o taken. Indeed most models predict an increase of k_{45} with temperature whereas most recent experimental data both on combination of methyl radicals[83] and the dissociation of ethane,[84] show that the reverse takes place.

As indicated earlier one of the interests in the methyl radical - ethane system is the fall-off of the rate constants as a function of pressure. Conventional studies can lead to data for the pressure dependence of k_{44}. For example Price et al.[99] studied the reaction of methyl radicals with toluene over a wide temperature range (611 to 883 K). The ratio R_{CH4}/R_{C2H6} is

$$Me + toluene \rightarrow CH_4 + benzyl \qquad\qquad (56)$$

clearly pressure dependent over the range 4.5 to 204 torr. By obtaining a limiting ratio and assuming $k_{45} = 10^{10.34} M^{-1}$ sec^{-1} they were able to determine k_{56} and hence k_{45} as a function of pressure. Although the chosen value for k_{45} is certainly in error, viable results are still obtained for k/k_∞ as a function of pressure. Equally, Trenwith[84] was able to obtain full-off data for the decomposition of ethane by the method already discussed. The most recent direct study comes from the photolysis of azomethane in the presence of argon. The pressure dependence of k_{44} was demonstrated over the temperature range 296 to 577 K and pressure range 5 to 500 torr by measuring the decay at 216.4 nm in absorption. In particular it was shown that both k_o and k_∞ fall as a function of temperature. The precision of these values is affected by the uncertainty in the absorption cross-section as a function of temperature. The data were analyzed by the semi-empirical method of Troe.[100] Trenwith[84] carried out RRKM extrapolations by fitting the data to a model activated complex. He tested several models including those of Waage and Rabinovitch,[4] Hase,[101] Bureat et al.,[92] Clark and Quinn,[94] and Growerek et al.[98] These models are essentially the same in assuming that the complex has a single active rotor although both ΔH°_o and the A factor vary from model to model. Hase's model differs in that the C–C bond length and hence the centrifugal correction is greater than in the other models. Trenwith calculated sums and densities of molecular quantum states using the Whitten - Rabinovitch approximation and centrifugal corrections using the Waage - Rabinovitch expression.[4] Skinner et al.[95] however preferred to use the more accurate exact summation. Trenwith found that the Hase model gave the best fit to his data but this model uses a value for E_o which is less than the accepted ΔH°_o. A useful study would be one where the decomposition is carried out under conditions as close as possible to a reliable reverse process study.

VI. Me + RADICALS OTHER THAN Me

In a system where methyl radicals can interact with a second type of radical R, three possible combination processes may take place:

Table 12
VALUES FOR THE RATIO
$$k_{57}/(k_{44} \ k_{58})^{1/2}$$

$$Me + Me \rightarrow C_2H_6 \ (44)$$
$$Me + R \rightarrow MeR \ (57)$$
$$R + R \rightarrow R_2 \ (58)$$

R	$k_{57}/(k_{44} \ k_{58})^{1/2}$	Ref.
Et	1.9 ± 0.1	105
n-Pr	2.1	106
i-Pr	2.0 ± 0.1	105
CF_3	2.7 ± 0.2	103
CF_3	2.0 ± 0.2	194
C_3F_7	2.7	105
CH_2COMe	2.0	107

Table 13
$P_{1/2}$ TORR VALUES FOR THE COMBINATION
OF METHYL RADICALS

Method	T/K	$P_{1/2}$/torr	Ref.
Sector	453	4.5	112
	513	2.7	
Cross-combination	288	0.012	
	333	0.048	105
	373	0.24	
Cross-combination	298	0.8 ± 0.2	
	393	1	103
Theory	298	0.21	4
	520	3.6	
	873	117	
	1058	240	

$$Me + Me \rightarrow C_2H_6$$

$$Me + R \rightarrow MeR \qquad (57)$$

$$R + R \rightarrow R_2 \qquad (58)$$

It has been assumed that in terms of simple collision theory[1] or the geometric mean rule,[102] the ratio $k_{57}/(k_{44}k_{58})^{1/2}$ should be 2 which is borne out by a number of results.[1] However, the difference between the molecular parameters of Me and R belie this simple relationship as pointed out by Kabrinsky et al.[103] Nevertheless a recent measurement of this ratio for methyl and trifluoromethyl radicals led to a value of 2.0 ± 0.2 over the temperature range 402 to 433 K.[194] Other values are shown in Table 12 and Reference 1. The interaction of different radicals R with the methyl radical in cross-combination provide another method of observing the fall-off of k_{44}[4,103,105] provided that Reactions 57 and 58 are always in their high pressure limits under the conditions studied. This has involved the determination of $P_{1/2}$,[4,103,105,111] the pressure at which the rate constant is equal to $1/2k_\infty$ (Table 13). Some discrepancies appeared in terms of a ''tight'' vs. a ''loose'' complex model for methyl radical combination but these were resolved by Kabrinsky et al.[103]

Although no disproportionation between methyl radicals has either been observed or expected, that with larger radicals occurs with high efficiency (Table 14) and little or no

Table 14
DISPROPORTIONATION COMBINATION RATIOS
(Δ) FOR THE RADICALS Me AND R

R	Δ	Ref.
Et	0.036 ± 0.005	105
n-Pr	0.05	1
i-Pr	0.15 ± 0.04	105
s-Bu	0.3	1
t-Bu	0.7	1

Note: For other values see References 113 and 114.

activation energy. This is in distinct contrast to the methyl hydrogen abstraction reactions with molecules which have A factors $= 10^{9 \pm 0.5}$ M^{-1} sec^{-1} and activation energies of 35 ± 15 kJ mol^{-1}. The very low activation energies for the disproportionation processes has been attributed to the high exothermicities involved whereas the high A factors are associated with the rearrangement of the highly energized transition state[108] or nascent molecule[1,107] formed in the initial combination process. An alternative mechanism involves a head-to-tail transition state which was thought untenable because the involved light transition state would imply a low A factor. Evidence for the head-to-tail mechanism comes from mass spectrometric analysis of the ethylene fraction arising from the disproportionation of 2MeCD$_2$ which turns out to be exclusively CH$_2$CD$_2$.[109] In the face of this evidence Benson[102,110] explored a very loose transition state which is essentially polar in nature. This gives a satisfactory interpretation of the experimental results in terms of both a high A factor and no activation energy. A comprehensive review of other radical - radical disproportionations has been given by Gibson and Corlay.[114]

As mentioned earlier, another chemical activation system occurs upon the combination of methyl and trifluoromethyl radicals. This is in terms of the

$$Me + CF_3 \rightleftarrows MeCF_3^* \rightarrow CH_2=CF_2 + HF$$

formation of a vibrationally excited trifluoreothane adduct which is only at the threshold of redissociation into two radicals, but contains considerable excess energy for the much lower energy path which leads to the elimination of hydrogen fluoride. Pritchard and Perona and other workers[115] analyzed their results in terms of RRK theory. A more satisfactory approach would have been to use a full RRKM analysis. Pritchard and Perona also studied the reaction of CD$_3$ with trifluoromethyl radicals and found a kinetic isotope effect $k_M/k_D = 3.1$ at 470 K.

Not much further work has been carried out on other methyl radical - radical reactions but some of these reactions are of considerable importance. In this category come the reactions with HO$_2$, OH, and CH$_2$.

The reaction with HO$_2$ is important both for atmospheric chemistry and combustion. There are two possible channels. Reaction 59 is a simple combination process in essence and would be expected to have a rate constant

$$Me + HO_2 \begin{cases} \nearrow (MeO_2H) \rightarrow MeO + OH & (59) \\ \searrow CH_4 + O_2 & (60) \end{cases}$$

of 10^{10} M^{-1} sec^{-1}. In fact the process is so exothermic $\sim$290 kJ mol^{-1} that the intermediate MeO$_2$H would not be observed since this would be expected to fly apart within a few vibrations. Reaction 60 is recognized both as the reverse of the initiation process in the combustion of methane and also as a simple disproportionation process. From arguements already discussed, this is also expected to have a rate constant in the range 10^9 to 10^{10} M^{-1} sec^{-1}. Reaction with OH may also be very important for combustion.[116,117] Here, only combination may take place. At the high temperatures involved other bonds may be broken in the nascent methanol molecule followed by dissociation of the methoxy or hydroxymethyl radicals.

$$Me + OH \rightleftharpoons MeOH \nearrow MeO + H \qquad (61)$$
$$\searrow CH_2OH + H \qquad (62)$$

$$MeO + M \rightarrow CH_2O + H + M \qquad (63)$$

$$CH_2OH + M \rightarrow CH_2O + H + M \qquad (64)$$

These sequence of reactions are preferred to direct formation of products such as Reaction 65.[116,118]

$$Me + OH \rightarrow CH_2O + H_2 \qquad (65)$$

One other reaction which may be important in combustion systems is reaction with CH$_2$ ($^3B_1$), which results in ethylene formation. The reaction has to proceed via an ethyl radical intermediate. The addition process is so exothermic $\sim$390 kJ mol^{-1} so elimination of hydrogen atom must occur within a few vibrations.

$$CH_2 + CH_3 \rightarrow C_2H_5^* \rightarrow C_2H_4 + H \qquad (66)$$

One would expect the addition process to have a rate constant $k_{66} \sim 10^{10}$ M^{-1} sec^{-1}. This is the kind of value that Gardiner and Olson[118] have used in their scheme for methane-rich combustion. Dees and Setser[119] also suggested this process as a source of ethylene in the reactions of methylene with butane. Setser[120] cited a similar mechanism in the cophotolysis of acetone with 1,1,3,3, tetrafluoroacetone.

$$CF_2 + CH_3 \rightarrow CH_3CF_2^* \rightarrow CH_2CF_2 + H \qquad (67)$$

Pilling and Robertson[121] attempted to measure Reaction 66 via the flash photolysis of mixture of ketene and azomethane. The best computer fit for product formation was given by $10^{10.5}$ M^{-1} sec^{-1} in reasonable agreement with estimates.

VII. H - ATOM ABSTRACTION REACTIONS

A. Me + H$_2$

The reaction of methyl radicals with hydrogen Reaction 68 is interesting for a

$$Me + H_2 \rightarrow CH_4 + H \qquad (68)$$

number of reasons. It is the simplest radical - molecule metathesis reaction. A considerable isotope effect is expected. There are two equally strong schools of thought for and against

Table 15
ARRHENIUS PARAMETERS FOR THE
REACTION OF Me + H_2/D_2

log (A/M^{-1} sec^{-1})	E_a/kJ mol^{-1}	H_2/D_2	Ref.
8.73 ± 0.06	43.51 ± 0.54	H_2	125
9.34	45.2	H_2[a]	128
8.7	44	H_2	127
8.77 ± 0.06	49.01 ± 0.58	D_2	125
9.204 ± 0.035	52.95 ± 0.34	D_2	129

[a] Calculated from the thermodynamics and the Arrhenius
parameters for the reverse process.

the importance of tunneling. The reverse step of Reaction 68, at high temperatures at least, constitutes a catalyzed

$$H + CH_4 \rightarrow Me + H_2 \qquad (69)$$

combination of hydrogen atoms. Using Walker's[122] evaluation for $k_{69} = 10^{11.10}$ exp ($-5989/$ T) M^{-1} sec^{-1} we find $k_{69} = 10^{8.5}$ M^{-1} sec^{-1} at 1000 K. According to Benson's[123] evaluation, $k_{70} = 10^{10}$ M^{-2} sec^{-1}. taking M in Reaction 70 to be

$$H + H + M \rightarrow H_2 + M \qquad (70)$$

CH_4, $R_{69}/R_{70} = 10^{-1.5}/[H]$ making Reaction 69 the preferred route for molecular hydrogen production for any reasonable concentration of hydrogen atoms. Reaction 68 has also been the subject of semi-empirical methods for the estimation of activation energies.[124]

There is no need for an in-depth survey of Reaction 68 because this has already recently been made.[125,126] It remains to consider the work that has been carried out since these two surveys. Marshall and Shahkar[127] studied Reaction 68 over the temperature range 584 to 671 K. Most studies involve the determination of k_{68} by measuring the rates of methane and ethane formation such that $R_{CH_4}/R_{C_2H_6} = k_{68}$ [H_2]/$k_{44}^{1/2}$. Marshall and Shahkar however base k_{68} on a rate of production of methyl radicals. Their value is in good agreement with a mean value compiled by Arthur, Donchi, and McDonell[125] (Table 15). One may also calculate a value for k_{68} from the reverse step (Reaction 69).[128] Unfortunately there appears to be a factor of 2 to 3 difference. The reason for this is not clear.

Whereas Clark and Dove[130] and Kobrinsky and Pacey[131] consider that tunneling is important for Reactions 68 and 69, Sepehrad et al.[128] and Arthur et al.[125] conclude that tunneling is not important. Pratt and Rogers[129] consider that tunneling is important for Me + H_2 but not for Me + D_2. Arthur et al.[125] used a modified bond energy - bond order method (BEBOA) to calculate Arrhenius parameters for Reaction 68. This involved replacing the Sato end-atom triplet repulsion term in the potential energy expression of the original BEBO method[132] by a function fitted to the potential energy values proposed by Hirschfelder and Linnett[133] for $^3\Sigma H_2$. Experimental and theoretical values of the Arrhenius parameters are compared at 450 K in Table 16. It is clear that the best agreement with the mean experimental results is obtained with BEBOA method without tunneling.

Arthur et al.[125] also considered the kinetic isotope effect for Me + H_2 and Me + D_2 over the temperature range 850 to 600 K. These results are summarized in Table 17. Once again best agreement with experiment was obtained without tunneling. Also, BEBOA was slightly superior to the BEBO method.

Table 16
EXPERIMENTAL AND THEORETICAL ARRHENIUS PARAMETERS FOR REACTION 68 AT 450 K

	Expt. (mean)	BEBO	BEBOA	BEBO	BEBOA
			Me + H_2		
log (A/M^{-1} sec^{-1})	8.73	8.97	8.97	8.29	8.44
E/kJ mol^{-1}	43.51	57.20	46.71	45.85	36.95
log (k/M^{-1} sec^{-1})	6.68	5.33	6.55	5.97	7.15
			Me + D_2		
log(A/M^{-1} sec^{-1})	8.77	8.84	8.85	8.49	8.57
E/kJ mol^{-1}	49.01	59.59	47.56	53.68	42.20
log (k/M^{-1} sec^{-1})	6.08	4.93	6.36	5.26	6.67

Table 17
KINETIC ISOTOPE EFFECTS FOR REACTION 68

		k_N/k_D		k_N/k_D (runs)	
T/K	k_N/k_D(expt)	BEBO	BEBOA	BEBO	BEBOA
350	5.99	6.15	6.09	19.94	15.73
400	4.74	5.19	5.14	12.63	10.93
450	3.95	4.51	4.47	9.07	8.27
500	3.41	4.02	3.99	7.06	6.64
550	3.02	3.64	3.61	5.80	5.57
600	2.74	3.34	3.32	4.95	4.81

B. Me + RH

As may be expected most data was found in this section. The rate constant k_{71} was usually determined with respect to the combination Reaction 44

$$Me + RH \rightarrow CH_4 + R \tag{71}$$

although other reference reactions have also been used. There are two notable exceptions which involve isobutane[140] and formaldehyde.[144] Here, Anastasi followed the reaction in real time using molecular modulation spectroscopy. Some data are given in Tables 18 to 22. This is by no means an exhaustive list since Kerr and Parsonage[125] have recently compiled a list of data. Other useful compilations have been made by Arthur and co-workers who in particular have made a very detailed analysis.

Where data has been determined over a wide temperature evidence has accrued for curved Arrhenius plots. Explanations for this curvature have ranged from planar vs. pyramidal forms of the methyl radical as a function of temperature, hydrogen abstraction at different sites in the molecule RH to tunneling. At this time since the consensus of opinion finds no evidence for tunneling in the reaction of methyl radicals with hydrogen or deuterium, it seems doubtful that a case could be made for tunneling with larger molecules. In terms of the simple collision theory one might expect the pre-exponential factor to vary as the absolute temperature — $T^{1/2}$. This is not enough to account for the observed curvature. One other explanation has been given by Benson[7] but this appears to have been overlooked. In the metathesis of a

Table 18
ARRHENIUS PARAMETERS FOR SOME METATHESIS REACTIONS
Me $\rightarrow$ RH $\rightarrow$ CH$_4$ + R (REACTION 71)

	log(A/M^{-1} sec^{-1})	E/kJ mol^{-1}	Temp. range (K)	Ref.
Hydrocarbons				
F$_2$ClH	9.0 ± 0.7	41.8 ± 6.7	430 — 480	134
C$_2$H$_6$	11.2	90 ± 20	920 — 1040	135
C$_2$H$_6$	9.24 ± 0.03	619 ± 1.3	385 — 770	136
C$_3$H$_8$	10.45 ± 0.52	47.3 ± 6.8	676 — 813	137
C$_4$H$_{10}$	11.4 ± 0.8	151.3 ± 26.6	886 — 951	138
C$_4$H$_{10}$	9.5	92.3 ± 16.6	980 — 1060	139
t-BuH	8.67 ± 0.07	36.6 ± 0.7	478 — 560	140
Neopentane	10.5 ± 0.1	67 ± 2	404 — 953	141
Toluene	8.07	66.5 ± 2.5	511 — 883	142

Table 19
ARRHENIUS PARAMETERS FOR SOME METATHESIS REACTIONS

	log (A/M^{-1} sec^{-1})	E/kJ mol^{-1}	Temp. range (K)	Ref.
Carbonyl compounds				
CH$_2$O	10.1 ± 0.2	45 ± 3	788 — 935	143
CH$_2$O	8.93 ± 0.03	29 ± 0.4	500 — 603	144
	8.2	27.5	385	
MeCHO-H	9.2	38.1	785	145
	19.7	47.3	1108	
	9.7	51.1	385	
H-CH$_2$CHO	10.8	63.0	785	145
	11.3	72.7	1108	
MeCOCHO		27.7 ± 2.8	3529 — 442	146
MeCOMe	8.54 ± 0.03	40.5 ± 0.28	299 — 748	147
CD$_3$ + CD$_3$COCD$_3$	8.78 ± 0.08	48.5 ± 0.69	373 — 798	147
3-pentanone				
CH$_3$CD$_2$COCD$_2$CH$_3$	8.10 ± 0.25	69.0 ± 3.3	513 — 572	148
CH$_3$CD$_2$COCD$_2$CH$_3$	8.3 ± 0.5	91.5 ± 7.5	513 — 572	148

Table 20
ARRHENIUS PARAMETERS FOR SOME METATHESIS REACTIONS

	log (A/M^{-1} sec^{-1})	E/kJ mol^{-1}	Temp. range (K)	Ref.
Oxygenated compounds other than carbonyl compounds				
MeOMe	9.0	44	373 — 573	149
	10.5 ± 0.4	63 ± 7	783 — 935	149
	9.55 ± 0.1	46.9 ± 0.8	373 — 473	150
HCOOMe	8.7 ± 0.1		400 — 513	151
CH$_2$ — CH$_2$	8.2			152
Oxetane	8.29 ± 0.19	35.22 ± 0.64	373.4 — 465.0	153
2-methylocetane	8.42 ± 0.18	3.404 ± 0.62	375.3 — 466.1	153
2,4-dimethyloxetane	8.51 ± 0.18	33.52 ± 0.69	375.3 — 466.1	153
3-methyloxetane	8.69 ± 0.20	38.23 ± 0.84	373.2 — 473.2	153
3,3-dimethyloxetane	8.72 ± 0.20	39.47 ± 1.0	373.2 — 473.2	153
2,2-dimethyloxetane	8.99 ± 0.23	41.24 ± 1.13	373.2 — 473.2	153
MeNO$_2$	8.85 ± 0.29		413 — 482	154
2-NO$_2$-propane	8.14 ± 0.24		413 — 479	155

Table 21
ARRHENIUS PARAMETERS FOR SOME METATHESIS REACTIONS

	$\log(A/M^{-1}\ \mathrm{sec}^{-1})$	$E/\mathrm{kJ\ mol}^{-1}$	Temp. range (K)	Ref.
Silanes				
SiH_4	8.89 ± 0.11	29.2 ± 0.8	$301 - 486$	156
SiD_4	8.96 ± 0.19	34.0 ± 1.3	$303 - 488$	157
Si_2H_6	8.9 ± 0.13	23.2 ± 0.9	$306 - 475$	157
Si_2D_6	9.19 ± 0.01	29.2 ± 0.7	$310 - 462$	157
$MeSiD_3$	9.25 ± 0.04	38.5 ± 0.3	$297 - 474$	158
$CD_3 + MeSiD_3$	9.49 ± 0.14	35.5 ± 0.9	$295 - 425$	158
$MeSiH_3$	9.28 ± 0.10	34.0 ± 0.7	$298 - 424$	158
$CD_3 + Me_2SiD_2$	9.92 ± 0.09	36.9 ± 0.6	$295 - 425$	158
Me_2SiH_2	9.07 ± 0.07	34.7 ± 0.5	$300 - 485$	158
$Me_3 SiH$	8.42 ± 0.10	33.3 ± 0.8	$345 - 526$	159
Me_3SiH	8.31 ± 0.10	32.6 ± 0.8	$345 - 526$	159
	8.02 ± 0.11	29.1 ± 0.8	$330 - 445$	160
	8.71 ± 0.11	34.9 ± 0.8	$302 - 486$	158
$CD_3 + Me_3SiD$	8.77 ± 0.07	38.5 ± 0.5	$297 - 474$	159
$SiHF_3$	9.21 ± 0.16	35.3 ± 1.2	$343 - 470$	160
Me_4Si	8.41 ± 0.30	41.5 ± 2.4	$331 - 526$	156
$CD_3 + Me_4Si$	8.87 ± 0.32	43.9 ± 2.8	$396 - 573$	156
$CD_3 + Me_3SiD$	7.98 ± 0.20	37.9 ± 1.5	$299 - 489$	158
$0\ SiD_3$	8.03 ± 0.38	24.6 ± 2.7	$310 - 470$	157
$MeSiCl_3$	9.90 ± 0.25	48.2 ± 2.0	$378 - 478$	161
Me_3SiCl	9.76 ± 0.16	44.1 ± 1.2	$337 - 471$	161
$CD_3 + MeSiF_3$	9.18 ± 0.10	53.0 ± 1.0	$448 - 573$	162
$CD_3 + Me_2SiF_2$	9.03 ± 0.04	50.6 ± 0.4	$448 - 573$	163
$CD_3 + Me_3SiF$	8.93 ± 0.08	48.2 ± 0.8	$448 - 573$	163

Table 22
ARRHENIUS PARAMETERS FOR SOME METATHESIS REACTIONS

	$\log(A/M^{-1}\ \mathrm{sec}^{-1})$	$E/\mathrm{kJ\ mol}^{-1}$	Temp. range (K)	Ref.
Miscellaneous organic compounds				
$MeCOMe$	8.54 ± 0.03	40.49 ± 0.28	$393 - 523$	163
CD_3COCD_3	8.78 ± 0.08	48.50 ± 0.69	$393 - 523$	163
MeN_2Me	8.05 ± 0.07	33.39 ± 0.56	$333 - 463$	163
MeN_2Me^a	8.21 ± 0.08	34.57 ± 0.59	$333 - 463$	163
Me_2S	8.62 ± 0.08	38.35 ± 0.68	$393 - 518$	164
$CD_3 + CH_2 - CH_2$	8.34 ± 0.60	39.9 ± 0.4	$303 - 477$	165
$MeCH - CH_2$	8.00 ± 0.48	34.6 ± 3.6	$339 - 435$	165

[a] Corrected for intramolecular formation of C_2H_6 from McN_2Me.

hydrogen atom between two polyatomic fragments, three new frequencies are added to the molecule being attacked, one stretching mode and two rocking modes. Three frequencies associated with the atom being transferred are also changed. Thus three old and six new frequencies must be considered, one of them being the reaction coordinate. It is generally assumed that a tight transition state is involved in the metathesis. Benson concludes that upon forming the transition state, the change in heat capacity ΔC_v can be as high as 25 J mol^{-1} K^{-1}. This will clearly lead to non-Arrhenius behavior when a wide temperature range is considered.

Table 23

ARRHENIUS PARAMETERS FOR SOME METATHESIS REACTIONS

	$\log(A/M^{-1}\ \text{sec}^{-1})$	$E/\text{kJ mol}^{-1}$	Temp. range (K)	Ref.
Inorganic hydrides				
NH_3	7.76 ± 0.30	40.7 ± 1.4	350 — 550	167
HN_3	8.0^a	17.6 ± 2.1	298	166
HCl	8.7	12.9	296 — 423	169
H_2S	8.00 ± 0.01	8.76 ± 0.08	334 — 432	168
H_2	8.73 ± 0.06	43.51 ± 0.54		125

^{a} Assumed value.

C. NH_3, HN_3, HCl, and H_2S

Table 23 lists the Arrhenius parameters for hydrogen atom abstraction from inorganic hydrides. The data for HN_3 is limited to one temperature such that an A factor was assumed to be $10^{8.0}\ M^{-1}\ \text{sec}^{-1}$ [166] in order to determine an activation energy. However the activation energy looks reasonable for such a weak H–N bond dissociation energy. No simple statistical relationship for the number of abstractable hydrogen atoms can be correlated with the pre-exponential factors. However these are higher for diatomic than for polyatomic species as is to be expected.[7] No correlation exists with the activation energies and the H–X bond dissociation energies of the hydrides but there is with the polar nature of HX. In other words, dipolar species have low activation energies.

VIII. OTHER ATOM ABSTRACTION REACTIONS

These sort of reactions have been studied for a number of interesting reasons. The first of these involving reactions of the type,

$$\text{Me} + X_2 \rightarrow \text{MeX} + X \tag{72}$$

where X_2 is a halogen molecule, is concerned with the field of molecular dynamics.[170] This has to do with crossed molecular beams and chemiluminescence, which provide evidence about energy distributions in reaction products. The second is concerned with a re-interpretation of the data of studies of the reaction of alkyl iodides (RI) with hydrogen iodide originally studied by Ogg.[171] Ogg had concluded that reaction took place as the result of a bimolecular step involving RI and HI:

$$\text{RI} + \text{HI} \rightarrow \text{RH} + I_2 \tag{73}$$

very similar to the classical reaction of 2MI

$$2\text{HI} \rightarrow H_2 + I_2 \tag{74}$$

The third system involves a radical buffer method for the determination of rate constants for the combination of alkyl radicals developed by Benson. The present author was involved in the early stages of this work.[172] Here it was realized that in the reaction of methyl radicals with ethyl iodide, the fastest reaction in the system was the iodine atom metathesis corresponding to the forward and reverse processes. Hiatt and Benson developed this

$$Me + EtI \rightleftharpoons MeI + Et \qquad (75)$$

system to determine rate constants for the combination of a number of alkyl radicals.[173]

A. Me + X_2

As mentioned in the introduction interest here centers on crossed molecular beam experiments in terms of single radical/molecule collisions. Analysis of the experimental results yields information about the distribution of energy and angular momentum among the reaction products, the dependence of the total reaction probability on the molecular energies, the lifetime of the collision complex, the quantum state of the products, and other important features. One important result that has emerged from such studies is that there are two types of reaction mechanisms, stripping and rebound mechanisms.

The first calculations on the dynamics of such processes involved the H + H_2 reaction. This moved on to H + Cl_2 and, therefore, the reactions Me + X_2 are a natural progression. Thus, Logan et al.[174] have looked at methyl radical reactions with a number of halogens. For MeI formation they considered the angular and velocity distributions for Me + I_2, IBr and ICl and for MeBr formation the same distributions for Me + Br_2. In all four reactions the methyl halide product recoils backward with respect to the incoming methyl radical. Although all reactions release similar fractions ($\sim$0.3) of the total energy into translational motion, the energy distribution of the products from the Br_2 and I_2 reactions indicate product repulsion, whereas those from the mixed halogen reactions indicate more statistical behavior.

As far as absolute rate constant determinations are concerned, Kovalenko et al.[175] studied the reactions of methyl radicals with Cl_2 and Br_2. Pulsed laser dissociation of methyl iodide was followed by time-resolved detection of I-R vibrational fluorescence from the C–H stretching modes of the methyl halide product. From the fluorescence intensity it was shown that about half of the exothermicity of the reaction goes into product vibration. A similarly large fraction of vibrational excitation was observed for MeF in the Me + F_2 reaction.[176] Kovalenko et al. found that at room temperature the rate constants determined were $k_{Cl_2} = (9.0 \pm 0.6) \times 10^8 \ M^{-1} \ sec^{-1}$ and $k_{Br_2} = (1.2 \pm 0.2) \times 10^{10} \ M^{-1} \ sec^{-1}$ such that the ratio $k_{Cl_2}/k_{Br} = 0.075$. Evans and Whittle[177] used a competitive method to measure the ratio of rate constants. Extrapolating to room temperature gives a value of 0.028 ± 0.015 about a factor of two different to the previous result.

B. RI + HI

Ogg interpreted his data[171] for the reaction of RI with HI as a four-center transition state implied by Reaction 73. However, a second scheme was needed because the apparent second order rate constant decreased with increasing initial concentration of HI. This second scheme involved a chain mechanism with initiation via the breaking of the weakest bond R-I and iodine atom termination. Benson and O'Neal realized that the slow initiation step could be replaced by the much faster iodine atom formation followed by an iodine atom catalyzed reaction. For methyl iodide this becomes:

$$I_2 + M \underset{b}{\overset{a}{\rightleftarrows}} 2I + M$$

$$I_2 + MeI \underset{d}{\overset{c}{\rightleftarrows}} Me + I_2 \qquad (76)$$

$$Me + HI \overset{e}{\longrightarrow} CH_4 + I \qquad (77)$$

Using the steady-state assumption, the rate of production of iodine becomes:

Table 24
RATE CONSTANTS FOR THE
COMBINATION OF ALKYL
RADICALS DETERMINED VIA
THE RADICAL BUFFER METHOD

R	$\log(k/M^{-1}\ sec^{-1})$	Ref.
Et	8.6 ± 0.8	173,181
i-Pr	8.6 ± 1.1	182
t-Bu	5.4	183
CF$_3$	9.7 ± 0.5	184

$$R_{I_2} = \frac{k_c k_e K_{a,b}^{1/2}\ [MeI][HI][I_2]^{1/2}}{k_d[I_2] + k_c[HI]}$$

Over the temperature range of the experiments — 543 to 593 K — the ratio k_d/k_e ~10 and so one can appreciate the roles played by the two terms in the denominator as the reaction proceeds. One may also appreciate the inhibiting role played by increasing amounts of HI as originally observed by Ogg.

It turns out that the pyrolysis of alkyl iodides are also iodine atom catalyzed reactions. An almost simultaneous, similar re-interpretation of Ogg's work was made by Sullivan.[177] Benson and co-workers restudied the reaction between MeI and HI[178] and measured the equilibrium:[179]

$$I_2 + CH_4 \rightleftarrows MeI + HI \tag{78}$$

which led to ΔH°_f (Me), D(Me − H) and D(Me − I). The work was subsequently extended to other iodides.[9] Although, as discussed earlier, the method for Me has now been superseded; this represents an important milestone for the experimental determination of the heats of formation and entropies of free radicals.

C. Me + RI

The corollary to the above work and the very fast setting up of the equilibrium such as Reaction 75 means that the equilibrium constant K_{75} gives a value for the ratio [Me]/[Et] and making use of the rate constant for the combination of methyl radicals gives a value for that of the ethyl radicals. Because ethane formation may also arise from ethyl radical hydrogen abstraction reactions in this system, the rate constant k_{79} had to be determined via the

$$Me + Et \rightarrow C_3H_8 \tag{79}$$

geometric mean rule for the combination of methyl and ethyl radicals. Studies were made using other alkyl iodides and extended by Whittle and co-workers to perfluoroalkyl radicals.[15] Values for the rate constants so determined are given in Table 24. These values apart from CF$_3$ were hotly contested by other workers[185,186] who judged those for *i*-Pr and *t*-Bu to be particularly low. The radical buffer method depends critically upon good precision for the equilibrium constant such as K_{75}. This, in turn, depends upon values for the entropies and heats of formation for the radicals concerned which may be considerably in error. The full mechanism in the buffer system may also complicate the issue as exampled for Et.

As far as iodine atom abstraction reactions were concerned, Benson concluded that the A factors would be ~10^9 M^{-1} sec^{-1} and the activation energies would be between 8 to 16

Table 25
ARRHENIUS PARAMETERS FOR THE METHASIS OF HALOGEN ATOMS

	Temp. range(K)	$\log(A/M^{-1}\ sec^{-1})$	$E/kJ\ mol^{-1}$	Ref.
Chlorine atom transfer				
CCl$_4$		10.4	54.0 ± 2.9	187,188
	293 — 423	8.6	38.1	189
	395 — 425	8.8 ± 0.3	42.3 ± 2.1	190
CFCl$_3$	293 — 423	9.4	35.2	189
CF$_2$Cl$_2$	293 — 423	9.1	35.2	189
CCl$_3$CN	363 — 418	9.9	43.5	188
C$_2$Cl$_6$	363 — 418	8.8	42.3 ± 3.8	187,188
C$_6$H$_5$CCl$_3$	363 — 418	7.3	31.8 ± 3.4	188
CCl$_3$COCCl$_3$	363 — 418	9.6	40.6 ± 3.4	187,188
Bromine atom transfer (363 — 418 K)				
CF$_3$Br		10.3	52.3 ± 4.2	188
CF$_2$Br		8.0	26.8 ± 4.2	188
CCl$_3$Br		10.2	29.7 ± 3.8	188
	(400 — 473)[a]	8.1 ± 0.3	14.6 ± 2.1	190
CCl$_2$Br$_2$		10.8	31.8 ± 4.6	188
CBr$_4$		11.2	33.1 ± 4.6	188
Iodine atom transfer				
CF$_3$I		10.8	31.4 ± 4.2	188

[a] Addition site.

kJ mol^{-1} for alkyl radicals. Table 25 lists Arrhenius parameters for methyl radical metathesis from halogenated species. For chlorine atom transfer the mean value for the A factor is $10^{9.0 \pm 0.9}$ although those values differing by more than ±0.5 log units should be treated with some suspicion. Similarly with one notable exception — which was ignored — all values for the activation energy lie in the range 38.6 ± 4.3 kJ mol^{-1} thus showing no real change in parameters with the nature of the substrate.

IX. ADDITION TO π-BOND SYSTEMS

I suppose that a primary reason for studying the addition of methyl radicals to doubly and to a lesser extent triply bonded molecules such as Reaction 80:

$$Me + C_2H_4 \rightarrow MeCHCH_2 \qquad (80)$$

is in terms of polymer formation. Here, ethene is present in excess. In the presence of small amounts of ethene, Marshall and Rahman[191] suggested that the system may be used to determine thermochemical information about the normal propyl radical. They pyrolyzed azomethane at 620 K in the presence of ethene. Argon (300 torr) was added to ensure that no unimolecular or bimolecular process was pressure dependent. Conditions were chosen such that the system (Reaction 81) was in equilibrium.

$$Me + C_2H_4 \rightleftarrows n\text{-Pr} \qquad (81)$$

It was assumed that ethane and propane were produced uniquely from Reactions 44 and 79. It was further assumed that $k_{79} < k_{44}$. Hence, the ratio $R_{C_2H_6}/R_{C_3H_8}$ gave the ratio [Me]/[n-Pr]. This allowed the equilibrium constant K_{81} to be determined from the relationship:

$$R_{C_2H_6}/R_{C_3H_8} = \frac{k_{44}}{k_{79}} \frac{[Me]}{[n\text{-}Pr]} = \frac{1}{K_{81}[C_2H_4]}$$

By measuring K_{81} over the temperature range 581 to 649 K[192] they found that ΔH°_f (n-Pr) = 94.6 ± 4.2 kJ mol^{-1} and S°_{300} = 282.0 ± 12.6 J mol^{-1} K^{-1}. In terms of the reverse of reaction 80, namely the decomposition of the n-propyl radical, Rabinovitch and co-workers[193] have studied the decomposition of vibrationally excited radicals via the addition of hydrogen atoms to olefins:

$$H + CH_2=CHMe \rightleftarrows n\text{-}Pr^* \rightarrow Me + C_2H_4 \tag{82}$$

In principal redissociation may take place, but since the C–C bond is the weaker, large extents of reaction to produce methyl radicals result. This reaction also constitutes a displacement process which is another important aspect of the addition of methyl radicals to π systems. This is for instance, the only sensible way to explain the formation of acetone in the pyrolysis of acetaldehyde (800 K):[35]

$$Me + MeCHO \rightleftarrows i\text{-}PrO \rightarrow H + Me_2CO \tag{83}$$

Here the addition and redissociation processes are in virtual equilibrium with a small amount of the isopropoxy radicals "leaking away" to form acetone and a hydrogen atom. Addition of methyl radicals to hexafluoroacetone followed by displacement provides a method to determine rate of addition to the C–O π-bond system:[194]

$$
Me + CF_3COCF_3 \rightleftarrows Me\!-\!\overset{\displaystyle CF_3}{\underset{\displaystyle CF_3}{C}}\!-\!O \rightarrow MeCOCF_3 + CF_3 \tag{84}
$$

It is pertinent to mention one other displacement process here which has nothing to do with π-bond systems. This is the displacement of hydrogen atom in CH_4 by a methyl radical:

$$Me + CH_4 \rightarrow C_2H_6 + H \tag{85}$$

This reaction may be important in high temperature pyrolysis and combustion systems. For example Tabajashi and Bauer found it necessary to invoke such a reaction for modeling of the early stages of the pyrolysis and oxidation of methane.[195] In order to calculate the entropy change $\Delta S^{\neq}_{85}$ upon achieving the transition state, the molecule ethane may be taken as a model:

The reaction coordinate corresponds to one of the four C–H breaking contribution R ln 4 to $\Delta S_{85}^{\neq}$. If a tight transition state is assumed, the translational and rotational entropies of ethane will not be seriously changed by the addition of a hydrogen atom. However one has to consider the extra vibrations in $C_6H^{\neq}$. In addition the contribution to symmetry for the transition state is 12 for $C_2H_7^{\neq}$ compared to 18 for C_2H_6. Also included is R ln 2 taking into account the fact that $C_2H_7^{\neq}$ is a radical. Assuming that the barrier to internal rotation in C_2H_6 is the same in $C_2H_7^{\neq}$, this makes $\Delta S_{85}^{\neq} = 128.5$ J mol^{-1} K^{-1}. In terms of athermochemical formulation of the transition state theory A_{85} is given by:

$$A_{85} = e^2 \frac{kT}{h} R'T \, e^{\Delta S\ddagger/R} M^{-1} \text{sec}^{-1}$$

where the symbols have their usual meaning and $R' = 0.082 \ell$ atm mol^{-1} K^{-1}. Substitution into this expression for $\Delta S_{85}^{\ddagger}$ etc., makes $A_{85} = 10^{8.4}$ M^{-1} sec^{-1} at 300 K.[196] The A factors for the forward (Reaction 85) and reverse (Reaction 86) steps

$$H + C_2H_6 \rightarrow CH_4 + Me \tag{86}$$

are related by the expression in $A_{85}/A_{86} = \Delta S^{\circ}_{55}/R$. A value for ΔS°_{85} of -36.0 J mol^{-1} K^{-1}[7] make $A_{86} = 10^{10.3}$ M^{-1} sec^{-1}. This may be compared with the value for the hydrogen atom abstraction process (Reaction 87)

$$H + C_2H_6 \rightarrow H_2 + Et \tag{87}$$

value $A_{87} = 10^{11.1}$ M^{-1} sec^{-1}.[197] Taking the upper limits for k_{85} determined by Back[198] from 802 to 983 K make E_{85} range from 150 to 175 kJ mol^{-1}. Using the higher value makes a chain contribution via Reaction 85 negligible in the decomposition of methane at 1000 K. Thus to date no evidence exists for the participation of Reaction 85 in any system.

A. C=C

One problem associated with such studies is the measuring of the rate of addition to π-bond systems in the absence of direct monitoring of the adduct. Cvetanovic and Irwin[199] used a mass balance technique which was originally used in solution. This involves the photolysis of biacetyl and a mass balance between the methyl radicals and the carbon monoxide yields, irrespective of the decomposition of the acetyl radical. In the presence of excess isobutane and small amounts of olefin or other unsaturated compound,

$$(MeCO)_2 + h\nu \rightarrow 2MeCO \tag{88}$$

$$MeCO + M \rightarrow Me + CO + M \tag{89}$$

it is assumed that either addition to olefin or abstraction from isobutane (*t*-BuH) takes place. The difference between methane and carbon monoxide yields

$$Me + t\text{-BuH} \rightarrow CH_4 + t\text{-Bu} \tag{90}$$

clearly gives the rate of the addition process. Hydrogen atom abstraction by methyl from the olefin may usually be neglected. This method has also been used by Holt and Kerr[200] and Sangster and Thynne.[201] However, Tedder and co-workers[202,203] used azomethane as a source of methyl radicals and excess methyl iodide which acted as a radical trap for the methyl adduct radical (MeE).

Table 26
THE ADDITION OF METHYL RADICALS TO OLEFINS

	$\log(A/M^{-1}\,sec^{-1})$	$E/kJ\,mol^{-1}$	Temp. range (K)	Ref.
Ethylenes				
C_2H_4	8.32 ± 0.5	30.5 ± 4.2	350 — 500	200
$CH_2^a = CHF$	8.20 ± 1.0	32.0 ± 1.3		203
$CHF^a = CH_2$	7.61 ± 0.2	33.3 ± 1.3		203
$CF_2^a = CH_2$	7.18 ± 0.2	40.0 ± 2.5		203
$CHF^a = CF_2$	7.95 ± 0.3	27.5 ± 1.7		203
$CF_2^a = CHF$	6.56 ± 0.3	13.5 ± 2.9		203
C_2F_4	8.38 ± 0.5	22.9 ± 1.1		203
C_2F_4	8.05 ± 0.23	21.8 ± 1.7	353 — 453	201, 203

Mean 8.02 ± 0.3

Other olefins				
$CH_2{=}CH_3Me$	7.9	39.7		202
	8.22	31.0	353 — 453	203
l-Butene	8.01	30.1	353 — 453	203
cis 2-Butane	7.65	30.5	353 — 453	203
trans-2-Butane	8.15	33.9	353 — 453	203
$CF_3CF = CF_2$	8.14	23.9	354 — 476	204
Allene	8.3	33.9	373 — 483	205
1,3-Butadene	7.91	17.2	353 — 533	3

Mean A factor 8.04 ± 0.21

| Benzene | 8.97 ± 0.5 | 31.8 ± 4.2 | 372 — 484 | 200 |

[a]   Addition site.

$$Me + E \rightarrow MeE \tag{91}$$

$$MeE + MeI \rightarrow MeEI + Me \tag{92}$$

In addition, a mixture of ethene + substituted ethane was used so that it also became a competitive technique. Data up to 1972 has been surveyed by Kerr and Parsonage.[3] The relative values of Tedder and co-workers have been normalized relative to the results of Holt and Kerr[200] for addition to ethylene shown in Table 26. The mean A factor for these addition processes is $10^{8.0}\,M^{-1}\,sec^{-1}$ in keeping with thermochemical kinetic estimates.[7] The value for $CF_2^* = CHF$ however, is much lower than this and, therefore, must be in error. As already mentioned, these A factors may be expected to have a large temperature coefficient.

Holt and Kerr studied the addition of methyl radicals to benzene by the biacetyl mass balance method. The Arrhenius parameters are given in Table 26.

No other data exist for the addition of methyl radicals to C=C aromatic systems but they do show some importance as a displacement process. An example would be the addition of a hydrogen atom to toluene where

$$\tag{93}$$

Table 27
ARRHENIUS PARAMETERS FOR THE
ADDITION OF METHYL RADICALS TO
KETONES

	$\log(A/M^{-1}\ sec^{-1})$	$E/kJ\ mol^{-1}$	Ref.
(MeCO)$_2$	7.3	26.4	207
(CD$_3$CO)$_2$	7.1 ± 0.6	27.6 ± 4.2	208
CD$_3$COCD$_3$	7.5 ± 0.4	48.1 ± 4.6	209
CF$_3$COCF$_3$	8.0 ± 0.4	21.3 ± 2.9	194
CF$_3$COCF$_3$	7.6 ± 0.1	19.0 ± 1.3	210
Me$_2$CO	7.4 ± 0.5	48.1 ± 4.2	

Note: Mean A factor $10^{7.5\ \pm\ 0.3}\ M^{-1}\ sec^{-1}$.

[a] Calculated from $\Delta H°$, $\Delta S°$, and the Arrhenius parameters for the reverse reaction.[2,11]

preferential dissociation of the adduct results in benzene formation.[206]

B. C=O

Interest in these laboratories on this type of reaction was in connection with our studies on the decomposition of alkoxy radicals.[212] It seemed worthwhile to consider the reverse process, e.g.,

$$Me\ +\ Me_2CO \rightarrow t\text{-BuO} \tag{94}$$

Clearly practical difficulties arise when one attempts to determine the rate constant k_{94}. Dainton, Ivin, and Wilkinson[213] studied the reaction using $^{14}CH_3$ radicals generated from the photolysis of the appropriately labeled acetone at 623 K. Varnerin[214] carried out a similar study at 823 K, using CD_3 radicals generated from the pyrolysis of deuterated acetaldehyde. Other studies have involved biacetyl and hexafluoroacetone (Table 27). Taking the latter reaction as an example;

$$Me\ +\ CF_3COCF_3 \rightarrow CF_3COCH_3\ +\ CF_3 \tag{95}$$

it is necessary to consider the reaction in detail:

$$
\begin{array}{ccc}
 & Me & \\
 & | & \\
a & | & c \\
Me\ +\ CF_3COCF_3 \underset{b}{\overset{a}{\rightleftarrows}} CF_3-C-CF_3 \underset{d}{\overset{c}{\rightleftarrows}} CF_3COCH_3\ +\ CF_3 \\
 & | & \\
 & O &
\end{array}
$$

The symbols a, b, c, and d refer to the rate constants k_a, k_b, k_c, and k_d, respectively for the reactions involved. Ignoring the reaction involving the addition of trifluoromethyl radicals to trifluoroacetone, a steady-state analysis of the system, together with the reference Reaction 44 leads to the result.

$$R_{CF_3COCH_3}/R_{C_2H_6}^{1/2}\ =\ k_ak_c[CF_3COCF_3]/k_{44}^{1/2}(k_b\ +\ k_c) \tag{A}$$

Table 28
ARRHENIUS PARAMETERS FOR ADDITION TO C≡C
BONDED COMPOUNDS

	log A/M^{-1} sec^{-1}	E/kJ mol^{-1}	Temp. range (K)	Ref.
HC≡CH	8.79 ± 0.8	32.2 ± 6.3	379 — 487	200
	8.4	32.2	371 — 479	217
MeC≡CH	8.7	36.8	379 — 465	218

This expression simplifies for two extreme conditions. First, if $K_c \ll k_b$,

$$R_{CF_3COCH_3}/R_{C_2H_6}^{1/2} = Kk_c[CF_3COCF_3]/k_{44}^{1/2} \tag{B}$$

The equilibrium constant K is equal to k_a/k_b. Second, if $k_b \ll K_c$,

$$R_{CF_3COCF_3}/R_{C_2H_6}^{1/2} = k_a[CF_3COCF_3]/k_{44}^{1/2} = k_{95}[CF_3COCF_3]/k_{44}^{1/2} \tag{C}$$

There are thus three possible interpretations of the data namely A, B, and C. In fact, interpretation of the data precludes Conditions A and B.[194] The mean A factor is $10^{7.5 \pm 0.3}$ M^{-1} sec^{-1} which is half a power of ten lower than that for the addition of methyl radicals to olefins. Benson and Choo[215] have pointed out that a lower A factor is expected because the transition state has an extra torsional motion which is absent in the case of ketones (see also Reference 180). This view is supported by transition state theory calculations of Köhler and Knoll.[216] They used MINDO/3-UHF calculations to show that the activation entropies for addition to the carbon atom of >C=O are less than that for addition to the oxygen atom.

C. C≡C and C≡N

Holt and Kerr[200] used the previously mentioned mass balance method to determine the rate of addition of methyl radicals to acetylene. There is a large discrepancy between the values of Garcia-Dominguez and Trotman-Dickenson[217] as quoted by Kerr and Parsonage.[3] The difference shows up in the A-factor as a factor of two. This would be resolved if addition of the methyl radical to acetylene was followed by secondary additions. In this method

$$Me + HC≡CH \rightarrow MeCH=CH \tag{96}$$

it was assumed that the adduct radical reacted with acetaldehyde followed by

$$MeCH=CH_2 + MeCNO \rightarrow MeCH=CH_2 + MeCO \tag{97}$$

decomposition of the acetyl radical. Thus, product analysis

$$MeCO + M \rightarrow Me + CO + M \tag{98}$$

gave the rate of addition process (Reaction 96). Arrhenius parameters are then determined relative to that for Me combination (Reaction 44) (Table 28). At this time the reason for the descrepancy is unresolved. A similar method was used for addition of methyl radicals to propyne.[218] Product analysis showed that addition occurred predominantly at the terminal C atom in the ratio $K_{99a}/k_{99b} = 8.3$.

$$Me + MeC≡CH \Big\langle {{MeC=CHMe} \atop {Me_2C=CH}}$$

(99a)

(99b)

Although the data is meager, it would appear that the A factors are the same as that for the addition of methyl radicals to benzene and higher than that for addition to olefins by a factor of $\sim$5.

No information is available for addition or displacement from C≡N triple bonds:

$$\text{Me} + \text{R} - \text{CN} \rightleftarrows \text{R}\overset{\displaystyle\overset{\text{Me}}{|}}{\text{C}}=\text{N} \rightarrow \text{MeCN} + \text{R} \tag{100}$$

It might be expected that such a reaction might occur where alkyl radicals are generated in the presence of ClCN. In fact only the chlorine atom exchange process (Reaction 101) was observed. Here R is CF_3[219] or C_2H_5.[220]

$$\text{R} + \text{ClCN} \rightarrow \text{RCl} + \text{CN} \tag{101}$$

For R $=$ CF_3, ΔH°_{101} is $+60$ kJ mol^{-1} whereas for the displacement process (Reaction 102) is $\sim$75kJ mol^{-1} exothermic.

$$\text{CF}_3 + \text{ClCN} \rightleftarrows \text{Cl}\overset{\displaystyle\overset{\text{CF}_3}{|}}{\text{C}}=\text{N} \rightarrow \text{CF}_3\text{CN} + \text{Cl} \tag{102}$$

It must mean that the addition process has a high activation energy.

ACKNOWLEDGMENT

The author is indebted to many people who sent preprints or reprints of their work, in particular, N. L. Arthur and D. Gutman.

REFERENCES

1. **Kerr, J. A. and Trotman-Dickenson, A. F.,** Reactions of alkyl radicals, in, *Progress in Reaction Kinetics* Vol. 1, Porter G., Ed., Pergamon, Oxford, 1961, 105.
2. **Gray, P., Herod, A. A., and Jones, A.,** Kinetic data for hydrogen and deuterium atom abstraction by methyl and trifluoromethyl radicals in the gaseous phase, *Chem, Rev.,* 71, 247, 1971.
3. **Kerr, J. A. and Parsonage, M. J.,** *Evaluated Kinetic Data on Gas Phase Addition Reactions,* Butterworths, London, 1972.
4. **Waage, E. V. and Rabinovitch, B. S.,** Some aspects of theory and experiment in the ethane-methyl radical system, *Int, J, Chem. Kinet.,* 3, 105, 1971.
5. **Cox, R. A., Derwent, R. G., Holt, P. M., and Kerr, J. A.,** Photo-oxidation of methane in the presence of nitric oxide and nitrogen dioxide, *J. Chem. Soc. Faraday Trans. I,* 72, 2044, 1976.
6. **Baldwin, R. R. Hopkins, D. E., Norris, A. C., and Walker, R. W.,** The addition of methane to slowly reacting hydrogen-oxygen mixtures: reactions of methyl radicals, *Combust. Flame,* 15, 33, 1970.
7. **Benson, S. W.,** *Thermochemical Kinetics,* 2nd ed., McGraw-Hill, New York, 1976.
8. **Stull, D. R., Prophet, H., et al.,** *Janaf Thermochemical Tables,* 2nd ed., NSRDS-NBS37, National Bureau of Standards, Washington D. C., 1971.
9. **Golden, D. M. and Benson, S. W.,** Free radical and molecule thermochemistry from studies of gas phase iodine atom reactions, *Chem. Rev.,* 69, 125, 1969.
10. **McCulloch, K. E. and Dibeler, V. H.,** Enthalphy of formation of methyl and methylene radicals from photoionisation studies of methane and ketene, *J. Chem. Phys.,* 64, 4445, 1976.
11. **Traeger, J. C. and McLoughlin, R. G.,** Absolute heats of formation for gas phase cations, *J. Am. Chem Soc.,* 103, 3647, 1981.

12. **Baghai-Vayjooee, M. H., Colussi, A. J., and Benson, S. W.,** Very low pressure reactor. A new technique for measuring rates and equilibria of radical-molecule reactions at low temperature. Heat of formation of the methyl radical, *J. Am. Chem. Soc.*, 100, 3214, 1978; *Int. J. Chem. Kinet.*, 11, 147, 1979.

13. **Golden, D. M., Spokes, G. N., and Benson, S. W.,** Very low pressure pyrolysis (VLPP); a versatile kinetic tool, *Angew. Chem.*, 12, 534, 1973.

14. **Paley, P. D. and Wimalasena, J. H.,** Kinetics and thermochemistry of the ethyl radical. The induction period in the pyrolysis of ethane, *J. Phys. Chem.*, 88, 5657, 1984.

15. **Whittle, E.,** private communication.

16. **Cheng, J-T, Lee, Y-S, and Yeh, C-T,** The ^3Hg photosensitised decomposition of C_2H_6 at high intensity, *J. Phys. Chem.*, 81, 687, 1977.

17. **Pilling M. J. and Robertson J. A.,** A rate constant for $CH_2(^3B_1)$ + Me, *Chem. Phys. Lett.*, 33, 336, 1975; Patrick, R., Pilling, M. J., and Rogers, G. J., A high pressure rate constant for methyl radical + atomic hydrogen and an analysis of the kinetics of the methyl radical + atomic hydrogen → methane reaction. *Chem. Phys.*, 53, 279, 1980; Pilling, M. J., private communication.

18. **Pilling, M. J., Robertson, J. A., and Rogers, G. J.,** The flash photolysis of azomethane. Minor products from the photolysis of methyl radicals and a rate constant for Me + NO, *Int. J. Chem. Kinet.*, 8, 883, 1976.

19. **Duchovic, R. J. and Hase, W. L.,** Sensitivity of the H + Me recombination rate constant to the shape of the C–H stretching potential, *Chem. Phys. Lett.*, 110, 474, 1984.

20. **Pratt, G. L. and Wood, S. W.,** Kinetics of the reaction of methyl radicals with oxygen, *J. Chem. Soc. Faraday Trans. 1*, 80, 3419, 1984.

21. **Bauschlicher, C. W., Jr.,** Barrier height for the abstraction reaction of methylene with hydrogen $CH_2(^3B_1)$ + H_2 → Me + H, *Chem. Phys. Lett.*, 56, 31, 1978.

22. **Tan, L. P.,** A partial open shell FSGO calculation on the triplet methylene abstraction reaction with the hydrogen molecule $^3CH_2$ + H_2 → Me + H, *Chem. Phys. Lett.*, 57, 239, (1978).

23. **Just, T.,** *Shock Waves in Chemistry,* Nifshitz, A., Ed., Marcel Dekker, New York, 1981, 279.

24. **Niki, H. and Morris, E. D., Jr.,** Reaction of methyl radicals with atomic oxygen, *Int. J. Chem. Kinet.*, 5, 47, 1972.

25. **Slagle, I. R., Prass, F. J., Jr., and Gutman D.,** Kinetics into the steady state. I. Study of the reaction of oxygen atoms with methyl radicals, *Int. J. Chem. Kinet.*, 6, 111, 1974.

26. **Washida, N. and Bayes, K. D.,** The reaction of methyl radicals with atomic and molecular oxygen, *Int. J. Chem. Kinet.*, 8, 777, 1976.

27. **Washida, N.,** Reaction of methyl radicals with $O(^3P)$, O_2 and NO, *J. Chem. Phys.*, 73, 1665, 1980.

28. **Batt, L., Burrows, J. P., and Robinson, G. N.,** On the isomerisation of methoxy radical: Relevance to atmospheric chemistry and combustion. *Chem Phys. Lett.*, 78, 467, 1981.

29. **Becker, K. J. and Klay, D.,** The formation of CH radicals in hydrocarbon atom flames, *Chem. Phys. Lett.*, 4, 62, 1969.

30. **Benson, S. W. and Shum, L. G. S.,** Review of the heat of formation of hydroperoxyl radical, *J. Phys. Chem.*, 87, 3479, 1983.

31. **Benson, S. W. and Shaw, R.,** Thermochemistry of organic peroxides, hydroperoxides, polyoxides, and their radicals, in, *Organic Peroxides* Vol. 1, Severa, D., Ed., Wiley-Interscience, New York, 1970, 106.

32. **Baker, G., Littlefair, J. H., Shaw, R., and Thynne, J. C. J.,** The heats of formation of dimethyl diethyl and di-t-butyl peroxide, *J. Chem. Soc.*, 1, 6970, 1965.

33. **Kondo, O. and Benson, S. W.,** Kinetics and equilibria in the system BR + $CH_3OOH \leftrightarrows HBr + CH_3OO$.

34. **Khachatryan, L.A., Niazyan, O.M., Mantashyan, A.A., Vedeneev, V.I., and Teitel'boim, M.A.,** Experimental determination of the equilibrium constant of the reaction Me + $O_2 \rightleftarrows MeO_2$ during the gas-phase oxidation of methane, *Int. J. Chem. Kinet.*, 14, 1231, 1982.

35. **Batt, L.,** The reactions of alkoxy and alkyl peroxy radical, *Int. Rev. Phys. Chem.*, 1986.

36. **Gutman, D. and Slagle, I. R.,** 1986, to be published.

37. **Hsu, D. S. Y., Shaub, W. M., Creaner, T., Gutman, D., and Lin, M. C.,** Kinetic modeling of CO production from the reaction of Me with O_2 in shock waves, *Ber. Bunsenges. Phys. Chem.*, 87, 909, 1983.

38. **Basco, N., James, D. G. L., and James, F. C.,** A quantitative study of alkyl radical reactions by kinetic spectroscopy, II. Combination of methyl radical with the oxygen molecule, *Int. J. Chem. Kinet.*, 4, 129, 1972; *Chem. Phys. Lett.*, 8, 265, 1971.

39. **Parkes, D. A.,** The oxidation of methyl radicals at room temperature, *Int. J. Chem. Kinet.*, 9, 451, 1977.

40. **Hochanadel, C. J., Ghormley, J. A., Boyle, J. W., and Ogren, P. J.,** Absorption spectrum and rates of formation and decay of the CH_3O_3 radical, *J. Phys. Chem.*, 81, 3, 1977.

41. **Washida, N. and Bayes, K. D.,** The reactions of methyl radicals with atomic and molecular oxygen, *Int. J. Chem. Kinet.*, 8, 777, 1976.

42. **Washida, N.,** Reactions of methyl radicals with $O(^3P)$, O_2 and NO, *J. Chem. Phys.*, 1665, 73, 1980.

43. **Selzer, E. A. and Bayes, K. D.,** Pressure dependence of the rate of reaction of methyl radicals with O_2, *J. Phys. Chem.*, 87, 392, 1983.

44. **Plumb, I. C. and Ryan, K. R.,** Kinetics of the reaction of Me with $O(^3P)$ and O_2 at 295 K, *Int. J. Chem. Kinet.*, 14, 861, 1982.

45. **Oref, I. and Rabinovitch, B. S.,** The experimental evaluation of k_∞ in unimolecular reaction systems, *J. Phys. Chem.*, 72, 4488 (1968).

46. **Baulch D. L., Cox R. A., Hamson R. F., Jr., Kerr, J. A., Troe J. and Watson R.,** Evaluated kinetic and photochemical data for atmospheric chemistry, *J. Phys. Chem. Ref. Data*, 9, 295, 1980.

47. **Clark, T. C., Izod, T. P. J., and Matsuda, S.,** Oxidation of methyl radicals studied in reflected shock waves using the time-of-flight mass spectrometer, *J. Chem. Phys.*, 55, 4644, 1971.

48. **Klais, O., Anderson, P. C., Laufer, A. H., and Kumylo, M. S.,** An upper limit for the rate constant of the bimolecular reaction Me $+ O_2 \rightarrow H_2CO + OH$ at 368 K, *Chem. Phys. Lett.* 66, 598, 1979.

49. **Baldwin, A. C. and Golden, D. M.,** Reactions of methyl radicals of importance in combustion systems, *Chem. Phys. Lett.*, 55, 350, 1978.

50. **Walker, R. W.,** A critical survey of rate constants for reactions in gas-phase hydrocarbon oxidation, in *Reaction Kinetics*, Vol. 1, Ashmore, P. G., Ed., Chemical Society of London, 1975, 161.

51. **Bhaskaran, K. A., Frank, P., and Just, Th.,** High temperature methyl radical reactions with atomic and molecular oxygen, *Proceedings 12th Internationl Symposium Shock Tubes and Waves*, Lifshitz, A. and Rom, J., Eds., The Magnes Press, Jerusalem, 1980, 503.

52. **Hinshelwood, C. N.,** *The Kinetics of Chemical Change*, Clarendon Press, Oxford, 1949.

53. **Gowenlock, B. G.,** *Progress in Reaction Kinetics*, Porter, G., Ed., Vol. 3, Pergamon, Oxford, 1966.

54. **Leathard, D. A. and Purnell, J. H.,** Paraffin pyrolysis, *Ann. Rev. Phys. Chem.*, 21, 197, 1970.

55. **Batt, L.,** *The Chemistry of Functional Groups, Supplement F*, Patai, S., Ed., Wiley, Chichester, 1982, 417.

56. **Batt, L. and Milne, R. T.,** The Me–NO bond dissociation energy, *Int. J. Chem. Kinet.*, 5, 1967, 1973.

57. **Basco, N., James, D. G. L., and Suart, R. D.,** A quantitative study of alkyl radical reactions by kinetic spectroscopy. I. Mutual combination of methyl radicals and combination of methyl radicals with nitric oxide, *Int. J. Chem. Kinet.*, 2, 215, 1970.

58. **Van den Bergh, H. E. and Callear, A. B.,** Spectroscopic measurements of the rate of the gas phase combination of methyl radicals with nitric oxide and oxygen at 295 K, *Trans. Faraday Soc.*, 67, 2017, 1971.

59. **Laufer, A. H. and Bass, A. M.,** Rate constants of the combination of methyl radicals with nitric oxide and oxygen, *Int. J. Chem. Kinet.*, 7, 639, 1975.

60. **Pilling, M. J., Robertson, J. A., and Rogers, G. J.,** The flash photolysis of azomethane, minor products from the photolysis of methyl radicals and a rate constant for Me $+$ NO, *Int. J. Chem. Kinet.*, 8, 883, 1976.

61. **Gorin, E.,** Photolysis of acetaldehyde in the presence of iodine, *Acta Physiochem. U.S.S.R.*, 9, 681, 1938.

62. **Yang, K. and Ree, T.,** Collision and activated complex theories for bimolecular reactions, *J. Chem. Phys.*, 35, 588, 1961.

63. **Pratt, G. and Veltman, I.,** Kinetics of addition of methyl and ethyl radicals to nitric oxide, *J. Chem. Soc. Trans. Faraday I*, 72, 2477, 1976.

64. **Cox, J. D. and Pilcher, G.,** *Thermochemistry of Organic and Organometallic Compounds*, Academic Press, London, 1970.

65. **Glanzer, K. and Troe, J.,** Thermal decomposition of nitro compounds: 1. Dissociation of nitromethane, *Helv. Chim. Acta*, 55, 2884, 1972.

66. **Phillips, L. and Shaw, R.,** Reactions of methyl and methoxy radicals with nitrogen dioxide and nitric oxide, *10th International Symposium on Combustion*, 1965, 453.

67. **Yamada, F., Slagle, I. R., and Gutman, D.,** Kinetics of the reactions of methyl radicals with nitrogen dioxide, *Chem. Phys. Lett.*, 83, 409, 1981.

68. **Rossi, M., Barker, J. R., and Golden, D. M.,** Infrared multiphoton dissociation yields via a versatile new technique. Intensity, fluence, and wavelength dependence for CF_3I, *Chem. Phys. Lett.*, 65, 523, 1979.

69. **Audley, G. J., Baulch, D. L., and Campbell, I. M.,** Formation of organic nitro-compounds in flowing $H_2O_2 + NO_2 + N_2 +$ organic vapour systems, *J. Chem. Soc. Faraday Trans. I*, 80, 599, 609, 1984.

70. **Watkins, K. W. and Ward, W. W.,** Addition of methyl radicals to carbon monoxide: chemically and thermally activated decomposition of acetyl radicals, *Int. J. Chem. Kinet.*, 6, 855, 1974.

71. **Parkes, D. A.,** The ultraviolet absorption spectrum of the acetyl radical and the kinetics of the Me $+$ CO reaction at room temperature, *Chem. Phys. Lett.*, 77, 527, 1981.

72. **Anastasi, C. and Maw, P. L.,** Reaction kinetics in acetyl chemistry over a wide range of temperature and pressure, *J. Chem. Soc. Faraday Trans. I*, 78, 2423, 1982.

73. **Szirovicza, L. and Walsh, R.,** Gas phase addition of HI to ketene and the kinetics of the decomposition of the acetyl radical, *J. Chem. Soc. Faraday Trans. I*, 70, 33, 1974.

74. **Lissi, E. A., Massiff, G., and Villa A. E.,** Addition of methoxy radicals to olefins, *Int. J. Chem. Kinet.*, 7, 625, 1975.

75. **James, F. C., Kerr, J. A., and Simons, J. P.,** Direct measurement of the rate of reaction of the methyl radical with sulphur dioxide, *J. Chem. Soc. Faraday Trans. 1,* 69, 2124, 1973.

76. **Simonaitis, R. and Heicklen, J.,** Reactions of Me, MeO and MeO$_2$ and O$_3$, *J. Phys. Chem.,* 79, 298, 1975.

77. **Washida, N., Akimoto, H., and Okuda, M.,** Reaction of methyl radicals with ozone, *J. Chem. Phys.,* 73, 1673, 1980.

78. **Falconer, J. W., Hoare, D. E., and Overend, R.,** The oxidation of methane by nitrous oxide, *Combust. Flame,* 21, 339, 1973.

79. **Borisov, A. A., Zamanskii, V. M., and Patmischel, K.,** The mechanism of methane oxidation with nitrous oxide, *Kinet. Katal.,* 19, 307, 1977.

80. **Glanzer, K., Quack, M., and Troe, J.,** A spectroscopic determination of the methyl radical recombination rate constant in shock waves, *Chem. Phys. Lett.,* 39, 304, 1976.

81. **Baulch, D. L. and Duxbury, J.,** Ethane decomposition and the reference rate constant for methyl radical recombination, *Combust. Flame,* 37, 313, 1980.

82. **Bass, A. M. and Laufer, A. H.,** The methyl radical combination rate constant as determined by kinetic spectroscopy, *Ber. Bunsenge Phys. Chem.,* 78, 198, 1974.

83. **Macpherson, M. T., Pilling, M. J., and Smith, M. J. C.,** The pressure and temperature dependence of the rate constant for methyl radical combination over the temperature range 296 - 577 K, *Chem. Phys. Lett.,* 94, 430, 1983.

84. **Trenwith, A. B.,** Re-examination of the thermal dissociation of ethane, *J. Chem. Soc. Faraday Trans. 1,* 75, 615, 1979.

85. **Clark, T. C., Izod, T. P. J., and Kistiakowsky, G. B.,** Reactions of methyl radicals produced by the pyrolysis of azomethane or ethane in reflected shock waves, *J. Chem. Phys.,* 54, 1295, 1971.

86. **Gomer, R. and Kistiakowsky, G. B.,** The rate constant of ethane formation, *J. Chem. Phys.,* 19, 85, 1951.

87. **Van den Bergh, H. E., Callear, A. B., and Norstrom, R. J.,** An experimental determination of the oscillator strength of the 2160 Å band of the free methyl radical and a spectroscopic measurement of the combination rate, *Chem. Phys. Lett.,* 4, 101, 1969.

88. **James, F. C., and Simons, J. P.,** Yet another direct measurement of the rate constant for the recombination of methyl radicals, *Int. J. Chem. Kinet.,* 6, 887, 1974.

89. **Truby, F. K. and Rice, S. K.,** Methyl radical association studied by time-resolved mass spectroscopy, *Int. J. Chem. Kinet.,* 5, 721, 1973.

90. **Parkes, D. A., Paul, D. M., and Quinn, C. P.,** Study of the spectra and recombination kinetics of alkyl radicals by molecular modulation spectroscopy. I. The spectrometer and a study of methyl recombination between 250 and 450 K and perdentero methyl recombination at room temperature, *J. Chem. Soc. Faraday Trans. 1,* 72, 1935, 1976.

91. **Adachi, H., Basco, N., and James, D. G. L.,** Mutual interactions of the methyl and methylperoxy radical studied by flash photolysis and kinetic spectroscopy, *Int. J. Chem. Kinet.,* 12, 949, 1980.

92. **Bureat, A., Skinner, G. B., Crossley, R. W., and Scheller, K.,** High temperature decomposition of ethane, *Int. J. Chem. Kinet.,* 5, 345, 1973.

93. **Quinn, C. P.,** The thermal dissociation and pyrolysis of ethane, *Proc. R. Soc. London,* A275, 190, 1963.

94. **Clark, J. A. and Quinn, C. P.,** Kinetic isotope effect in the thermal dissociation of ethane, *J. Chem. Soc. Faraday Trans. 1,* 72, 706, 1976.

95. **Skinner, G. B., Rogers, D., and Patel, K. B.,** Consistency of theory and experiment in the ethane-methyl radical system, *Int. J. Chem. Kinet.,* 13, 481, 1981.

96. **Quack, M. and Troe, J.,** Specific rate constants of unimolecular processes. II. Adiabatic channel model, *Ber. Bunsenges. Phys. Chem.,* 78, 240, 1970.

97. **Roth, P. and Just, T.,** Measurements of the high temperature pyrolysis of ethane, *Ber. Bunsenges. Phys. Chem.,* 83, 577, 1979.

98. **Crowcock, F. B., Hase, W. L., and Simons, J. W.,** Kinetics of chemically activated ethane, *Int. J. Chem. Kinet.,* 5, 77, 1973.

99. **Dunlop, A. N., Kominer, R. J., and Price, S. J. W.,** Hydrogen abstraction from toluene by methyl radicals and the pressure dependence of the recombination of methyl radicals, *Can. J. Chem.,* 48, 1269, 1970.

100. **Troe, J.,** Theory of thermal unimolecular reactions in the fall-off range. I. Strong collision rate constants, *Ber. Bunsenges. Phys. Chem.,* 87, 161, 1983.

101. **Hase, W. L.,** Theoretical critical configuration for ethane decomposition and methyl radical recombination, *J. Chem. Phys.,* 57, 730, 1972.

102. **Benson, S. W.,** Molecular models for recombination and disproportionation of radicals, *Can. J. Chem.,* 61, 881, 1983.

103. **Kabrinsky, P. C., Pritchard, G. O., and Toby S.,** Pressure dependence of the cross-combination ratio for CF$_3$ and Me radicals, *J. Phys. Chem.,* 75, 2225, 1971.

104. **Chupka, W. A.,** Mass-spectrometric study of the photoinisation of methane, *J. Chem. Phys.,* 48, 2337, 1968.

105. **Grotewold, J., Lissi, E. A., and Neumann, M. G.,** Pressure dependence of the methyl radical combination reaction, *J. Chem. Soc. A,* 375, 1968.

106. **Nicholson, A. J. C.,** The photochemical decomposition of the aliphatic methyl ketones, *Trans. Faraday Soc.,* 50, 1067, 1954.

107. **Bradley, J. N.,** Mechanism of disproportionation in alkyl radical reactions, *J. Chem. Phys.,* 35, 748, 1961.

108. **Bradley, J. N. and Rabinovitch, B. S.,** On the mechanism of disproportionation of alkyl radicals, *J. Chem. Phys.,* 36, 3498, 1962.

109. **McNesby, J. R., Drew, C. M., and Gordon, A. S.,** Synthesis of n-butane-2-3-d$_4$ by the photolysis of $\alpha\alpha'$-diethyl ketone-d$_4$, *J. Phys. Chem.,* 59, 988, 1955.

110. **Benson, S. W.,** Some problems of structure and reactivity in free radical and molecule reactions in the gas phase, *Advances in Photochemistry,* Noyes, W. A., Jr., Hammond, G. S., and Pitts, J. N., Jr., Eds., Vol. 2, J. Wiley and Sons, New York, 1964.

111. **Rabinovitch, B. S. and Setser, D. W.,** Unimolecular decomposition and some isotope effects of simple alkanes and alkyl radicals, *Advances in Photochemistry,* Noyes, W. A., Jr., Hammond, G. S., and Pitts, J. N., Jr., Eds., Vol. 3, J. Wiley and Sons, New York, 1964.

112. **Kistiakowsky, G. B. and Roberts, E. K.,** Rate of association of methyl radicals, *J. Chem. Phys.,* 21, 1637, 1953.

113. **Kerr, J. A. and Moss, S. J.,** *CRC Handbook of Biomolecular and Termolecular gas reactions,* Vol. 2, CRC Press, Boca Raton, Fla., 1981.

114. **Gibson, M. J. and Corley, R. C.,** Organic radical - radical reactions, disproportionation vs. combination, *Chem. Rev.,* 73, 441, 1973.

115. **Pritchard, G. O. and Perona, M. J.,** The elimination of HF from vibrationally excited fluoroethanes. The decomposition of 1,1,1-trifluoroethane-d$_0$ and d$_3$, *Int. J. Chem. Kinet.,* 2, 281, 1970.

116. **Tabayashi, K. and Bauer, S. H.,** The early stages of pyrolysis and oxidation of methane, *Combust. Flame,* 34, 63, 1979.

117. **Just, T.,** Atomic resonance absorption spectrometry in shock tubes in *Shock Waves in Chemistry,* Lifshitz, A., Ed., Marcel Dekker, New York, 1981.

118. **Olson, D. B. and Gardiner, W. C., Jr.,** Combustion of methane in fuel-rich mixtures, *Combust. Flame,* 32, 151, 1978.

119. **Dees, K. and Setser, D. W.,** The photolysis of ketene - butane mixtures with and without added CO, *J. Phys. Chem.,* 75, 2240, 1971.

120. **Kim, K. C., Setser, D. W., and Holmes, B. E.,** Hydrogen fluoride and deuterium fluoride elimination reactions of chemically activated 1,1,1-trideuterio-2,2-difluoroethane, 1,1-difluoroethane and 1,1,1-trideuterio-2-fluoroethane, *J. Phys. Chem.,* 77, 725, 1973.

121. **Pilling, M. J. and Robertson, J. A.,** A rate constant for CH$_2$(^{3}B$_1$) + CH$_3 \rightarrow$ C$_2$H$_5^* \rightarrow$ C$_2$H$_4$ + H, *Chem. Phys. Lett.,* 33, 336, 1975.

122. **Walker, R. W.,** Activation energies of the reversible reaction between hydrogen atoms and methane to give hydrogen and methyl radicals, *J. Chem. Soc. A,* 2391, 1968.

123. **Benson, S. W.,** *The Foundations of Chemical Kinetics,* McGraw-Hill, New York, 1960.

124. **Alfassi, Z. B. and Benson, S. W.,** A simple empirical method for the estimation of activation energies in radical molecule methatesis reactions, *Int. J. Chem. Kinet.,* 5, 879, 1973.

125. **Arthur, N. L., Donchi, K. F., and McDonell, J. A.,** BEBO calculations. IV. Arrhenius parameters and kinetic isotope effects for the reactions of CH$_3$ and CF$_3$ radicals with H$_2$ and D$_2$, *J. Chem. Soc. Faraday Trans. I,* 71, 2431, 1975.

126. **Kerr, J. A. and Parsonage, M. S.,** Evaluated kinetic data on gas phase transfer reactions of methyl radicals, Butterworths, London, 1976.

127. **Marshall, R. M. and Shahkar, G.,** Rate parameters for CH$_3$ + H$_2$ › CH$_3$ + H in the temperature range 584 — 671 K, *J. Chem. Soc. Faraday Trans. I,* 77, 2271, 1981.

128. **Sepehrad, A., Marshall, R. M., and Purnell, J. H.,** Reaction between hydrogen atoms and methane, *J. Chem. Soc. Trans. Faraday I,* 75, 835, 1979.

129. **Pratt, G. and Rogers, D.,** Homogenous isotope exchange reactions II. CH$_4$/D$_2$, *J. Chem. Soc. Trans. Faraday I,* 72, 2769, 1976.

130. **Clark, T. C. and Dove, J. E.,** The rate coefficients for CH$_3$ + H$_2 \rightarrow$ CH$_4$ + H measured in reflected shock waves; a non-Arrhenius reaction, *Can. J. Chem.,* 51, 2155, 1973.

131. **Kabrinsky, P. C. and Pacey, P. D.,** The reaction of methyl radicals with molecular hydrogen, *Can. J. Chem.,* 52, 3665, 1974.

132. **Johnston, H. S.,** *Gas Phase Reaction Rate Theory,* Ronald Press, New York, 1966.

133. **Hirschfelder, S. O., Curtiss, C. F., and Bird, R. B.,** *Molecular Theory of Gases and Liquids,* Wiley, New York, 1954.

134. **Nikisha, L. V., Polyak, S. S., and Sakolova, N. A.,** Rate constant of the reaction of methyl radicals with difluorochloromethane, *Kinet. Katal.,* 17, 1074, 1976.

135. **Pacey, P. D. and Purnell, J. H.,** Arrhenius parameters of the reaction, $CH_3 + C_2H_6 \rightarrow CH_4 + C_2H_5$, *J. Chem. Soc. Trans. Faraday I,* 68, 1462, 1972.

136. **McNesby, J. R. and Gordon, A. S.,** The photolysis and pryolysis of acetone -d_6 in the presence of ethane and of acetone in the presence of ethane -d_6, *J. Am. Chem. Soc.,* 77, 4719, 1955.

137. **Camilleri, P., Marshall, R. M., and Purnell, J. H.,** Arrhenius parameters for the unimolecular decompositions of azomethane and n-propyl and isopropyl radicals and for methyl radical attack on propane, *J. Chem. Soc. Trans. Faraday I,* 71, 1491, 1975.

138. **Pacey, P. D. and Purnell, J. H.,** Arrhenius parameters for the reactions $CH_3 + C_4H_{10} \rightarrow CH_4 + C_4H_9$ and $C_2H_5 + C_4H_{10} \rightarrow C_2H_6 + C_4H_9$, *Int. J. Chem. Kinet.,* 4, 657, 1972.

139. **Yampol'skii Yu, P. and Nametkin, N. S.,** Rate constants of reactions of CH_3, C_2H_5, and atomic hydrogen with butane at high temperatures, *Kinet. Katal.,* 17, 57, 1976.

140. **Anastasi, C.,** Study of the methyl - isobutane reaction in the range 478 < T/K < 560, *J. Chem. Soc. Trans. Faraday I,* 79, 741, 1983.

141. **Pacey, P. D.,** The reactions of methyl radicals with neopentane, *Can. J. Chem.,* 51, 2415, 1973.

142. **Dunlop, A. N., Kominar, R. J., and Price, S. J. W.,** Hydrogen abstraction from toluene by methyl radicals and the pressure dependence of the recombination of methyl radicals, *Can. J. Chem.,* 48, 1269, 1970.

143. **Manthorne, K. C. and Pacey, P. D.,** The reaction of methyl radicals with formaldehyde in dimethyl ether prolysis, *Can. J. Chem.,* 56, 1307, 1978.

144. **Anastasi, C.,** Reaction of methyl radicals with formaldehyde in the range 500 < T/K < 603, *J. Chem. Soc. Faraday Trans. I,* 79, 749, 1983.

145. **Bardi, I., Berces, T., Marta, F., and Szilagyi, I.,** Reactions of methyl radicals with acetaldehyde and acetaldehyde-d_1, *Int. J. Chem. Kinet.,* 8, 285, 295, 1976.

146. **Kyle, E. and Orchard, S. W.,** The photolysis of methyl glyoxal vapour at 436 nm, *J. Photochem.,* 7, 305, 1977.

147. **Arthur, N. L. and Newitt, P. J.,** An evaluation of the kinetic data for the hydrogen transfer reactions: $CH_3 + CH_3COCH_3$, $CD_3 + CD_3COCD_3$ and $CH_3 + CH_3NNCH_3$, to be published, 1986.

148. **Scherzer, K. and Wolf, R.,** Reaction of methyl radicals with 2,2,4,4-d_4-diethyl ketone, *Z. Phys. Chem. Leipzig,* 250, 97, 1972.

149. **Pacey, P. D.,** The initial stages of the pyrolysis of dimethyl ether, *Can. J. Chem.,* 53, 2742, 1975.

150. **Batt, L., Alvarado-Salinas, G., Reid, I. A. B., Robinson, C., and Smith, D. B.,** *19th International Symposium on Combustion,* The Combustion Institute, 1982, 81.

151. **Donovan, T. R., Dorko, W., and Harrison, A. G.,** Hydrogen abstraction from methyl formate by methyl radicals, *Can. J. Chem.,* 79, 828, 1970.

152. **Jones, S. H. and Whittle, E.,** Reactions of trifluoromethyl and methyl radicals with ethylene oxide, *Can. J. Chem.,* 48, 3601, 1970.

153. **Duke, M. G. and Holbrook, K. A.,** Reactions of methyl radicals with oxetanes, *J. Chem. Soc. Faraday Trans. I,* 76, 1232, 1980; 80, 3391, 1984.

154. **Ballod, A. P., Fedorova, T. V., Chikvaidze, N., Titarchuk, T. A., and Kumova, A. A.,** Rate constants for abstraction of hydrogen atoms from nitromethane by methyl radicals, *Kinet. Katal.,* 21, 1095, 1980.

155. **Ballod, A. P. and Titarchuk, T. A.,** Rate constants of cleavage of a hydrogen atom by a methyl radical from 2-nitropropane, *Kinet. Katal.,* 18, 1359, 1977.

156. **Arthur, N. L. and Bell, T. N.,** An evaluation of the kinetic data for hydrogen abstraction from silanes in the gas phase, *Rev. Chem. Intermed.,* 2, 37, 1978.

157. **Strausz, O. P., Jakubowski, E., Sandhu, H. S., and Gunning, H. E.,** Arrhenius parameters for the abstraction reactions of methyl radicals with silanes, *J. Chem. Phys.,* 51, 552, 1969.

158. **Berkeley, R. E., Safarik, I., Gunning, H. E. and Strausz, O. P.,** Arrhenius parameters for the reactions of methyl radicals with silane and methylsilanes, *J. Phys. Chem.,* 77, 1734, 1973.

159. **Morris, E. R. and Thynne, J. C. J.,** Hydrogen atom abstraction from silane, trimethyl silane and tetramethylsilane by methyl radicals, *J. Phys. Chem.,* 73, 3294, 1969.

160. **Kerr, J. A., Slater, D. H., and Young, J. C.,** Hydrogen - abstraction reactions for silanes. II. The reactions of methyl radicals with trifluorosilane, trimethylsilane, Dimethyl dichlorosilane and trimethyl chlorosilane, *J. Chem. Soc. A.,* P. 134, 1967.

161. **Kerr, J. A., Slater, D. H., and Young, J. C.,** Hydrogen abstraction reactions from silanes. I. The reaction of methyl radicals with trichlorosilane, methyl dichlorosilane and methyltrichlorosilane, *J. Chem. Soc. A.,* P. 104, 1966.

162. **Bell, T. N. and Platt, A. E.,** Reactions of deuterated radicals with methylfluorosilanes. Polar effects in hydrogen abstraction, *J. Phys. Chem.,* 75, 603, 1971.

163. **Arthur, N. L. and Newitt, P. S.,** submitted for publication, 1986.

164. **Arthur, N. L. and Lee, M.-S.,** Reactions of methyl radicals. I. Hydrogen abstraction from dimethyl sulphide, *Aust. J. Chem.,* 29, 1483, 1976.

165. **Jakubowski, E., Ahmed, M. G., Lown, E. M., Sandhu, H. S., Gosavi, R. K., and Strausz O. P.,** Sulphur atom abstraction from episulphides and carbonyl sulphide by methyl radicals, *J. Am. Chem. Soc.,* 94, 4094, 1971.

166. **Konar, R. S. and Darwent, B.DeB.,** The photolysis of azomethane in the presence of HN_3, *Can. J. Chem.,* 48, 2280, 1970.

167. **Arthur, N. L. and McDonell, J. A.,** BEBO calculations. II. Arrhenius parameters and kinetic isotope effects for the reactions of CH_3 and CF_3 radicals with NH_3, *J. Chem. Phys.,* 57, 3228, 1972.

168. **Arthur, N. L. and Arican, H.,** Reactions of methyl radicals. V. Hydrogen abstraction from hydrogen sulphide, *Aust. J. Chem.,* 36, 2195, 1983.

169. **Cvetanovic, R. S. and Steacie, E. W. R.,** Photolysis of acetone - hydrogen chloride mixtures, *Can. J. Chem.,* 31, 158, 1953.

170. **Levine, R. D. and Bernstein, R. B.,** *Molecular Reaction Dynamics,* Clarendon Press, Oxford, 1974.

171. **Ogg, R. A., Jr.,** Kinetics of the thermal reaction of gaseous alkyl iodides with hydrogen iodide, *J. Am. Chem. Soc.,* 56, 526, 1934.

172. **Batt, L. and Benson, S. W.,** unpublished work, University of Southern California, 1961.

173. **Hiatt, R. and Benson, S. W.,** A new method for measuring the ratios of rate constants for radical recombination, *J. Am. Chem. Soc.,* 94, 25, 1972.

174. **Logan, J. A., Mims, C. A., Stewart, G. W., and Ross, J.,** Molecular beam studied of methyl radical reactions with halogen molecules: product angular and velocity distributions., *J. Chem. Phys.,* 64, 1804, 1976.

175. **Kovalenko, L. J. and Leone, S. R.,** Laser studies of methyl radical reactions with Cl_2 and Br_2: absolute rate constants, product vibrational excitation and hot radical reactions, *J. Chem. Phys.,* 80, 3656, 1984.

176. **Moss, M. G., Hudgens, J. W., and McDonald, J. D.,** Infrared chemiluminescence investigation of the reactions of methyl radicals with oxygen and fluorine, *J. Chem. Phys.,* 72, 3486, 1980.

177. **Sullivan, J. H.,** The thermal reactions of hydrogen iodide with alkyl iodides, *J. Phys. Chem.,* 65, 722, 1961.

178. **Flowers, M. C. and Benson, S. W.,** Kinetics of the gas-phase reaction of MeI and HI, *J. Chem. Phys.,* 38, 882, 1963.

179. **Golden, D. M., Walsh, R., and Benson, S. W.,** The thermochemistry of the gas phase equilibrium I_2 + CH_4. MeI + HI and the heat of formation of the methyl radical, *J. Am. Chem. Soc.,* 87, 4053, 1965.

180. **Batt, L., McCulloch, R. D., and Milne, R. T.,** Thermochemical and kinetic studies of alkyl nitrites (RONO) - D(RO-NO), the reactions between RO and NO and the decomposition of RO, *Int. J. Chem. Kinet.* S1, 359, 1975.

181. **Hiatt, R. and Benson, S. W.,** Rate constants for radical recombination. IV. The activation energy for ethyl radical recombination, *J. Am. Chem. Soc.,* 94, 686, 1972.

182. **Hiatt, R. and Benson, S. W.,** Rate constants for alkyl radical recombination. II. The isopropyl radical, *Int. J. Chem. Kinet.,* 4, 151, 1972.

183. **Hiatt, R. and Benson, S. W.,** Rates of recombination of free radicals. V. The tert-butyl radical, *Int. J. Chem. Kinet.,* 5, 385, 1973.

184. **Hiatt, R. and Benson, S. W.,** Rate constants for radical recombination. III. The trifluoromethyl radical, *Int. J. Chem. Kinet.,* 4, 479, 1972.

185. **Parkes, D. A. and Quinn, C. P.,** Study of the spectra and recombination kinetics of alkyl radicals by molecular modulation spectrometry. II. The recombination of ethyl, isopropyl and t-butyl radicals at room temperature and t-butyl radicals between 250 and 450 K, *J. Chem. Soc. Faraday Trans. I,* 72, 1952, 1976.

186. **Golden, D. M., Piszkiewicz, L. W., Perona, M. J., and Beadle, P. C.,** An absolute measurement of the rate constant for isopropyl radical combination, *J. Am. Chem. Soc.,* 96, 1645, 1974.

187. **Tomkinson, D. M. and Pritchard, H. P.,** Abstraction of halogen atoms by methyl radicals, *J. Phys. Chem.,* 70, 1579, 1966.

188. **Tomkinson, D. M., Garvin, J.P., and Pritchard, H. P.,** Abstraction of chlorine atoms by methyl radicals, *J. Phys. Chem.,* 68, 541, 1964.

189. **Hautecloque, S.,** Reactions of methyl radicals with the molecules CCl_4, $CFCl_3$ and CF_2Cl_2, *J. Photochem.,* 7, 83, 1977.

190. **Macken, K. V. and Sidebottom, H. W.,** The reactions of methyl radicals with chloromethanes, *Int. J. Chem. Kinet.,* 11, 511, 1979.

191. **Marshall, R. M. and Rahman, L.,** New technique for investigation of the thermochemistry of free radicals; the n-propyl radical, *J. Chem. Soc.* Chem. Commun. 614, 1976.

192. **Marshall, R. M. and Rahman, L.,** Radical equilibrium studies, the thermodynamic parameters of n-propyl, *Int. J. Chem. Kinet.,* 9, 705, 1977.

193. **Oref, I. and Rabinovitch, B. S.,** Do highly excited polyatomic molecules behave ergodically?, *Acct. Chem. Res.,* 12, 166, 1979.

194. **Batt, L. and Mowat, S. I.,** The addition of methyl radicals to hexafluoroacetone, *Int. J. Chem. Kinet.,* 16, 603, 1984.
195. **Tabayashi, K. and Bauer, S. H.,** The early stages of pyrolysis and oxidation of methane, *Combust. Flame,* 34, 63, 1979.
196. **Batt, L., MacKay, M., and Stewart, P.,** unpublished results.
197. **Baldwin, R. R. and Melvin, A. J.,** The reaction of hydrogen atoms with oxygen and with ethane, *J. Chem. Soc.,* 1785, 1964.
198. **Back, R. A.,** A search for a gas-phase free-radical inversion displacement reaction at a saturated carbon atom, *Can. J. Chem.,* 61, 916, 1983.
199. **Cvetanovie, R. J. and Irwin, R. S.,** Rates of addition of methyl radicals to olefin in the gas phase, *J. Chem. Phys.,* 46, 1694, 1967.
200. **Holt, P. M. and Kerr, J. A.,** Kinetics of gas phase addition reactions of methyl radical. I. Addition to ethylene, acetylene and benzene, *Int. J. Chem. Kinet.,* 9, 185, 1977.
201. **Sangster, J. M. and Thynne, J. C. J.,** The addition of methyl radicals to tetrafluoroethylene, *Int. J. Chem. Kinet.,* 1, 571, 1969.
202. **Tedder, J. M., Walton, J. C., and Winton, K. D. R.,** Free radical addition to olefins. IX. Addition of methyl radicals to fluoro-ethylenes, *J. Chem. Soc. Faraday Trans. I,* 68, 1866, 1972.
203. **Low, H. C., Tedder, J. M., and Waltson, J. C.,** Free radical addition to olefins. XX. A reinvestigation of the addition of methyl radicals to fluoroethylenes, *J. Chem. Soc., Faraday Trans. I,* 72, 1707, 1976.
204. **Thynne, J. C. J.,** The addition of methyl radicals to hexafluoropropylene, *Int. J. Chem. Kinet.,* 3, 155, 1970.
205. **Getty, R. R., Kerr, J. A., and Trotman-Dickenson, A. F.,** The reactions of alkyl radicals. XIII. The additions of methyl, isopropyl and t-butyl radicals to propylene and the isomerisation of alkenyl radicals, *J. Chem. Soc. A,* 1360, 1967.
206. **Benson, S. W. and Shaw, R.,** Kinetics and mechanism of the prolysis of 1,3-cyclohexadiene. A thermal source of cyclohexadrenyl radicals and hydrogen atoms. The addition of hydrogen atoms to benzene and toluene, *J. Am. Chem. Soc.,* 89, 5351, 1967.
207. **Hole, K. J. and Mulcahy, M. F. R.,** The pyrolysis of biacetyl and the third body effect on the combination of methyl radicals, *J. Phys. Chem.,* 73, 177, 1969.
208. **Knoll, H., Scherzer, K., and Schliebs, R.,** The thermal and initiated thermal decomposition of biacetyl-d_6, *Int. J. Chem. Kinet.,* 9, 349, 1977.
209. **Knoll, H., Richter, G., and Schliebs, R.,** On the gas-phase free radical displacement reaction Me + $CD_3COCD_3 \rightarrow CD_3$ + $MeCOCD_3$, *Int. J. Chem. Kinet.,* 12, 623, 1980.
210. **Drew, R. M. and Kerr, J. A.,** Kinetics of the gas phase addition of methyl radicals to hexafluoroacetone: an extension of the temperature range, *J. Chem. Res. P.* 254, 1983.
211. **Batt, L. and Robinson, G. N.,** Arrhenius parameters for the decomposition of the t-butoxy radical, *Int. J. Chem. Kinet.,* 14, 1053, 1982.
212. **Batt, L.,** The gas-phase decomposition of alkoxy radicals, *Int. J. Chem. Kinet.,* 11, 977, 1979.
213. **Dainton, F. S., Ivin, K. J., and Wilkinson, F.,** The kinetics of the exchange reaction Me + $CH_4 \rightarrow CH_4$ + Me using ^{14}C as a tracer, *Trans. Faraday Soc.,* 55, 929, 1959.
214. **Varnerin, R. E.,** The metathetical reactions of methyl radicals with ethane, dimethyl ether, acetone and propylene, *J. Am. Chem. Soc.,* 77, 1426, 1955.
215. **Benson, S. W. and Choo, K. Y.,** Arrhenius parameters for the alkoxy radical decomposition reactions, *Int. J. Chem. Kinet.,* 13, 833, 1981.
216. **Kohler, H.-J. and Knoll, H.,** Activation entropies of free radical addition reactions via MINDO/3-UHF calculations, *J. Prakt. Chem.,* 323, 466, 1981.
217. **Gareca-Dominguez, J. A. and Trotman-Dickenson, A. F.,** The reactions of alkyl radicals. IX. The addition of methyl, ethyl, isopropyl and t-butyl radicals to acetylene and the isomerisation of alkenyl radicals, *J. Chem. Soc., P.* 940, 1862.
218. **Getty, R. R., Kerr, J. A., and Trotman-Dickenson, A. F.,** The reactions of alkyl radicals. XIII. The additions of methyl, isopropyl and t-butyl radicals to propyne and the isomerisation of alkenyl radicals, *J. Chem. Soc. A, P.* 1360, 1967.
219. **Cosa, F., Oexler, E. V., and Staricco, E. H.,** Arrhenius parameters for the reaction of trifluoromethyl radicals with cyanogen chloride, *J. Chem. Soc. Faraday Trans. I,* 77, 253, 1981.
220. **De Vohringer, C. M. and Staricco, E. H.,** Reaction of pentafluoroethyl radicals with cyanogen chloride, *J. Chem. Soc. Faraday Trans. I,* 78, 3493, 1982.

Chapter 12

THE REACTIONS OF CCl$_3$ AND CClF$_2$ RADICALS

Zeev B. Alfassi and Shlomo Mosseri

TABLE OF CONTENTS

I. INTRODUCTION

CCl_xF_{3-x} (x = 0 to 3) radicals are very important in several fields: in lipid peroxidation induced by CCl_4[1-4] (the biological importance is given with more details in the chapter concerning the reaction with biologically important molecules); in atmospheric tropospheric and stratospheric chemistry,[5-14] in the radiolysis of freons used for extraction of noble gases in the reprocessing of used nuclear fuel,[15-19] or for extracting heat from nuclear waste in order to produce electricity using low temperature turbines, besides being important to the study of chemical kinetics. Large amounts of freons (Mtons/year[5]) are produced and added to the natural environment due to their use as aerosol propellants and as refrigerants. In the stratosphere these compounds are photolyzed producing mainly Cl atoms and CCl_xF_{3-x} radicals. Several studies[5-12] suggested that the chlorine atoms produced in this photolysis are leading to the destruction of the ozone in the stratosphere, which will eventually expose the earth to UV radiation, eliminated now by the ozone layer in the atmosphere. The destruction of the ozone occurs mainly via the chain

$$Cl + O_3 \rightarrow ClO + O_2$$

$$ClO + O \rightarrow Cl + O_2$$

$$\overline{}$$

$$O_3 + O \rightarrow 2O_2 \text{ (net)}$$

Taylor et al.[9] suggested that CF_2ClBr (which are used for fire retarding) might be photolyzed already in the tropospheric level due to the relative weakness of the C–Br bond, which leads to a considerable absorption cross-section for wavelengths prevalent in this level. In the presence of OH radicals the following reaction[10] may be another source for CF_2Cl radicals:

$$OH + CF_2ClH \rightarrow H_2O + CF_2Cl$$

There are several estimations for the concentration of CF_2ClH[11,12] in the troposphere and an average value is about 10% of its atmosphere concentration.

In the reprocessing of burnt nuclear fuel, the noble gases krypton and xenon are released. From this point of view of health physics the extraction of ^{85}Kr is essential for prevention of radioactive pollution of the atmosphere. Several techniques are used for this extraction: low temperature adsorption, cryogenic distillation, permeation membranes, and solvent absorption. From these processes, the solvent absorption appears to offer the best advantages in terms of economy and in operating safety and convenience, from all methods. This process, on the other hand, is susceptible to radiation damage of the solvent by the β-rays of ^{85}Kr and to corrosion of equipment by degradation products. The low halocarbons were found to be the most advantageous solvents in terms of selectivity of extraction of noble gases rather than nitrogen and oxygen.

II. PRODUCTION OF CCl_3 AND CF_2Cl RADICALS

CCl_3 radicals were produced by photolysis or radiolysis of CCl_4,[20] CCl_3Br,[21-23] or $(CCl_3)_2CO$, by the pyrolysis of C_2Cl_6[24] and by the reaction of radicals with CCl_4[25-27] or chlorine atom with $CHCl_3$.[28]

$$R + CCl_4 \rightarrow RCl + CCl_3$$

$$Cl + CHCl_3 \rightarrow HCl + CCl_3$$

CF_2Cl radicals were formed by photolysis or radiolysis of CF_2Cl_2,[29-31] CF_3COCF_2Cl,[32] $(CF_2Cl)_2CO$[33]. $CF_2ClH + Hg$,[34] $CF_2ClCOCF_2H$,[35] and $(CF_2ClCO)_2O$.[36]

In the last few years, large numbers of studies concerning the hydrogen abstraction by CCl_3 radicals from RH substrates were done in the liquid phase. In these studies, induced by radiolysis, two systems were used:

1. RH as a solvent and small concentration of CCl_4
2. CCl_4 as a solvent and low concentration of RH

The only difference between those two systems is in the initiating step.

RH as a solvent: $RH \rightarrow R + H$ or $R' + R''$

$$R + CCl_4 \rightarrow RCl + CCl_3$$

$$H + CCl_4 \rightarrow HCl + CCl_3$$

CCl_4 as a solvent: $CCl_4 \rightarrow Cl + CCl_3$

$$Cl + RH \rightarrow HCl + R$$

$$R + CCl_4 \rightarrow RCl + CCl_3$$

The propagation and the termination steps in the two systems are the same:

Propagation: 1. $CCl_3 + RH \rightarrow CHCl_3 + R$

2. $R + CCl_4 \rightarrow RCl + CCl_3$

Termination: 3. $CCl_3 + CCl_3 \rightarrow C_2Cl_6$

4. $R + CCl_3 \rightarrow RCCl_3$ or disproportionation products

5. $R + R \rightarrow R_2$ or disproportionation products

Assuming that $CHCl_3$ is formed only via Reaction 1 and not at all through Reaction 4, e.g.,

$$C_6H_{11} + CCl_3 \xrightarrow{k_{4d}} C_6H_{10} + CHCl_3$$

the above mechanism leads through steady-state analysis to the following equation:

$$\frac{[CHCl_3]}{[C_2Cl_6]^{1/2}} = \frac{k_1}{k_3^{1/2}} [RH].t^{1/2} \qquad (1)$$

where t is the time of irradiation. This equation indicates how k_1 can be measured if k_3 is known independently. The advantages of each method can be summarized as follows:
RH as a solvent:

1. Since the concentration of RH is large, the length of the chain (number of propagation steps per each termination step, i.e., $[CHCl_3]/[C_2Cl_6]$) will be large and the contribution of $CHCl_3$ formed by reactions other than Reaction 1 is negligible compared to the yield from Reaction 1.

2. Since most of the interaction of the photons is with RH molecules, the termination product C_2Cl_6 will be formed only through Reaction 3, with no contribution from primary processes as was found for CCl_4 irradiation.[37,38]
3. The use of RH as a solvent enables formation of high yield of $CHCl_3$ without a considerable change in the concentration of RH during the reaction, thus leaving Equation 1 correct for all situations. The change of the concentration of CCl_4 does not change the yields of the products according to Equation 1.

CCl_4 as a solvent

1. Using RH as a solvent, allows the examination of the proposed mechanism only through the proportionality of $[CHCl_3]/[C_2Cl_6]^{1/2}$ with $t^{1/2}$. The use of only a small concentration of RH allows large variations in RH concentration, which enable the examination of the mechanism by the dependence of $[CHCl_3]/[C_2Cl_6]^{1/2}$ on RH concentration.
2. The ratio of CCl_3 and R radicals in the steady-state condition is given by

$$\frac{[R]}{[CCl_3]} = \frac{k_1[RH]}{k_2[CCl_4]}$$

and thus at high $[RH]/[CCl_4]$ ratio there will be a considerable concentration of R radicals which can lead to $CHCl_3$ formation from Reaction 4. In addition, this will decrease the yield of C_2Cl_6, the product with the lower yield and lower sensitivity to detection. The results of Katz et al.[39,40] show that at high value of $[RH]/[CCl_4]$ the yield of C_2Cl_6 was only one forth of the maximal value, obtained when Reaction 3 is the only termination pathway. This effect will be more pronounced the larger is k_1.

3. When RH is a branched alkane, its radiolysis will lead to several radicals and consequently to several halocarbons which may complicate the analysis of the products.
4. When RH is an olefin, its radiolysis leads to polymerization of the olefin, which will be inhibited only at high concentration of CCl_4.
5. In liquid phase the rate constants and Arrhenius' parameters may be solvent dependent. Using CCl_4 as a standard solvent, enables the comparison of the rate constants and Arrhenius' parameters of various compounds through the ratio $k_1/k_3^{1/2}$ and the standard rate constant k_3.
6. The product whose yield determines the extent of the reaction is C_2Cl_6, since its yield is usually lower and mainly due its lower detection efficiency with flame ionization detectors. Very long chains will lead to high yield of $CHCl_3$ in order to have sufficient C_2Cl_6 for determination. This high concentration of $CHCl_3$, while it is advantageous in neglecting production of $CHCl_3$ in reactions other than reaction 1, may lead to other reactions, such as:

$$R + CHCl_3 \rightarrow RCl + CHCl_2$$

$$CHCl_2 + R–H \rightarrow CH_2Cl_2 + R$$

Although the yields of these reactions are expected to be very low, it can be a source of errors when more than 50% of the initial CCl_4 is consumed.

III. TERMINATION REACTIONS OF CCl_3 AND CF_2Cl RADICALS

A. The Combination of Two CCl_3 Radicals

During the last 20 years various studies were done on the rate constant for combination of two CCl_3 radicals, k_3

$$CCl_3 + CCl_3 \xrightarrow{k_3} C_2Cl_6$$

Most of the studies were done on various systems by applying the rotating sector method. The first reported data show disagreement between themselves. Goldfinger and his co-workers[41] estimated, from the study of gas phase chlorination of $CHCl_3$ and alkanes, k_3 to be $10^{8.8}$ M^{-1} sec^{-1}. Tedder and Walton[22] applied the rotating sector technique to the photochemical addition of bromotrichloromethane to ethylene in the gas phase and determined k_3 to be $10^{10.9}$ M^{-1} sec^{-2} independent of temperature in the range 77 to 173°C. De Mare and Huybrechts[42] studied the gas-phase photochlorination of $CHCl_3$ and by applying the rotating sector technique they found k_3 to be $10^{9.66}$ M^{-1} sec^{-1}, independent of temperature in the range 30 to 152°C. This conflict was solved by White and Kuntz,[43] who determined the combination rate constant using the rotating sector technique applied to the photoinitiated chain reaction between CCl_4 and cyclohexane between 112 to 183°C. They obtained log k_3 = 9.6, in quite a good agreement with De Mare and Huybrecht. In addition, they repeated De Mare and Huybrecht's experiment at 40°C and obtained an agreement with their results, log k_3 = 9.5. However, when they studied the photoaddition of CCl_3Br to ethylene (Tedder and Walton's system) at 130°C, they obtained k_3 = 1.3×10^9 M^{-1} sec^{-1} (which is three times lower than the value for the other systems and 60 times lower than Tedder and Walton's results). Moreover, although the termination product rate of formation was found to be time independent, the formation of CH_2BrCH_2Br was found to have an induction period, which indicates that this system is more complex than originally assumed and that the use of the simple rotating sector equations is not justified in this system. Additional work was done also by Matheson et al.[44] who studied the effect of the addition of HCl on the photolysis of carbon tetrachloride in the presence of cyclohexane. By evaluating the rate constant for the following reaction:

$$CCl_3 + HCl \rightarrow CHCl_3 + Cl$$

from known thermochemical data and from the known rate of the reverse process, they found k_3 = 7.25×10^9 M^{-1} sec^{-1}.

While the rate constant for combination of CCl_3 in the gaseous phase was found to be the same in the last studies, there is still controversy about the value in the liquid phase. The first measurement of the rate of trichloromethyl radical combination was made by Melville et al.[45] who applied the rotating sector technique to the liquid phase addition of bromotrichloromethane to cyclohexene. Melville et al. found the combination rate constant to be temperature independent (30 to 50°C) within experimental error and equal 5×10^7 M^{-1} sec^{-1}. Later they extended their studies in cyclohexene and made also studies in vinyl acetate[46] and found k_3 to be in both solvents 5×10^7 M^{-1} sec^{-1} with low activation energy, if at all. Benough and Thomson[47] studied the combination of two CCl_3 radicals in vinyl acetate and obtained k_3 to be 2.65×10^7, 5.3×10^7 and 8.05×10^7 M^{-1} sec^{-1} depending on the method of measuring the rate of initiation. In all these experiments, the rate constants were determined by dilatometry and no products determination has been done. Later, Carlsson et al.[48,49] studied the kinetics of the photochemically initiated reaction of t-butyl hypochlorite with chloroform in carbon tetrachloride solution at 24°C,[48] and the kinetics of photoinitiated reductions of methyl iodide and carbon tetrachloride by tri-*n*-butylgermanium hydride in cyclohexane at 25°C.[49] The rate constants k_3 in both studies were determined by measurements of temperature changes under adiabatic conditions and by using the rotating sector intermittent illumination. The first study[48] leads to $k_3 = (3.5 + 1.5) \times 10^7 \, M^{-1} \sec^{-1}$ while the second study [49] leads to $k_3 = (5.5 \pm 0.5) \times 10^8 \, M^{-1} \sec^{-1}$. The authors[49] preferred the second value only due to its resemblence to the rate of combination of other small radicals in nonviscous solvents. Another reason for this preference might be that in another paper they found the chloroform + *t*-butylhypochlorite system to be a complicated one, involving the combination of both CCl_3 and *t*-butoxy radicals.[50] Three years later, Ingold and his co-workers[51] studied again the combination reaction of two CCl_3 radicals, this time with direct observation of the radicals by electron spin resonance spectroscopy in solution of *t*-butyl peroxide. The signal of CCl_3 was studied as a function of time and was found to be second order with a rate constant of $k_3 = (5 \pm 1) \times 10^7 \, M^{-1} \sec^{-1}$, independent of temperature in the range of -75 to $-15°C$. The authors themselves remarked that they cannot explain why the termination rate constant obtained by the rotating sector method for the chain reaction of tributyl-germanium hydride with CCl_4 is so different from all the other determined values for $CCl_3 + CCl_3$. In the same year another study from the same group[52] yields $(k_c + k_d)^{CCl_3 + CCl_3}/(k_c + k_d)^{B+B} = 0.4 \pm 0.3$ and $(k_c + k_d)^{Et + Et}/(k_c + k_d)^{B+B} = 1.0 \pm 0.5$, in pentane solution ($k_c$ and k_d stand for the rate constants of combination and disproportionation respectively, B and Et stand for *t*-butyl radical and ethyl radical, respectively). Thus, the ratio between the self-combination rate constant of CCl_3 and that of C_2H_5 radicals (or *t*-butyl radicals) is 1:2.5 as compared to the direct measurement[51] which shows that the rate constant for CCl_3 combination is about two orders of magnitude less than that for methyl or *t*-butyl radicals combination in *tert*-butylperoxide solution. No explanation was given for this discrepancy by the authors. Moreover, although their last paper[52] gives higher values for the rate of combination of CCl_3 radicals, it is still lower than the rate constants for combination of methyl and ethyl radicals. As CCl_3 is not resonance stabilized and since the steric hinderance in CCl_3 is less than that in *t*-butyl radical, the low rate of self-combination was explained as due to polar factors. While Ingold and his co-workers[51,52] applied time resolved ESR experiments, where the ESR signal was recorded as a function of time after some sudden changes in radical initiation, Paul[53] used ESR spectroscopy for radical detection and either a steady-state UV initiation or harmonically modulated initiation technique which enables measurement of radical lifetimes. Paul used methanol solution containing *tert*-butyl peroxide for radical formation with varying concentrations of CCl_4. In steady-state studies the addition of CCl_4 leads to a decrease in the intensity of the signal of CH_2OH radicals and to formation of a signal of CCl_3 radicals. The concentrations of the radicals are given by the equation:

$$\frac{[M]_o^2 - [M]^2}{[C]^2} = \frac{k_X}{k_M} \cdot \frac{[M]}{[C]} + \frac{k_C}{k_M}$$

where [M] and [C] are the steady-state concentrations of the hydroxymethyl and trichloromethyl radicals (the subscript o denotes zero concentration of CCl_4), $2k_M$ and $2k_c$ are the rate constants of self-combination of the two radicals, respectively, and k_x stands for the rate constant for cross combination. Paul obtained that within his experimental error of $\pm 50\%$, $k_c = k_M = 1/2 \, k_x$ at the temperature range of $-55°$ to $17°C$. The absolute value of k_M was studied by the modulation ESR method[54] and found to be: $k_M \, (M^{-1} \sec^{-1}) = 5.2 \times 10^{11} . \exp(-2.67/RT \text{ kcal.mol}^{-1})$. This activation energy is very close to that of the self-duffusion of methanol, leading to the conclusion that there is no indication for an

Table 1
PHASE DEPENDENCE ON ARRHENIUS PARAMETERS FOR HYDROGEN ABSTRACTION (IN THE LIQUID PHASE THE SOLVENT IS THE SUBSTRATE)

	Liquid phase		Gaseous phase	
Substrate	$\log A_1 - {}^1/_2 \log A_3$ $(M^{-1/2} \sec^{-1/2})$	$\Delta E_1 - {}^1/_2 \Delta E_3$ (kcal mol^{-1})	$\log A_1 - {}^1/_2 \log A_3$ $(M^{-1/2} \sec^{-1/2})$	$\Delta E_1 - {}^1/_2 \Delta E_3$ (kcal mol^{-1})
Cyclohexane	3.28 ± 0.08	8.81 ± 0.12[55]	4.01 ± 0.13	10.0 ± 1.0[56]
			4.0	10.7[43]
n-Hexane	3.69 ± 0.05	9.62 ± 0.10[25]	4.00 ± 0.18	10.8 ± 0.3[57]

activation energy of termination reaction step and the activation energy of the combination rate constant is completely due to diffusion. Furthermore, the observed rate constant agrees very well with the calculated value for diffusion controlled rate from Smoluchowski's equation which predicts $k_c = 0.9 \, k_M$ over the whole temperature range of the experiments. This measurement of k_c was further supported by Paul's experiments of modulated radical initiation studies. From his results Paul[53] concluded that the combination of CCl_3 in methanol is diffusion controlled. However he could not explain the discrepancies between his results and the previous studies of Watts and Ingold.[51] This might be due to a wrongly assumed mechanism in some of the studies, or to a possible difference between the different solvents, although it seems unreasonable.

Katz et al.[25,55] measured the abstraction of a hydrogen atom by a CCl_3 radical from both cyclohexane and *n*-hexane and compared their results in the liquid phase to the gas phase results of White and Kuntz[43] and Currie et al.[56] for cyclohexane and of Wampler and Kuntz[57] for *n*-hexane. From these comparisons and by assuming that the Arrhenius' parameters for the hydrogen abstraction should be the same for both phases, they concluded that the combination of two CCl_3 radicals is a diffusion controlled reaction. In these studies, in both phases, the rates of formation of $CHCl_3$ (or alternatively RC1) and C_2Cl_6 were measured and $k_1/k_3^{1/2}$ were calculated from Equation 1. The temperature dependence gives $\log A_1 - {}^1/_2 \log A_3$, and $E_1 - 1/2 E_3$. The observed data are given in Table 1. Katz et al. argued that the difference of the activation energies between the gas phase and the liquid phase of about 2.0 and 1.0 kcal mol^{-1}, for cyclohexane and *n*-hexane, respectively, agree with half of the activation energies for the self-duffusion of cyclohexane and *n*-hexane (4.54 and 2.07 kcal mol^{-1}, respectively[58]). However, it seems that Katz et al. maybe pushed the experimental results a little too much. According to Currie et al.[56] results for the gas phase, the difference between the liquid phase and the gas phase is 1.2 kcal mol^{-1} for both solvents, in spite of the large difference in the activation energies for self-diffusion. Moreover, this phase difference of 1.2 kcal mol^{-1} is not much more than the difference of 0.7 kcal mol^{-1} found between two laboratories studying both the gas phase.

Katz et al. found also an agreement between the A factors for hydrogen abstraction in the gaseous phase and the liquid phase. However, this required that $\log A_3$ in the liquid phase is higher than in the gas phase by 0.6 units in the case of *n*-hexane or 1.4 units in the case of cyclohexane. They rather did not mention this unreasonable assumption of higher A_3 in the liquid phase (in their paper there is even a difference of 1.0 units in $\log A_3$ between cyclohexane and *n*-hexane) and explain only how to calculate A_3, from the observed rate constants for combination of CCl_3 in solution and assuming that this is due to activation energy of diffusion. However, they used the activation energy of diffusion of *n*-hexane or cyclohexane rather than that of the solvent in which the rate of combination was measured and they took different values for k_3 at ambient temperature. For cyclohexane they used k_3 $5 \times 10^7 M^{-1} \sec^{-1}$ and for *n*-hexane they used $k_3 = 4 \times 10^8 \, M^{-1} \sec^{-1}$. Thus, it seems that these comparisons cannot bring an answer to the question of the right constant for CCl_3 combination, specially that the basic assumption of equal rates for hydrogen abstraction in the gaseous and liquid phases was not proven since the rate of collision is influenced differently by the temperature in the gas and liquid phase.

The combination of CCl_3 radicals in aqueous solution was studied by Lesigne et al.[59] who radiolyzed aqueous solution of CCl_4 and followed the absorption of 230 mm as a function of time. They attributed this absorption to CCl_3 radical and found that at concentrations of the order of 10^{-4} M and at pH range 5.5 to 12.4 the absorption disappears by a second-order process with a rate constant of $(7.4 \pm 1.8) \times 10^8$ M^{-1} sec^{-1}.

To summarize, the value of the rate constant for the combination of two CCl_3 radicals in the liquid phase is still one of the unsolved problems of chemical kinetics. Most of the experimental results show on lower rate constant in the liquid phase than in the gas phase, however the observed rate constant in the liquid phase is almost independent of temperature and solvent, which is very hard to reconcile. The explanation of activation energy of diffusion which decreases the rate constant for liquid phase reaction cannot explain why no temperature dependence was found in *tert*-butylperoxide solutions and also why the other radicals react in the same solvent with a rate constant very close to that of the gas phase.

B. Cross Combination of CCl_3 and Other Radicals

The cross combination of CCl_3 with other radicals was studied in the gaseous phase mainly by Wijnen and his co-workers.[60-64] In these works they studied the photolysis of $CHCl_3$ or CCl_4 in the presence of ethane, ethylene ethyl chloride, and diethyl ketone and from the ratio of the yields of the various combination products they found the ratio of the rate constants of combination. Thus from the study of the photolysis of a mixture of CCl_4 with ethane[60] they studied the following rate constants:

$$CCl_3 + CCl_3 \xrightarrow{k_3} C_2Cl_6$$

$$C_2H_5 + C_2H_5 \xrightarrow{k_5} C_4H_{10}$$

$$C_2H_5 + CCl_3 \xrightarrow{k_{4c}} C_2H_5CCl_3$$

$$C_2H_5 + CCl_3 \xrightarrow{k_{4d}} C_2H_4 + CHCl_3$$

and found $k_{4c}/(k_3^{1/2} k_5^{1/2}) = 2.0 \pm 0.15$ in accordance with classical collision theory and $k_{4d}/k_{4c} = 0.22 \pm 0.14$[60], or 0.24 ± 0.04.[62] They also found the disproportionation to combination ratio of CCl_3 radicals with CH_2ClCH_2 radical to be 0.14 ± 0.03.[60-61] The ratio of disproportionation over combination for CCl_3 and C_2H_5 or C_2H_4Cl radicals is considerably higher than that for CH_3 and C_2H_5 radicals (0.05 ± 0.01). A high ratio of disproportionation to combination was found[62] for $CCl_3 + CH_3CHCOC_2H_5$, where this ratio was found to be 0.9. These ratios were found by the authors to be in accordance with Bradley's curve (a linear plot of log D/C vs. $\triangle$S, where D/C is the ratio of disproportionation over combination and $\triangle$S is the entropy differences between disproprotionation and combination products) for non substituted alkyl radicals. The ratio of disproportionation over combination of $CHCl_2 + C_2H_5$ was found to be [63] 0.07 indicating that $CHCl_2$ is closer to CH_3 than to CCl_3. For $CCl_3 + CH_3CHCl$, Yu and Wijnen found[64] D/C $= 0.25 \pm 0.03$. Matheson et al.[65] studied the photolysis of carbon tetrachloride in the presence of alkanes in the gaseous phase. In the presence of methane and in the range of 57 to 250°C, their results give $k_{4c}^2/(k_3 \cdot k_5)$, =

4.1 $\pm$ 0.3 for the CCl_3 and CH_3 radicals; similar values were obtained for the ethyl radical (CCl_4 in the presence of ethane). In the case of CCl_4 + propane, the ratio of $k_{4c}^2/(k_3 \times k_5)$, for CCl_3 and isopropyl radical was found to fluctuate very much in the different experiments so that no conclusion could be made. However, it can be concluded that at least for the radicals CH_3, C_2H_5, CH_3CHCl, and CH_2ClCH_2 the rate of cross combination with CCl_3 behaves according to classical collision theory and all the rate constants are about 10^{10} M^{-1} sec^{-1}.

Gaumann and his co-workers[65,66] studied the radiolysis of mixtures of CCl_4 with alkanes in the liquid phase and within quite large ranges of temperature (-90 to $110°C$). The authors assumed that the contribution of Reaction 1 to the formation of $CHCl_3$ is negligible at temperatures below $-30°C$ and assumed that its only source at these temperatures is the disproportionation reaction k_{4d}. From this they accepted $k_{4d}/k_{4c} = R[CHCl_3]/R[C_2Cl_6]$ where R is the rate of formation. For n-hexane[66] the values of k_{4d}/k_{4c} were found to be 2.9 $\pm$ 0.2($-90°C$), 2.4 $\pm$ 0.1 ($-70°C$), 2.4 $\pm$ 0.1 ($-35°C$) with 2.6 as the recommended average value. For cyclohexyl radical Gaumann and co-workers used[67] the same value however without any measurement and for the perdeuterated cyclohexyl radical they used a ratio of 1.8. As this is the only value measured for CCl_3 in the liquid phase, and since it was not measured in the gaseous phase, the effect of the phase cannot be assessed.

C. Disproportionation of CF$_2$Cl Radicals

In several works[9,32-36] on the photolysis of systems giving CF_2Cl radicals in the gaseous phase, it was found that one of the products is CF_2Cl_2. One of its formation routes was suggested as:

$$CF_2Cl + CF_2Cl \xrightarrow{k_{4d}} CF_2Cl_2 + CF_2$$

If CF_2Cl_2 is formed only by this reaction, than the ratio of the yields of CF_2Cl_2 and $(CF_2Cl)_2$ should give the ratio of the rate constants for disproportionation and combination k_{4d}/k_{4c}. However, none of these works proved that this reaction is the sole source for CF_2Cl_2 and other sources, such as the following reactions, were also suggested.

$$CF_2Cl + Cl \rightarrow CF_2Cl_2$$

$$CF_2 + Cl_2 \rightarrow CF_2Cl_2$$

Majer et al.[33] found that the ratio of the yields of $CF_2Cl_2/C_2F_4Cl_2$ in the photolysis of $(CF_2Cl)_2CO$ decreased with increasing the wavelength of the photolyzing light at $25°C$. Thus, it is clear that CF_2Cl_2 is formed also in other reactions. At wavelengths larger than 3000 Å this ratio was found to reach a plateau which gives an upper limit for k_d/k_c for CF_2Cl + CF_2Cl of 0.06, this is lower than all other values. For example, Pritchard and Perona[35] in photolysis of $CF_2ClCOCF_2H$ by 3130 Å light obtain $CF_2Cl_2/C_2F_4Cl_2 = 0.17 \pm 0.04$. Bellas et al. suggested[34] 0.5 as the upper limit and Watkin and Whittle[36] found $k_d/k_c = 0.13$.

IV. ADDITION OF CCl$_3$ TO DOUBLE BONDS

A. Addition vs. Abstraction of Hydrogen Atoms

When CCl_3 radical reacts with an olefinic compound it can either add to the double bond or abstract a hydrogen atom. The abstraction reaction takes place usually only if there is an allylic hydrogen atom. Huyser[68] studied the ratio of addition to the double bond vs. abstraction of an allylic hydrogen, for CCl_3 radicals which is equal to the ratio of the appropriate rate

Table 2
THE RELATIVE YIELDS OF ADDITION TO DOUBLE BOND TO HYDROGEN ABSTRACTION BY TRICHLOROMETHYL RADICAL[67] (k_{ad}/k_{ab}) AND THE FACTOR OF TEMPERATURE (R)

Olefin	77.8°C	40.0°C	R[b]
cis-2-Butene	34[a]		
trans-2-Butene	26[a]		
2-Pentene	5.7		
4-Methyl-2-pentene	1.26	1.68	1.33
3-Heptene	3.5	5.0	1.43
1-Octene	43		
1-Decene	44		
Cyclopentene	5.4	6.5	1.20
Cyclohexene	1.20	1.85	1.54
Cycloheptene	5.5	8.3	1.51

[a]   At 99°C.

[b] $(k_{ad}/k_{ab})_{78.8°C} / (k_{ad}/k_{ab})_{40.0°C}$.

constants ($k_{ad}/k_{ab} = k_{addition}/k_{abstraction}$). The reactions were studied by photolysis of CCl_3Br (as a source for CCl_3 radicals) in the presence of different olefins. The abstraction was measured as the amount of $CHCl_3$ formed, and the addition product yield was deduced from the difference of the consumption of CCl_3Br and the formation of $CHCl_3$, as shown by the equation below. Huyser ignored completely the possibility of disappearance of CCl_3 by termination reaction, e.g., the formation of C_2Cl_6, which is true only for long chains.

$$\frac{k_{ad}}{k_{ab}} = \frac{\text{moles of } CCl_3Br \text{ reacted} - \text{moles of } CHCl_3 \text{ formed}}{\text{moles of } CHCl_3 \text{ formed}}$$

Huyser's results which are given in Table 2 show a decrease of k_{ad}/k_{ab} with increasing temperature, indicating the higher activation energy required for the hydrogen abstraction reaction relative to that of the addition reaction. The effect of temperature is between 1.20, which indicates an activation energy of 1.06 kcal mol^{-1}, and 1.54 correlated with ΔE_a of 2.51 kcal mol^{-1}. Thus, it can be generalized that the hydrogen abstraction reaction has an activation energy higher by 2 ± 1 kcal mol^{-1}, than that of addition of the CCl_3 to the double bond.

Huyser's results show the importance of three structural factors: terminal vs. nonterminal olefins, cyclic vs. open-chain olefins, and geometrical isomers.

1.   The higher addition to abstraction ratio found for terminal olefins (1-octene and 1-decene) relative to nonterminal olefins (2-pentene and 3-heptene) indicates sterical hinderance to addition of CCl_3 radical to nonterminal double bonds. In the case of 2-butenes, in which the steric hinderance could not be much different than that of 2-pentene, the high k_{ad}/k_{ab} is probably due to lower value of k_{ab} rather than to low k_{ad}. The 2-butenes having only primary allylic hydrogens have lower rate constants for hydrogen abstraction than 2-pentene which has two secondary allylic hydrogens. This also explains why k_{ad}/k_{ab} is even lower for 4-methyl-2-pentene which has a tertiary allylic hydrogen atom which should lead to higher k_{ab}.

2. k_{ad}/k_{ab} for cyclohexene is smaller than could be expected from the results for 2-heptene or 2-pentene, probably due to larger hinderance for CCl_3 addition to more rigid cyclic

molecule than to open-chain molecules. The difference between the cyclohexene and cyclopentene or cycloheptene is probably due to differences in k_{ab} as was found for the saturated cyclanes. However, there the trend was in the opposite direction, cyclohexane being the least reactive toward H atom abstraction.[57,70]

3. Huyser found higher k_{ad}/k_{ab} ratio for the *cis*-butene than for the *trans*-isomer. As there is no evidence for higher reactivity of *cis*-2-butene over *trans*-2-butene toward abstraction of allylic hydrogens by other radicals the higher k_{ad}/k_{ab} ratio, in the case of *cis*-isomer should be attributed to higher k_{ad}. This agrees with Skell and Woodworth[69] results who found that at 0°C, CCl$_3$ adds 2.4 times faster to *cis*-butene than to *trans*-2-butene. Skell and Woodworth[69] studied the photoinitiated reaction of CCl$_3$Br with either pure *cis*- or *trans*-2-butene or with a mixture of them. With pure isomers they found no isomerization, indicating that the decomposition of the radical CH$_3$-CH(CCl$_3$)-CH-CH$_3$ is slow compared to its reaction with CCl$_4$. The same products were found for the addition of CCl$_3$Br to both isomers, indicating that the rate of interconversion of the diastereoisomers of this radical, by rotation about the C_2-C_3 bond, must be larger than the reaction of this radical with CCl$_3$Br.

B. Addition of CCl$_3$ Radicals to Double Bonds

Tedder, Walton, and co-workers[71-74] studied the kinetics of addition of trichloromethyl to various olefins in the gas phase, by the photolysis of CCl$_3$Br in the presence of various olefins with wavelengths of light longer than 280 nm. The following mechanism was suggested:

$$CCl_3Br + h\nu \rightarrow CCl_3 + Br \tag{2}$$

$$CCl_3 + Ol \rightarrow CCl_3Ol \qquad (Ol = Olefin) \tag{3}$$

$$CCl_3Ol + CCl_3Br \rightarrow CCl_3OlBr + CCl_3 \tag{4}$$

$$2CCl_3 \rightarrow C_2Cl_6 \tag{5}$$

Similar reactions are taking place with the Br atom. In the case of excess of CCl$_3$Br Reaction 5 was established to be the main termination reaction. The rate constant for the addition reaction was measured relative to the combination reaction by applying steady-state treatment, giving:

$$R_{CCl3OlBr}/R_{C2Cl6} = k_7[Ol]/(k_3\phi I_a)^{1/2}$$

where I_a is the absorbed dose, ϕ is the quantum yield of production of CCl$_3$ and ϕI_a is the rate of CCl$_3$ formation (Reaction 2). Tedder et al. did not make any actinometry and derived ϕI_a from the measured rate of formation of C$_2$Cl$_6$, usually only at one temperature. However, Kerr and Parsonage[75] pointed out that the same group results[76] showed that for constant initial concentration of CCl$_3$Br the quantum yield of C$_2$Cl$_6$ formation in the photolysis of CCl$_3$Br increase from 0.0222 to 0.336 over the temperature range 365 to 457K. Consequently, Kerr and Parsonage recalculated k_7 whenever it was possible by using the same equation as Equation 1. Sidebottom et al.[77] agreed with the re-analysis of their data by Kerr and Parsonage and gave a more complete recalculation of these data. They also found that ϕI_a can be represented by an Arrhenius type expression: $\phi I_a = 6.1 \times 10^{-3} \exp(-6,260 \pm 900 \text{ cal}/RT) \text{ sec}^{-1}$, which shows that their earlier data should be corrected by adding 3.1 kcal mol^{-1} to the activation energy. According to Kerr and Parsonage[75] findings, the Arrhenius parameters for addition of CCl$_3$ and CH$_3$ to ethylene are very close ([log $A(M^{-1} \text{ sec}^{-1})$] = 8.3

Table 3
THE FRACTIONAL YIELDS OF
CCl$_3$ ADDITION TO THE DOUBLE
BONDS OF SEVERAL
HALOETHENES (150°C, GAS
PHASE)

CH$_2$=CHF	CF$_2$=CHF
1:0.077	1:3.5
CH$_2$=CHCl	CF$_2$=CHCl
1:not observable	1:0.04
CH$_2$=CHCH$_3$	CF$_2$=CFBr
1:0.071	1:0.03
CH$_2$=CFCH$_3$	CCl$_2$=CHCl
1:0.0071	1:30
CH$_2$=CF$_2$	CF$_2$=CCl$_2$
1:0.071	1:0.2

and 8.5 and E = 7.5 and 7.7 kcal mol^{-1}, respectively), while for CF$_3$ radical the A factor is about the same but the activation energy is much lower (2 kcal mol^{-1}).

Johari et al.[74] who studied the addition of CCl$_3$ radicals to chloro olefins, using photolysis of CCl$_3$Br, in the gas phase, found also olefinic products which were accounted for by chlorine atom elimination from the radical produced by CCl$_3$ addition to the chlorine substituted carbon atom of the olefin. They found that the rate of addition of CCl$_3$ radical to a CHCl= group or a CCl$_2$= group is slower than the rate of addition to a CF$_2$= group and much slower than addition to a CH$_2$= group. Thus, in the case of vinyl chloride, CCl$_3$ is added only to the CH$_2$ group. This is the reason why no chlorine elimination occurs in this molecule (as was proven by the absence of olefinic products in CCl$_3$Br + CH$_2$ = CHCl) while in the case of CHCl=CHCl, 57% of the products were found to be olefinic. Comparison of the activities of various olefins shows that the reactivity of the olefins decrease with increasing chlorine content with the exception of trichloroethylene. The fraction of CCl$_3$ adding to the various sides of the double bond at 150°C in the gas phase is given in Table 3.

Horowitz and Baruch[26,78] studied the addition reaction of CCl$_3$ radicals in liquid phase (mixtures of cyclohexane with CCl$_4$) to C$_2$Cl$_4$, C$_2$HCl$_3$, and C$_2$H$_2$Cl$_2$. The Arrhenius parameters (within the temperature range 50 to 175°C) were determined competitively vs. H-atom transfer from cyclohexane. Since H-atom transfer rate constants were those derived from competition with CCl$_3$ self-combination, the absolute rate constants for addition depend on the value used for k$_3$. The authors found that the rate constants for liquid phase addition agree quite well with the gas-phase results of Johari et al.[74] Comparing their results with gas-phase results for the addition of CF$_3$ radicals to chloroethylenes, they found that CF$_3$ radicals are markedly more reactive than CCl$_3$ radicals. However, in spite of the fact that at 423 K CF$_3$ addition is faster by about three orders of magnitude than the addition of CCl$_3$ radicals, it was found that the selectivity of the two radicals for reaction with different chloroehylenes is about the same. This lack of difference in selectivity was shown quantitatively by the linear relation found for the logarithms of k$_{CF_3}$ and k$_{CCl_3}$ (log k$_{CF_3}$ = 0.97 log k$_{CCl_3}$ + 2.96). This correlation shows that Hammond's postulate of inverse relation between reactivity and selectivity is not obeyed. In addition, Horowith and Baruch found that the linear free energy relationship used by Tedder and Walton[79] does not describe adequately the reactivity trends observed in the addition reaction of CF$_3$ and CCl$_3$ radicals with ethylene and chloroethylenes. Horowith and Baruch pointed out that the A factors for the addition of CF$_3$ and CCl$_3$ radicals are almost identical so that the difference in the addition rate constant of the two radicals can be ascribed almost entirely to the difference in the addition activation energies E$_{CCl_3}$ − E$_{CF_3}$, which is about 5.6 kcal mol^{-1} for all chloro-

Table 4
ARRHENIUS PARAMETERS FOR THE ADDITION OF
TRICHLOROMETHYL RADICALS TO OLEFINS[77,78]

Phase	Addition to	Olefin	log A (M^{-1} sec^{-1})	E (kcal mol^{-1})
Gas	$CH_2=$	$CH_2=CH_2$	8.3	6.3
		$CH_2=CHF$	8.1	6.4
		$CH_2=CHCH_3$	9.6	6.5
		$CH_2=CFCH_3$	8.8	6.3
		$CH_2=CHCl$	9.3	6.5
		$CH_2=CHCF_3$	9.1	8.0
		$CH_2=CF_2$	8.4	7.7
	$CHF=$	$CHF=CH_2$	8.1	8.4
		$CHF=CF_2$	9.1	9.2
	$CF_2=$	$CF_2=CH_2$	8.4	11.4
		$CF_2=CHF$	9.1	10.2
		$CF_2=CF_2$	9.7	9.2
		$CF_2=CFCF_3$	9.3	9.3
Liquid	$CHCl=$	*cis*-$CHCl=CHCl$	7.8	8.5
		trans-$CHCl=CHCl$	8.2	8.5
	$CHCl=$ and $CCl_2=$	$CHCl=CCl_2$	8.0	8.1
	$CCl_2=$	$CCl_2=CCl_2$	7.6	9.5

ethylenes. The observation that $E_{CCl_3} - E_{CF_3}$ is constant for the chloroethylenes suggests that the rate of addition reaction is primarily controlled by the strength of the C–C being formed, and that the C–C bond in CF_3 - ethylene is considerably stronger than the C–C bond in CCl_3 - ethylene. The values of the Arrhenius parameters for CCl_3 addition to ethylenes in both gas and liquid phase is given in Table 4, which shows that reactivity in the addition reaction is affected to a greater extent by Cl substitution at the site of the attack than by substitution at the other end of the double bond. Horowith and Baruch also noted that the trend of the CCl_3 radicals reactivity toward chloroethylenes is not only the same as that of CF_3 radicals (both radicals being electrophilic radicals), but also the same as that of the cyclohexyl radicals, which are nucleophilic radicals. This observation can be rationalized by assuming that the order of reactivity in this series of chloroethylenes is mainly determined by steric factors.

Kim et al.[80] studied the effect of the media of varying polarity on the kinetics of the addition of trichloromethyl radicals to 1-hexene. The authors found that the rate constants can be described quite well by Kirkwood equation.

El Soueni et al.[81] studied the reactions of trichloromethyl radicals with conjugated diene (penta-1,3-diene) and methylene interrupted enyne (penta-1-en-4-yn). Photochemical reaction of CCl_3Br with *cis*-penta-1,3-diene in the liquid phase gave a mixture of 80% *trans*- and *cis*-1,4- adducts (*trans*-to-*cis* ratio 18:1) and 20% 1,2 adduct. In this system, hydrogen abstraction appears to be a minor process. The photochemical reaction of CCl_3Br with penta-1-en-4-yn in the liquid phase at ambient temperature produced both the 1,2 dibromide $BrCH=CBrCH_2CH=CH_2$ from bromine atom addition to the triple bond, and the 1,2-adduct $CCl_3CH_2CHBrCH_2=CCH$ from addition of the CCl_3 radicals to the double bond which was found to be the main product. No adducts from CCl_3 radicals attack on the triple bond were detected. A small amount of $CHCl_3$ was found, indicating that some hydrogen abstraction had taken place. The CCl_3 radicals add to the conjugated double bond to form substituted allyl radicals which exist in the two resonant forms a and b shown below.

$$CCl_3 + C=C-C=C \rightarrow CCl_3-C-\overset{\cdot}{C}-C=C \leftrightarrow CCl_3-C-C=C-\overset{\cdot}{C}$$

(a)   (b)

The substituted allylic radical abstracts a bromine atom from CCl_3Br to form the adduct and a CCl_3 radical which continues the chain. Two different adducts can be formed according to which resonant form reacted with CCl_3Br (1,2 is formed from a and 1,4 is formed from b). It is most probable that steric factors govern the relative proportions of the two types of adduct. The preponderance of the *trans*-1,4-adduct on the *cis*-isomer is probably due to the well-established fact that the conjugated dienes exist predominantly in the s-*trans* conformation.

Addition to double bond occurred more rapidly than addition to triple bond. For CF_3 a ratio of 2:4 at 437 K was found. CCl_3 radical addition to penta-1-en-4-yn occurred exclusively at the double bond. This behavior of the two radicals may be a reflection of the greater selectivity of CCl_3 as compared with CF_3 radicals.

C. Addition to Aromatic Hydrocarbons

Kooyman and Farenhorst who studied the addition of trichloromethyl radicals to *n*-hexadecene and styrene,[82] found that these reactions were retarded by added aromatic hydrocarbon. From the extent of retardation they were able to measure the relative rates of reactivities of the trichloromethyl radical toward the various aromatics. The authors found that there is an approximate linear relationship between the logarithms of the relative rate constants and the maximum free valence energies (F_{max}, as calculated by a Hückel molecular orbital approach).

Arnold et al.[83] studied the relative rates (relative to *trans*- stilbene) of addition of trichloromethyl radical to a series of 9-substituted anthracenes at 70°C. Using benzene-bromo-trichloro-methane (3:2 molar ratio) as a solvent, and benzoyl peroxide as a radical initiator, the authors found that there is a substituent dependence for the rate of addition to the ring and reported on a good correlation (r = 0.97) between the logarithms of the relative rates and Brown and Okamoto's σ_p^+ substituent parameter with $\rho^+ = 0.83$. The deviation of the nitro- and methyl-substituted compounds from this correlation was explained as due to steric inhibition of resonance in the case of the nitro substituted and as due to side-chain reaction of hydrogen abstraction in the case of the methyl-substitution as a result of its readily abstractable hydrogen atoms.

V. ABSTRACTION REACTIONS

The only abstraction reaction studied for CCl_3 radicals is the abstraction of an hydrogen atom from various compounds.

A. Selectivity in Abstraction Reactions — Structural Effects

Pryor et al.,[84] in their paper on reactivity patterns of the methyl radical, compared the reactivity of several radicals with primary, secondary, and tertiary hydrogens. They found that from all studied radicals, CCl_3 was the most selective one (except bromine atom). The ratios of reactivities toward tertiary:secondary:primary hydrogens are 50:7:1 for CH_3 radical, 24:7.8:1 for CF_3 radical, and 2300:80:1 for CCl_3 radical, all in the gas phase and for 182 to 190°C. The high selectivity of the CCl_3 radical (and the bromine atom) in abstracting hydrogen atoms seems to be due to the relative low strength of the CCl_3–H bond[85,86] (and the Br–H bond[86]). All the other radicals studied have a stronger bond with hydrogen. The bond between primary carbon atom and hydrogen is 98 kcal mol^{-1} while the CCl_3–H bond dissociation enthalpy is 95 kcal mol^{-2} [85] (similar to the bond strength of secondary carbon

Table 5
BOND DISSOCIATION ENTHALPIES IN KCAL MOL^{-1} FOR CCl$_3$ WITH SEVERAL ATOMS AND RADICALS

Bond	Enthalpy	Bond	Enthalpy	Bond	Enthalpy
CCl$_3$–H	95.2	CCl$_3$–CH$_3$	87.7	CCl$_3$–NO	32
CCl$_3$–F	102	CCl$_3$–CH$_2$Cl	85.3		
CCl$_3$–Cl	70.3	CCl$_3$–CHCl$_2$	79.0		
CCl$_3$–Br	56	CCl$_3$–CCl$_3$	70.7		

Table 6
RELATIVE REACTIVITIES OF TRICHLOROMETHYL RADICAL TOWARD VARIOUS ARALKYL HYDROCARBONS AT 40°C (TOLUENE REACTIVITY ASSUMED TO BE UNITY[87])

Hydrocarbon	Reactivity
C$_6$H$_5$CH$_2$–H	1.0
C$_6$H$_5$CH(CH$_3$)–H	50
C$_6$H$_5$C(CH$_3$)$_2$–H	260
(C$_6$H$_5$)$_2$CH–H	50
(C$_6$H$_5$)$_2$C(CH$_3$)–H	650
(C$_6$H$_5$)$_3$C–H	160

atom with hydrogen). The bromine atom which is even more selective than CCl$_3$, has even a weaker bond with hydrogen (87.4 kcal mol^{-1}). Table 5 gives the bond energy of CCl$_3$ with various atoms and radicals.[85,86] Pryor et al.[84] based their selectivity values on rate constants measured in the gas phase at one temperature by McGrath and Tedder. Wampler and Kuntz[57] who studied the temperature dependence of the rate of hydrogen abstraction by CCl$_3$ radicals in the gas phase, found that the activation energy for abstraction of a tertiary hydrogen is lower by 2.1 to 2.4 kcal mol^{-1} than that of a secondary hydrogen and by 3.1 kcal mol^{-1} from that of a primary hydrogen. Alfassi and Feldman[20,70] studied the temperature dependence of the rate constants for abstraction of hydrogen atoms by CCl$_3$ in the liquid phase from 2,3-dimethyl butane and cyclohexane and show from these studies that also in the liquid phase the activation energy for abstraction of secondary hydrogen by CCl$_3$ is lower by about 2.5 kcal mol^{-1} than that of tertiary hydrogen. Comparison of abstraction of hydrogen atoms from cyclohexane and *n*-hexane in the liquid phase by CCl$_3$ radicals[25,55] showed that primary hydrogens are considerably less reactive than secondary hydrogens.

Selectivity for CCl$_3$ radical abstracting an hydrogen atom was also found in the case of aralkyl hydrocarbons[87] although the abstraction of a benzylic hydrogen is always exothermic [D(C$_6$H$_5$ CH$_2$–H) = 85 kcal mol^{-1} [86]]. The relative reactivity per an hydrogen atom for various aralkyl hydrocarbons is given in Table 6.

B. Polar Effects
Matheson et al.[65] reported on similar activation energies for abstraction of an hydrogen atom from various alkanes by either CH$_3$ or CCl$_3$ radicals. The authors pointed out the similar activation energies as compared to the large difference between the bond dissociation enthalpies of CH$_3$-H (104 kcal mol^{-1}) and CCl$_3$-H (95.2 kcal mol^{-1}) and explained this as

due to polar effects induced by the electronegativity of the CCl_3 radical. Baruch and Horowitz[88] pointed out that hydrogen abstraction reaction of CCl_3 radicals with CCl_3SiH and $(CH_3)_3$ SiH proceed less readily than the reactions of both CH_3 and CF_3 radicals, in accordance with bond dissociation enthalpies of CCl_3H, CH_4, and CF_3H (95, 104, and 106 kcal mol^{-1}, respectively). However, whereas for CCl_3SiH this claim is clear, the situation is not so clear for $(CH_3)_3SiH$ where the rate for CH_3 is much closer to CCl_3 than to CF_3 although the order of reactivities fits the bond dissociation energies [log k (M^{-1} sec^{-1}) at 400 K is 3.74, 4.10, and 6.23 for CCl_3, CH_3, and CF_3, respectively[88]]. Baruch and Horowitz explained the higher reactivity of CCl_3 toward $(CH_3)_3SiH$ as compared to that toward Cl_3SiH in terms of polar effects. CCl_3 as an electrophilic radical is expected to be less reactive when there is electron-withdrawing substituent (Cl atoms) than when the substituent is increasing the electron density in the Si–H bond (alkyl substitution). The higher inductive effect of the ethyl group compared to the methyl group was explained as the cause for the higher reactivity of CCl_3 toward $(C_2H_5)_3$ SiH as compared to $(CH_3)_3SiH$. Ormson et al.,[89] on the other hand, deduced from the activation energy of hydrogen abstraction from Cl_3SiH by CCl_3 and CF_3CCl_2 that these reactions are not affected by polar effects.

Huyser[90] studied the relative reactivities of ring-substituted toluenes toward trichloromethyl radicals by carrying out competitive photochemically induced bromination with bromotrichloromethane as the brominating agent. The author suggested the following free radical chain sequence as the bromination mechanism:

$$CCl_3 + ArCH_3 \rightarrow CHCl_3 + ArCH_2$$

$$ArCH_2 + BrCCl_3 \rightarrow ArCH_2Br + CCl_3$$

Using a mixture of toluene with substituted toluene, the ratio of rate constants of the first reaction (above) for a substituted toluene, k_s, and toluene, k_t, is given by the equation

$$k_s/k_t = \frac{\log S_0/S_f}{\log T_0/T_f}$$

where S and T stand for the concentrations of the substituted toluene and toluene, respectively, and the subscripts o and f assigned the concentration at the beginning and at the end of the reaction. Huyser found that log k_s/k_t do not correlate well with Hammett's σ values (parameter which indicate the electron releasing or withdrawing ability for a nuclear substituent) but is correlated well with the σ^+ value of the nuclear substituent, which is a measure of the effect of the substituent on the ease of formation of the transition state.

$$ArCH_3 + CCl_3 \rightarrow [ArCH_2.....H.....CCl_3] \rightarrow ArCH_2 + CHCl_3$$

The later correlation yields the value $\rho = -1.46$ which is abnormally high for a free-radical substitution reaction. The values for ρ for various reactants correlated with σ^+ are given in Table 7, together with the relative reactivity per α-hydrogen atom. For most of the reactants the correlation with σ^+ was found to be better than the correlation with σ. For cumenes, the same agreement (similar correlation coefficient $- r$) was found,[93] ρ (σ) = -0.89. For allyl-benzenes and benzaldehyde a better agreement was found for σ than for σ^+, ρ (σ) = -0.58,[95] for allylbenzene and ρ (σ) = -0.74[96] for benzaldehydes.

Tanner et al.[91] reinvestigated Huyser's studies[90] of the relative rates of the photoinitiated bromination of substituted toluenes with bromotrichloromethane. The authors obtained similar results to those of Huyser and reported on ρ (σ^+) = -1.24. However, they found that under those experimental conditions mixtures of toluene-α-d_3 and toluene undergo extensive

Table 7
RELATIVE REACTIVITIES OF BENZYLIC HYDROGENS TOWARD
ABSTRACTION BY CCl$_3$ RADICALS (PER α-HYDROGEN ATOM) AND
HAMMET'S ρ (σ$^+$) PARAMETER FOR SUBSTITUENT EFFECT.

Substrate	Temperature (°C)	Solvent	Relative reactivity	ρ	Ref.
PhCH$_3$	50	Chloroform	1.00	− 1.46	90
PhCH$_3$	50	Freon-112 + CBrCCl$_3$		− 1.24	91
PhCH$_3$	50	Freon-112 + CBrCl$_3$ + ethylene oxide		− 0.69	91
PhCH–C(CH$_3$)$_3$	80	CCl$_4$		− 0.94	92
PhCH(CH$_3$)$_2$	70	CCl$_4$	66	− 0.67	93
PhCH$_2$CH$_3$	80	CCl$_4$	24	− 0.53	94
PhCH(OCH$_3$)$_2$	80	CCl$_4$	39	− 0.18	94
PhCHO	80	CCl$_4$	58	− 0.67	23,93
PhCH$_2$CH=CH$_2$	80	CCl$_4$		− 0.58	95
PhCH$_2$OCH$_2$Ph	80	CCl$_4$	89		94
PhCH$_2$OCH$_3$	80	CCl$_4$	29	− 0.36	96
Ph$_2$CHOCH$_3$	80	CCl$_4$	80	− 0.12	94

exchange of protium and deuterium during the brominations. The exchange reaction is ascribed to the presence of hydrogen bromide, which is formed in the process by the following sequence of reactions:

$$CCl_3Br \xrightarrow{h\nu} CCl_3 + Br$$

$$Br + RH \rightarrow HBr + R$$

$$R + CCl_3Br \rightarrow RBr + CCl_3$$

$$CCl_3 + RH \rightarrow CHCl_3 + R$$

The HBr present takes part in the following two reversible reactions:

$$CCl_3 + HBr \rightarrow CHCl_3 + Br$$

$$R + HBr \rightarrow RH + Br$$

The last reaction is the one leading to the deuterium-hydrogen exchange. This mechanism points out the possibility that the rate determining step will be

$$Br + RH \rightarrow HBr + R$$

In order to get rid of this possibility, they add either potassium carbonate, which eliminated most of the scrambling of toluene-α-d$_3$ and toluene, or ethylene oxide, which eliminated it completely. Significant differences in the values of the relative reactivities were observed with and without the additives. Under conditions where reversible abstraction with HBr is reduced, the sensitivity to polar substituents in the aromatic nucleus is very much reduced. This effect can be quantitated through the ρ (σ$^+$) which for substituted toluenes in the presence of 20% ethylene oxide (condition where no scrambling was found) has a value of only − 0.69 as compared to − 1.24 found in the absence of ethylene oxide.

The findings that substituted benzaldehydes are correlated with σ while substituted toluenes, ethyl benzenes, and benzyl methyl ethers are correlated with σ^+ were explained by Lee[23] as due to the resonance of benzaldehyde which reduces the electron availability on the aldehydic H atom.

C. Cylic Molecules

Huyser et al.[97] studied the bromination of cycloalkanes by bromotrichloromethane in the liquid phase at 80°C, induced by thermal decomposition of benzoyl peroxide. The reactivities of various cyclanes relative to a unit reactivity of cyclohexane were found to be $C_5 = 1.57$, $C_7 = 3.30$, and $C_8 = 9.20$. Kuntz et al.[43,57] studied the rate of abstraction of a hydrogen atom by CCl_3 radicals from cycloalkanes (C_5 - C_8) as a function of temperature.

In a different study, Alfassi and Feldman[70] studied the reaction of CCl_3 in the liquid phase with cyclohexane, cyclooctane, and cyclododecane. In both studies cyclohexane was found to be the least reactive one, in both cases due to the highest activation energy. The order of activation energies $E(C_6) > E(C_{12}) > E(C_8)$ found by Alfassi and Feldman[70] was explained as due to the stability of the cyclic rings, as it is measured by the excess enthalpy relative to cyclohexane.[70] The trend $C_5 > C_6 < C_7 < C_8$ was found also for various radicals and the decrease in the reactivity toward c-$C_{12}H_{24}$ is in accordance with the increasing stabilization of cyclic molecules with 12 or more carbon atoms in the ring.

Wong and Gleicher[98] who studied the relative rate of hydrogen abstraction from various benzocycloalkenes by trichloromethyl radicals in the liquid phase at 70°C found a good linearity between the logarithms of the relative rates and calculated strain energy (correlation coefficient -0.974). The calculation of the strain energy is relatively simple for this series of benzocycloalkene due to the annelation of the benzene ring which imparts a high degree of rigidity to the hydrocarbons and consequently reduces the number of conformers. The presence of the aromatic ring also causes the reaction to take place exclusively in the benzylic position. The strain energies were calculated assuming the transition states structures to resemble strongly the intermediate radicals. Although cyclic radicals may assume a pyramidal structure to relieve potential angle strain, Wong and Gleicher maintained the planarity in order to maximize the favorable resonance interaction.

Unruh and Gleicher[99] studied the relative rates of hydrogen abstraction from a series of 13 arylmethanes by the trichloromethyl radical at 70°C. The authors found a very good correlation between the logarithms of the rate constants and the change in the π-binding energies between the incipient radicals and the arylmethanes (correlation coefficient = 0.977) when the π-binding energies were calculated by the SCF method. However, when the π-binding energies were calculated by the HMO method the correlation was much worse. A better correlation was found when the data were divided into two sets consisting of α-methylnaphtyl-type and β-methyl naphthyl-type compounds. The tremendous improvement obtained when the data were correlated by SCF calculation rather than HMO calculation was explained as due to the neglect of spin polarization interaction in HMO calculation.

D. H/D Isotope Effect

Lee[23] studied the ratio of hydrogen abstraction rate vs. deuterium abstraction rate from benzaldehyde (C_6H_5CHO and C_6H_5CDO) by CCl_3 radicals in the liquid phase. At 80°C this ratio was found to be $k_H/k_D = 2.41 \pm 0.09$.

Most of the works in the gas phase on H/D isotope effect in hydrogen abstraction by CCl_3 radicals was done by Hautecloque.[100-103] Hautecloque[100] studied the reaction of CCl_3, from photolysis of CCl_3Br, with H_2 and D_2 at the temperature range of 190 to 300°C and found the same A factor for $CCl_3 + H_2$ and $CCl_3 + D_2$ ($10^{9.7} M^{-1}$ sec^{-1}) while the activation energy for deuterium was higher by 1.2 kcal mol^{-1} than that of hydrogen abstraction (15.2 and 14.3 kcal mol^{-1}, respectively). Calculation of the isotope effect done by the BEBO method

Table 8
ISOTOPE EFFECT k_H/k_D FOR ABSTRACTION OF
HYDROGEN OR DEUTERIUM BY CCl_3 RADICALS

Phase	Substrates	Temperature range (°C)	k_H/k_D	Ref.
Gas	H_2/D_2	200 — 300	3.7 — 3.1	100
	CH_4/CD_4	278 — 330	4.0 — 3.6	100
	C_2H_6/C_2D_6	248 — 304	2.1 — 2.8	101
	CH_3Br/CD_3Br	215 — 300	2.3 — 1.3	102
	CH_2Br_2/CD_2Br_2	215 — 300	3.0 — 2.8	102
	$CHBr_3/CDBr_3$	215 — 300	2.5 — 1.4	102
	$C_2H_4Br_2/C_2D_4Br_2$	248 — 304	6.3 — 4.0	103
	$C_2H_2Br_4/C_2D_2Br_4$	248 — 304	8.0 — 5.0	103
Liquid	C_6H_5CHO/C_6H_5CDO	80	2.4	23
	$c\text{-}C_6H_{12}/c\text{-}C_6D_{12}$	28.5 — 140.0	23.5 — 7.93	67

of Johnston[105] gave very good agreement when the tunneling effect was included in the calculations. The k_H/k_d ratio was found to vary with the temperature between 3.1 and 3.7. A similar work was done also for CCl_3 + CH_4 (CD_4) and a similar isotope effect (k_H/k_D: 3.6 to 4.0) was found. In addition the activation energy for abstraction of an hydrogen atom from CH_4 was found to be 7 kcal mol^{-1} [100] higher than that from H_2. The authors attributed these findings to the repulsion between the attacking radical and the methane molecule. However, this result might be wrong as the A factor found for CCl_3 + CH_4, $10^{12.1}M^{-1}$ sec^{-1}, is too high compared to similar reactions and so it seems that both A and E_a were wrongly determined. Assigning a similar A value for both reactions CCl_3 + H_2 and CCl_3 + CH_4, yields a difference of less than 2 kcal mol^{-1} between the two reactions.

Hautecloque studied also the H/D isotope effect in the reaction of CCl_3 in the gas phase with $C_2H_6(C_2D_6)$,[101,103] $CH_3Br(CD_3Br)$, $CH_2Br_2(CD_2Br_2)$, $CHBr_3(CDBr_3)$,[102] $C_2H_2Br_4$ ($C_2D_2Br_4$), and $C_2H_4Br_2$ ($C_2D_4Br_2$).[103] The k_H/k_D values given in Table 8 show that although the isotope effects vary with the temperature, these changes are quite small. In all cases k_H/K_D decreases with increasing the temperature. The theoretical transition state calculations for the isotope effect was found to agree with the experimental results for CCl_3 + $H_2(D_2)$ and CCl_3 + CH_4 (CD_4) only when the calculation includes the tunneling effect. An opposite finding was found for CCl_3 + C_2H_6 (C_2D_6)[101] and for CCl_3 + $CH_x Br_{4-x}$ (CD_xBr_{4-x}),[102] where the calculation agreed with the experimental results only when the tunneling effect was neglected. Table 8 also shows that for the serie CH_xBr_{4-x} (x = 1,2,3,4) k_H/k_D is higher for the symmetric cases CH_4 and CH_2Br_2. Hautecloque explained this as due to polar effects which predict that the least polar molecule CH_4 will have the highest k_H/k_D. However a different situation was found in bromoethanes where the least polar molecule, C_2H_6, has the lowest k_H/k_D.

Comparing CBr_3 to CCl_3, the isotope effect of CBr_3 + H_2 (D_2) was found to be 2.75 to 2.25 in the temperature range of 140 to 275°C,[104] quite a lower value than found for CCl_3. Nguyen et al.[67] studied the H/D isotope effect in the reaction CCl_3 + cyclohexane (perdeuterated cyclohexane) and found $\log_{10} (k_H/k_D) = -0.38 + 2.40/2.303 RT$ kcal mol^{-1}. The high value for the difference was suggested to be due to the tunneling in the proton abstraction by CCl_3. The tunneling effect can also explain the rather low ratio of A factor, 0.42. However it should be kept in mind as Nguyen et al. mentioned that an underestimation of this ratio could be compensated for by an overestimation in the difference of the activation energies.

E. Blocking Effect

Wampler and Kuntz[57] who studied the Arrhenius parameters for the reaction of CCl_3 with

Table 9
ARRHENIUS PARAMETERS FOR HYDROGEN ABSTRACTION FROM SUBSTITUTED METHANES BY CCl_3 RADICALS IN THE GAS PHASE AND THE VALUES OF LOG k AT 400 K

Substrate	log A $(M^{-1} sec^{-1})$	E (kcal mol^{-1})	log k (400 K)	Ref.
CH_4	12.3	21.6	0.58	100
CH_2Cl_2	7.8	7.3	3.84	107
CH_2ClF	9.4	11.9	2.94	107
CH_2F_2	8.6	12.4	1.87	107
$CHFCl_2$	7.8	8.8	3.02	107
CH_3Br	13.38	25.4	0.41	102
CH_2Br_2	12.17	19.7	1.48	102
$CHBr_3$	12.70	21.3	1.14	102

various hydrocarbons in the gaseous phase found that the A factor for hydrogen abstraction by CCl_3 radical from 2,3 dimethylbutane, in which the attacked hydrogen is blocked by a methyl group, is lower by one order of magnitude than for other hydrocarbons. Similar results were found also for CF_3 and CH_3 radicals, however the effect is larger in the case of CCl_3, possibly due to its larger size. For example, the ratio of the A factors for abstraction of hydrogen from cyclohexane and 2,3 dimethylbutane is $10^{0.91}$, $10^{0.6}$, and $10^{0.8}$ for CCl_3, CF_3, and CH_3, respectively.

F. Polar Effects in Halogenated Hydrocarbons

In 1968, Tedder and Watson[106] studied the reaction of trichloromethyl radicals with 1-fluorobutane and 1,1,1-trifluoropentane in the gas phase. The Arrhenius parameters reported by Tedder and Watson for those reactions were corrected 4 years later by Sidebottom et al.,[77] who recommended the following Arrhenius parameters for hydrogen abstraction by trichloromethyl radicals from the various carbon atoms in these molecules:

	XCH_2		CH_2		CH_2		CH_3	
X	log A	E	log A	E	log a	E	log A	E
H	9.7	14.3	9.5	10.7	9.5	10.7	9.7	14.3
F	9.8	12.6	—	—	9.5	10.7	10.1	14.5
CF_3	9.1	12.8	9.7	12.3	9.5	10.7	9.9	14.8

where A is in the M^{-1} sec^{-1} and E is in kcal mol^{-1}. These results show that there is no substituent effect beyond the β-carbon atom. The presence of electron attracting groups decreases the activation energy for hydrogen abstraction from α carbon atoms. The abstraction of hydrogen from several chlorofluoromethanes and bromomethanes by CCl_3 radicals in the gas phase was studied by Copp and Tedder[107] and by Hautecloque,[102] respectively. The Arrhenius parameters and the calculated rate constant at 400 K are given in Table 9 which shows that the A factors obtained by Hautecloque[100,102] are too high (higher than the diffusion rate constants), and that the difference between the A factors of the dichloromethane and chlorofluoromethane are too large (1.6). According to this the reported data should be treated cautiously. Table 9 also shows that for the fluorochloromethanes there is an inverse correlation between the C—H bond dissociation energy and the rate constant for hydrogen abstraction. By contrast, the bromomethanes do not fit this scheme.

Table 10
ARRHENIUS PARAMETERS FOR THE ABSTRACTION OF HYDROGEN BY CCl₃ IN THE LIQUID PHASE, RELATIVE TO CCl₃ COMBINATION

$$(CCl_3 + RH \xrightarrow{k_1} CHCl_3 + R, \ 2CCl_3 \xrightarrow{k_3} C_2Cl_6)$$

RH	$E_1 - E_3/2$	Log $A_1/A_3^{1/2}$	Log $k_1/k_3^{1/2}$ at 400 K	Solvent	Ref.
Alkanes					
n-Hexane	9.62 ± 0.10	3.69 ± 0.05	0.0272	n-Hexane	25
n-Hexane	9.08 ± 0.23	3.23	0.0186	n-Hexane	65
2,3 Dimethyl-butane	6.96 ± 0.26	2.40 ± 0.17	0.0396	CCl₄	20
Cyclanes					
Cyclohexane	8.81 ± 0.12	3.28 ± 0.08	0.0293	Cyclohexane	55
Cyclohexane	9.20 ± 0.50	3.53 ± 0.34	0.0319	CCl₄	70
Cyclooctane	7.90 ± 0.20	3.42 ± 0.14	0.1271	CCl₄	70
Cyclododecane	8.20 ± 0.30	3.33 ± 0.49	0.0708	CCl₄	70
Alkenes					
Cyclohexene	4.71 ± 0.29	1.93 ± 0.18	0.2274	CCl₄	38
Primary alcohols					
Ethanol	7.70 ± 0.19	2.51 ± 0.21	0.0201	CCl₄	110
n-Pentanol	8.86 + 0.62	3.23 ± 0.37	0.0245	CCl₄	110
Secondary alcohols					
2-Propanol	7.63 ± 0.21	2.99 ± 0.13	0.0663	CCl₄	111
Cyclohexanol	7.26 ± 0.43	2.85 ± 0.26	0.0765	CCl₄	111
Others					
Cholestanic esters	5.50 ± 0.40	1.70 ± 0.40	0.0495	CCl₄	112
Cholesteric esters	4.70 ± 0.50	2.00 ± 0.40	0.2706	CCl₄	112
Cholesta-4n-3-one	3.90 ± 0.30	1.00 ± 0.10	0.0740	CCl₄	112
Cholesteryl-iso-propyl ether	5.00 ± 0.60	2.40 ± 0.50	0.4660	CCl₄	112
Me₃SiH	6.43	2.97	0.2865	Cyclohexane	87
Et₃SiH	5.79	3.11	0.8847	Cyclohexane	87

G. Arrhenius Parameters for Hydrogen Abstraction

The Arrhenius parameters for hydrogen abstraction by CCl₃ radicals in the gas phase were summarized by Kerr and Moss.[108] Few results for the liquid phase were complied by Hendry et al.[109] who compiled mainly rate constants measured for one temperature and by assuming the A factors, they calculated the activation energies. From that time several new systems were studied in the liquid phase. The Arrhenius' parameters of the later studies in the liquid phase are summarized in Table 10. As all of these parameters were studied by competition with the combination of two CCl₃ radicals and as there are still questions about the rate of CCl₃ + CCl₃ → C₂Cl₆ in the liquid phase, the results are given relative to this reaction. Table 11 summarizes recent results on the abstraction of hydrogen from silanes in the gas phase which do not appear in the above-mentioned compilation. Here, the values were calculated assuming log $A_3(M^{-1} \ sec^{-1}) = 9.7$ and $E_3 = 0.0$ kcal mol⁻¹.

VI. THE ADDITION REACTIONS: CCl₃ + O₂ → CCl₃O₂ AND CF₂Cl + O₂ → CF₂ClO₂

In 1961, McCarthy and MacLachlan[114] attributed the near UV absorportion spectra which was detected during absorption spectrometry of cyclohexane and cyclohexanol to the peroxy radicals derived from those substrates. Further studies[115] have shown that in the liquid phase,

Table 11
ARRHENIUS PARAMETERS FOR HYDROGEN ABSTRACTION FROM R_3SiH AND R_3SiCH_3 BY CCl_3 RADICALS IN THE GAS PHASE.[113]

R_3SiH	log A	E	R_3SiCH_3	log A	E
$(CH_3)_3SiH$	8.7 ± 0.3	7.9 ± 0.5	$(CH_3)_3SiCH_3$	9.0 ± 0.3	12.9 ± 0.6
$(CH_3)_2ClSiH$	8.7 ± 0.3	9.1 ± 0.5	$(CH_3)_2ClSiCH_3$	8.8 ± 0.7	13.2 ± 1.6
CH_3Cl_2SiH	8.7 ± 0.3	9.5 ± 0.6	$CH_3Cl_2SiCH_3$	8.9 ± 0.2	14.1 ± 0.6
Cl_3SiH	8.6 ± 0.3	9.7 ± 0.5	Cl_3SiCH_3	8.7 ± 0.3	14.2 ± 0.6
$(C_2H_5O)_3SiH$	8.4 ± 0.4	10.1 ± 0.8	$CH_3F_2SiCH_3$	8.4 ± 0.4	14.3 ± 1.0
			$(CH_3)_3BrSiCH_3$	8.9 ± 0.3	13.0 ± 0.6

peroxy radicals generally possess structureless absorption spectra with λ_{max} in the range of 230 to 300 nm.

The major product of radiolysis and photoinduced oxidation of halocarbons is the corresponding carboxyl halide which is produced, probably, when the haloradical reacts with oxygen.[116-118] The exact process, however, is still unclear.

In order to explain the appearance of COF_2 and $COCl_2$ during the photoxidation of some halocarbons, Heicklen suggested the following one-step reaction mechanism (X = Cl, F).[119,120]

$$\text{Mechanism A:} \quad CX_3 + O_2 \rightarrow COX_2 + CXO$$

Several years later retreated Heicklen et al. from the above mechanism and suggested the following three step reaction mechanism for the production of COX_2 (X = Cl).[121-124]

$$\text{Mechanism B:} \quad CX_3 + O_2 \rightarrow CX_3O_2$$

$$2CX_3O_2 \rightarrow 2CX_3O + O_2$$

$$CX_3O \rightarrow COX_2 + X$$

Russell[125] and others, on the other hand, favored the following mechanism upon which the carboxyl halide is formed during recombination and/or wall reactions (X = F):

$$\text{Mechanism C:} \quad CX_3 + O_2 \rightarrow CX_3O_2$$

$$2CX_3O_2 \rightarrow 2CX_3O + O_2$$

$$2CX_3O \rightarrow CX_2O + CX_3OX$$

$$CX_3O + wall \rightarrow CX_2O + \ldots$$

Mechanism A was favored earlier by Heicklen, mainly because of the exothermicity of the reaction involved on that mechanism (e.g., for X = Cl, $\triangle H = -46.9$ Kcal/mol) and the failure of detection of products derived from CCl_3O (such as CCl_3OOCCl_3). Recently, Heicklen and co-workers favored mechanism B over A, based on the kinetic data obtained from chlorine-atom-sensitized oxidation study, the ozonolysis of C_2Cl_4[121,122] and the photolysis of CCl_4 and chlorofluoromethanes. Gillespie et al.[126] showed the appearance of ClO during the flash photolysis of $CFCl_3$ and CCl_3Br in the presence of oxygen. The authors, however, favored mechanism B over A by attributing the formation of ClO to secondary reaction of Cl with O_2. Otha and Mizoguchi confirmed mechanism B on the basis of kinetic data obtained from the gas phase photooxidation of chloral[127] (which was used as a source for CCl_3 radical and as a scavenger for the Cl atoms which may react with O_2 to produce ClO). The kinetic data observed at room temperature in different initial pressures of the reactants and in the presence or the absence of inert gases (such as N_2) showed the following relation among reactants and products, leading to the reaction mechanism given below.[127]

$$-R_{CCl_3CHO} = R_{Cl} = R_{COCl_2} = R_{CO} + R_{CO_2}$$

$$CCl_3CHO + h\nu \rightarrow CCl_3 + HCO$$

$$CCl_3 + O_2 \rightarrow CCl_3O_2$$

$$2CCl_3O_2 \rightarrow 2CCl_3O + O_2$$

$$CCl_3O \rightarrow COCl_2 + Cl$$

$$Cl + CCl_3CHO \rightarrow CCl_3CO + HCl$$

$$CCl_3CO \rightarrow CCl_3 + CO$$

$$CCl_3CO + O_2 \rightarrow CCl_3COO_2$$

$$2CCl_3COO_2 \rightarrow 2CCl_3CO_2 + O_2$$

$$CCl_3CO_2 \rightarrow CCl_3 + CO_2$$

Both mechanisms (B and C) involve the formation of CCl_3O_2 in the first stage. Simonaitis and Heicklen[128] obtained evidences which support the existence of CCl_3O_2 from steady-state photolysis of Cl_2-$HCCl_3$-O_2-N_2-NO-NO_2 mixtures. Niki et al.[28] identified that species during FTIR spectroscopic observation of haloalkyl peroxynitrates formed via ROO + $NO_2 \rightarrow$ ROONO$_2$ (R = CCl_3, $CFCl_2$, and CF_2Cl). Cooper et al.[129] have shown that gas-phase pulse radiolysis of $CCl_4/O_2/Ar$ mixtures yields intense absorption of light in the near UV (which was attributed to the ClO) together with broad underlying absorption which was attributed to CCl_3O_2 and which appeared only when oxygen was present on the irradiated mixtures. Similar absorption spectra were recorded with CF_3Cl and CH_4 as the radical sources and have been attributed to the peroxy radicals CF_3O_2 and CH_3O_2, respectively. Recently, Symons et al.[130] proved the existence of the CCl_3O_2 radical by using spin traps.

The kinetic data concerning the addition reaction of oxygen to halomethyl radicals are quite few. Basco et al.[131] suggested the following mechanism for CX_3O_2 production (X = H):

$$CX_3 + O_2 \rightarrow CX_3O_2^*$$

$$CX_3O_2^* + M \rightarrow CX_3O_2 + M$$

where M stands for third body stabilizing molecule.

The authors who used N_2 as M found the methyl radical decay to be third order at low pressure and second order at ~200 torr of nitrogen (the high pressure limit) in accordance with the following kinetic expressions:

$$\frac{d[CX_3O_2]}{dt} = \frac{k_1 k_2 [CX_3][O_2][M]}{k_{-1} + k_2[M]}$$

$$\frac{d[CX_3O_2]}{dt} = k_1[CX_3][O_2] \qquad \text{(if } k_2[M] \gg k_{-1})$$

$$\frac{d[CX_3O_2]}{dt} = \frac{k_1 k_2}{k_{-2}} [CX_3][O_2][M] \qquad \text{(if } k_{-1} \gg k_2[M])$$

The rate constant k_1 for the addition reaction was found by Basco et al. to be 3.1×10^8 M^{-1} sec^{-1} as compared to 1.1×10^9 M^{-1} sec^{-1} and 1.3×10^9 M^{-1} sec^{-1} reported by Hochanadel et al.[132] and Van den Bergh and Caller, respectively.[133]

The value of the rate constant for the addition of oxygen to CCl_3 was evaluated by Cooper et al.[129] who exposed mixtures of $CCl_4/O_2/Ar/C_2H_6$ to pulse radiolysis in the gas phase (in this sytem the C_2H_6 was used as a scavanger for the Cl atoms). The authors found k' — the pseudo first-order rate constant — to be linearly dependent on the oxygen pressure and concluded from the slope of the curve of k' vs. oxygen pressure the value of the rate constant for the reaction $CCl_2 + O_2 \rightarrow CCl_3O_2$ to be 3.25×10^9 M^{-1} sec^{-1}. This value was corrected to 3.09×10^9 M^{-1} sec^{-1} after substraction of the value 0.2×10^9 M^{-1} sec^{-1} attributed to C_2H_6 which was found to be an efficient third body.

VII. THE ADDITION OF NO_2 TO CCl_3, CF_2Cl^+, AND CCl_3O_2

The high yields of CCl_3 and Cl species produced in the radiolysis of CCl_4 and other halocarbons have been used to study the reactions of these species with Br_2,[134] NO,[135] olefins,[74] and other radical scavengers. The gas-phase reaction between NO_2 and CCl_3 (yielding CCl_3NO_2) was studied in 1979 by Cumming et al.[136] The authors exposed mixtures of CCl_4, NO_2, and argon to pulse radiolysis and suggested the following mechanisms in order to explain the kinetics data and the appearance of CCl_3NO_2.

Energy transfer mechanism:

$$CCl_3 + NO_2 \rightarrow CCl_3NO_2^*$$

$$CCl_3NO_2^* + M \rightarrow CCl_3NO_2 + M$$

Radical-molecule complex mechanism:

$$CCl_3 + M \rightarrow CCl_3M^*$$

$$CCl_3M^* + M \rightarrow CCl_3M + M$$

$$CCl_3M + NO_2 \rightarrow CCl_3NO_2 + M$$

Although both mechanisms were found to be in accordance with the kinetic data, the authors favored the energy transfer mechanism, which was found effective in many addition reactions (unless the third body has an affinity for the radical and cases in which the reacting radical is very simple, i.e., atoms and some diatoms).

The energy transfer mechanism was proved as an effective one also in describing the addition of oxygen to the CCl_3 radical. However, in contrast to the later system in which the CCl_3O_2 concentration was found to be independent of the concentration of both, CCl_4 and the inert gas argon (probably because of the unequality $k_2[M] \gg k_1$), the CCl_3NO_2

produced on the $CCl_4/NO_2/Ar$ system was found to be linearly dependent (by concentration) on the argon concentration (at constant NO_2 and CCl_4 concentration) and on the CCl_4 concentration (at constant Ar, NO_2). Substituting NO_2 instead of O_2 in the kinetic expressions introduced for the system produced CCl_3O_2, the ratio k_1k_2/k_{-1} for the addition of NO_2 to CCl_3 was found to be

$$\frac{k_1}{k_{-1}} k_2 = (1.2 \pm 0.3) \times 10^{11} \, M^{-2}sec^{-1} \quad (at\ 300\ K\ M = Ar)$$

$$\frac{k_1}{k_{-1}} k_2 = (1.2 \pm 0.5) \times 10^{12} \, M^{-2}sec^{-1} \quad (at\ 300\ K\ M = CCl_4)$$

The rate and the mechanism of the reaction of CF_2Cl (produced by infrared multiple-photon-induced decomposition [IRMPD] of CF_2Cl_2) with NO_2 were studied by Slagle and Gutman.[29] The kinetic data show an agreement with the following one-stage reaction which leads to the kinetic expression given below:

$$CF_2Cl + NO_2 \rightarrow CF_2ClO + NO$$

$$\frac{[CF_2ClO]}{[CF_2Cl]t} = \frac{[NO]}{[CF_2Cl]t} = k[NO_2]$$

where t stands for the reaction time.

The rate constant derived from the above expression was found to be $5.8 \times 10^9 \, M^{-1}$ sec^{-1}.

The addition of NO_2 to CCl_3O_2 in the gas phase was studied by Niki et al.[28] who exposed mixtures of $RH/NO_2/Cl_2$ ($R = CF_2Cl, CCl_3, CFCl_2$) to IR irradiation. From FTIR spectroscopic observation data (of the haloalkyl peroxynitrates produced) the following mechanism for the appearance of $CCl_3O_2NO_2$ and other species was suggested:

$$Cl + NO_2 \rightarrow ClNO_2 \quad or \quad ClONO$$

$$Cl + CHCl_3 \rightarrow HCl + CCl_3$$

$$CCl_3 + O_2 \rightarrow CCl_3O_2$$

$$CCl_3O_2 + NO_2 \rightarrow CCl_3O_2NO_2$$

$$2CCl_3O_2 \rightarrow 2CCl_3O + O_2$$

$$CCl_3O + O_2 \rightarrow CCl_2O + ClO_2$$

$$ClO_2 \rightarrow Cl + O_2$$

The last two reactions were suggested by the authors in order to explain the appearance of CCl_2O during the irradiation. Alternatively, unimolecular formation of CCl_2O from CCl_3O was also suggested:

$$CCl_3O \rightarrow CCl_2O + Cl$$

Several studies have been done in order to determine the products produced when CF_2Cl

and oxygen are involved together. Milstein and Rowland[137] photolyzed CF_2Cl_2 in the gas phase in the presence of oxygen and reported, from IR spectra data, on the appearance of COF_2 as the product of the reaction (with no indication of COFCl). Alfassi and Heusinger,[17] on the other hand, radiolyzed a similar system in the liquid phase, and reported from gas chromatograph-mass-spectra combination data on the appearance of CO_2 and COFCl. Suong and Carr[10] exposed CF_2ClOCF_2Cl in the gas phase to photolysis in the presence of oxygen and reported from gas chromatograph-IR spectra combination data on the appearance of CO_2 and COF_2 as the product of the reaction. The authors who reported on the CF_2O-to-CO_2 ratio to be 2 with no dependence on the pressures of the oxygen and the inert gas helium presented in the reaction mixtures and with no dependence on the photolysis time, found their result to be in accordance with the following steps initiated by the reaction of O_2 and CF_2Cl as suggested by Jayanty et al.[124]

$$CF_2Cl + O_2 \rightarrow CF_2ClO_2$$

$$2CF_2ClO_2 \rightarrow 2CF_2ClO + O_2$$

$$CF_2ClO \rightarrow CF_2O + Cl$$

The one step reaction $Cl + CF_2ClO_2 \rightarrow ClO + CF_2ClO$, suggested by Gillespie et al.[126] to explain the appearance of ClO during photolysis of halomethanes in the presence of oxygen, was rejected by Suong and Carr (with respect to their system) on the basis of low rate constant compared to that of the reaction $Cl + O_2 + M \rightarrow ClO_2 + M$.

Further studies concerning the reaction taking place in the presence of CF_2Cl and oxygen were done also by Francis[117] and Haszeldine and by Atkinson et al.[138]

REFERENCES

1. **Link, B., Duerk, H., Thiel, D., and Frank, H.,** Binding of trichloromethyl radicals to lipids of the hepatic endoplasmic reticulum during tetrachloromethane metabolism, *Biochem. J.,* 223, 577, 1984.
2. **Ansari, G. A. S., Moslen, M. T., and Reynolds, E. S.,** Evidence for *in vivo* covalent binding of CCl_3 derived from carbon tetrachloride to cholesterol of rat liver, *Biochem. Pharmacol.,* 31, 3509, 1982.
3. **Lai, E. K., McCay, P. B., Noguchi, T., Toshikazu, N., and Fong, K.,** *in vivo* spin-trapping of trichloromethyl radicals formed from carbon tetrachloride, *Biochem. Pharmacol.,* 28, 2231, 1979.
4. **Trudell, J. R., Boesterling, B., and Trevor, A. J.,** Reductive metabolism of carbon tetrachloride by human cytochromes P-450 reconstituted in phospholipid vesicles: mass spectral identification of trichloromethyl radical bound to dioleoyl phosphatidylcholine, *Proc. Natl. Acad. Sci. U.S.A.,* 79, 2678, 1982.
5. **Molina, M. J. and Rowland, F. S.,** Stratospheric sink for chlorofluoromethanes chlorine atom-catalyzed destruction of ozone, *Nature (London),* 249, 810, 1974.
6. **Stolarski, R. S. and Cicerone, R. J.,** Stratospheric chlorine: a possible sink for ozone, *Can. J. Chem.,* 52, 1610, 1974.
7. **Wofsi, S. C. and McElroy, M. B.,** HO_x, NO_x and ClO_x: their role in atmospheric photochemistry, *Can. J. Chem.,* 52, 1582, 1974.
8. **Crutzen, P.,** A review of upper atmospheric photochemistry, *Can. J. Chem.,* 52, 1569, 1974.
9. **Taylor, P. D., Tuckerman, R. T., and Whittle, E.,** U. V. absorption spectrum and photochemistry of CF_2ClBr, *J. Photochem.,* 19, 277, 1982.
10. **Suong, J. Y. and Carr, R. W., Jr.,** The photooxidation of 1,3-dichlorotetrafluoroacetone: mechanism of the reaction of CF_2Cl with oxygen, *J. Photochem.,* 19, 295, 1982.
11. **Seigneur, C., Caram, H., and Carr, R. W., Jr.,** Atmospheric diffusion and chemical reaction of the chlorofluoromethanes $CHFCl_2$ and CHF_2Cl, *Atmos. Environ.,* 11, 205, 1977.
12. **Rowland, F. S.,** Atmospheric chlorine and stratospheric ozone, *J. Photochem.,* 17, 413, 1981.
13. **Lovelock, J. E., Maggs, R. J., and Wade, R. J.,** Halogenated hydrocarbons in and over the Atlantic, *Nature (London),* 241, 194, 1973.

14. **Lovelock, J. E.**, Atmospheric halocarbons and stratospheric ozone, *Nature (London)*, 252, 292, 1974.
15. **Yamamoto, T. and Ootsuka, N.**, Radiation damage of fluorocarbon by krypton-85 beta rays: radiolysis of trichlorofluoromethane by cobalt-60 gamma rays, *J. Nucl. Sci. Tech.*, 17, 913, 1980.
16. **Yamamoto, J. and Ootsuka, N.**, Radiation damage of fluorocarbon by krypton-85 beta rays: radiolysis of some fluorocarbons by cobalt-60 gamma rays, *J. Nucl. Sci. Tech.*, 18, 913, 1981.
17. **Alfassi, Z. B. and Heusinger, H.**, The radiation chemistry of CF_2Cl_2 in the liquid phase, *Radiat. Phys. Chem.*, 22, 995, 1983.
18. **Mosseri, S., Alfassi, Z. B., Furst, W., and Heusinger, H.**, The radiolysis of mixtures of CF_2Cl_2 with RH substrates, *Radiat. Phys. Chem.*, 26, 89, 1985.
19. **Furst, W. and Heusinger, H.**, Effect of oxygen, nitrogen dioxide, water, tetrachloroethylene and cyclohexane on the radiolysis of dichlorodifluoromethane, 1,2 dichloro1,1,2,2 tetrafluoroethane and 1,1,2-trichloro-1,2,2-trifluoroethane in the liquid phase, *Atomkernenerg. Kerntech.*, 45, 111, 1984.
20. **Alfassi, Z. B. and Feldman, L.**, Kinetics of radiation induced abstraction of hydrogen atoms by trichloromethyl radicals in the liquid phase: 2,3-dimethylbutane, *Int. J. Chem. Kinet.*, 12, 379, 1980.
21. **Cadogan, J. I. G. and Sadler, I. H.**, Quantitative aspects of radical addition: rate constant ratios for addition of trichloromethyl and thiyl radicals of olefins, *J. Chem. Soc. B*, 1191, 1966.
22. **Tedder, J. M. and Walton, J. C.**, Rate of combination of trichloromethyl radicals, *Trans. Faraday Soc.*, 63, 2464, 1967.
23. **Lee, K. H.**, Polar effects in hydrogen abstraction of benzaldehydes. Abstraction by the trichloromethyl radical, *Tetrahedron*, 24, 4793, 1968.
24. **White, M. L. and Kuntz, R. R.**, The pyrolysis of hexachloroethane, *Int. J. Chem. Kinet* 5, 187, 1973.
25. **Katz, M. G., Baruch, G., and Rajbenbach, L. A.**, Radiation induced dechlorination of carbon tetrachloride: the kinetics of liquid phase reactions of trichloromethyl radicals in n-hexane solutions, *Int. J. Chem. Kinet.*, 8, 599, 1976.
26. **Horowitz, A. and Baruch, G.**, Addition of trichloromethyl radicals to tetrachloro-ethylene, *Int. J. Chem. Kinet.*, 11, 1263, 1979.
27. **Kuruc, J.**, Quantum yields of photosubstitution of the nitro group in nitrobenzene-tetrachloromethane mixtures, *J. Radioanal. Nucl. Chem.*, 82, 93, 1984.
28. **Niki, H., Maker, P. D., Savage, C. H., and Breitenbach, L. P.**, FTIR spectroscopy observation of haloalkyl peroxynitrates formed via $ROO + NO_2$ $ROONO_2$ ($R = CCl_3$, $CFCl_2$ and CF_2Cl), *Chem. Phys. Lett.*, 61, 100, 1979.
29. **Slagle, I. R. and Gutman, D.**, Kinetics of free radicals produced by infrared multiphoton-induced decomposition: formation of acetyl and chlorodifluoromethyl radicals and their reactions with nitrogen dioxide, *J. Am. Chem. Soc.*, 104, 4741, 1982.
30. **Rebbert, R. E.**, Gas phasephotolysis of CF_2Cl_2, $CFCl_3$ and CCl_4 in the presence of bromine at 213.9, 163.3, 147.0 and 123.6 nm. *J. Photochem.*, 8, 363, 1978.
31. **Hautecloque, S. and Bernas, A.**, Photolyse UV de $CFCl_3$ et CF_2Cl_2 en presence de divers capteurs de radicaux, *J. Photochem.*, 7, 73, 1977.
32. **Majer, J. R., Olavesen, C., and Robb, J. C.**, Photolysis of chloropentafluoroacetone (CF_3COCF_2Cl), *J. Chem. Soc. A*, 893, 1969.
33. **Majer, J. R., Olavesen, C., and Robb, J. C.**, Wavelength effect in the photolysis of halogenated ketones, *J. Chem. Soc. B*, 48, 1971.
34. **Bellas, M. G., Strausz, O. P., and Gunning, H. E.**, The Hg $6(^3P_1)$ photosensitization of chlorodifluoromethane, *Can. J. Chem.*, 43, 1022, 1965.
35. **Pritchard, G. O. and Perona, M. J.**, The decomposition of vibrationally excited 1,1,2,2-tetrafluoro-1-chloroethane, *J. Phys. Chem.* 73, 2944, 1969.
36. **Watkins, P. A. and Whittle, E.**, unpublished work.
37. **Bibler, N. E.**, The radiolysis of carbon tetrachloride, radical yields and the formation of tetrachloroethylene as an initial product, *J. Phys. Chem.*, 75, 24, 1971.
38. **Alfassi, Z. B. and Feldman, L.**, The kinetics of radiation induced hydrogen abstractions by trichloromethyl radicals in the liquid phase: cyclohexene, *Int. J. Chem. Kinet.*, 13, 771, 1981.
39. **Katz, M. G., Horowitz, A., and Rajbenbach L. A.**, Abstraction of chlorine atoms from chloromethanes by the cyclohexyl radical, *Int. J. Chem. Kinet.*, 7, 183, 1975.
40. **Alfassi, Z. B.**, The rate constant for combination of trichloromethyl radical with cyclohexyl or n-hexyl radicals. An evidence for the value of the rate constant for a chlorine atom transfer, *Int. J. Chem. Kinet.*, 12, 217, 1980.
41. **Eckling, R., Goldfinger, P., Huybrechts, G., Martens, G., Meyers, L., and Smoes, S.**, Kinetische Studien uber die Konkurrenz bei Photo-chlorierungsreaktionen, *Chem. Ber.*, 93, 3104, 1960.
42. **DeMare, G. R. and Huybrechts, G.**, Rate constants for the recombination of CCl_3 radicals and for their reactions with Cl, Cl_2 and HCl in the gas phase, *Trans. Faraday Soc.*, 64, 1311, 1968.
43. **White, M. L. and Kuntz, R. R.**, Rotating sector study of the gas phase photolysis of the carbon tetrachloride-cyclohexane systems, *Int. J. Chem. Kinet.*, 3, 127, 1971.

44. **Matheson, I. A., Sidebottom, H. W., and Tedder, J. M.,** The reaction of trichloromethyl radicals with hydrogen chloride and a new estimation of the rate of combination of trichloromethyl radicals, *Int. J. Chem. Kinet.,* 6, 493, 1974.
45. **Melville, H. W., Robb, J. C., and Tutton, R. C.,** The kinetics of the interaction of trichloromethyl radicals with cyclohexene, *Disc. Faraday Soc.,* 10, 154, 1951.
46. **Melville, H. W., Robb, J. C., and Tutton, R. C.,** The kinetics of the addition of bromochloromethane to unsaturated compounds; cyclohexene and vinyl acetate, *Disc. Faraday Soc.,* 14, 150, 1953.
47. **Bengough, W. I. and Thomson, R. A. M.,** Addition of bromotrichloromethane to vinyl acetate: kinetics of the photochemical reaction, *Trans. Faraday Soc.,* 10, 162, 1951.
48. **Carlsson, D. J., Howard, J. A., and Ingold, K. U.,** Absolute rate constants for the combination of trichloromethyl radicals and for their reaction with t-butyl hypochlorite, *J. Am. Chem. Soc.,* 88, 47726, 1966.
49. **Carlsson, D. J., Ingold, K. U., and Bray, L. C.,** Kinetics and rate constants for the reaction of tri-n-butylgermanium hydride with methyl iodide and carbon tetrachloride. The combination of methyl and trichloromethyl radicals in solution, *Int. J. Chem. Kinet.,* 1, 315, 1969.
50. **Carlsson, D. J. and Ingold, K. U.,** Reactions of alkoxy radicals: the kinetics and absolute rate constants for some t-butyl hypochlorite chlorinations, *J. Am. Chem. Soc.,* 89, 4891, 1967.
51. **Watts, G. B. and Ingold, K. U.,** Kinetic applications of electron paramagnetic resonance spectroscopy: self reactions of some group IV radicals, *J. Am. Chem. Soc.,* 94, 491, 1972.
52. **Griller, D. and Ingold, K. U.,** Rate constants for the bimolecular self-reactions of ethyl, isopropyl and tert-butyl radicals in solution, a direct comparison, *J. Am. Chem. Soc.,* 6, 453, 1974.
53. **Paul, H.,** Rate constant for the combination of trichloromethyl radicals and the electron transfer reaction between hydroxymethyl radicals and carbon tetrachloride in solution, *J. Am. Chem. Soc.,* 11, 495, 1976.
54. **Paul, H.,** Second order rate constants and CIDEP enhancements of transient radicals in solution by modulation ESR spectroscopy. *Chem. Phys.* 15, 115, 1976.
55. **Katz, M. G., Baruch, G., and Rajbenbach, L. A.,** Radiation induced dechlorination of carbon tetrachloride in cyclohexane solutions. The kinetics of liquid phase reactions of trichloromethyl radicals. *Int. J. Chem. Kinet.,* 8, 131, 1976.
56. **Currie. J., Sidebottom, H., and Tedder, J.,** The reaction of cyclohexyl radicals with carbon tetrachloride, *Int. J. Chem. Kinet.,* 6, 481, 1974.
57. **Wampler, F. B. and Kuntz, R. R.,** Some hydrogen abstraction reactions of the trichloromethyl radical, *Int. J. Chem. Kinet.,* 3, 283, 1971.
58. **McCall, D. W., Douglass, D. C., and Anderson, E. W.,** Self-diffusion by means of nuclear magnetic resonance spin-echo techniques, *Ber. Bunsenges. Phys. Chem.,* 67, 336, 1963.
59. **Lesigne, B., Gilles, L., and Woods, R. J.,** Spectra and decay of trichloromethyl radicals in aqueous solution, *Can. J. Chem.,* 52, 1135, 1974.
60. **Roquitte, B. C. and Wijnen, M. H. J.,** The photolysis of carbon tetrachloride in the presence of ethane and ethylene, *J. Am. Chem. Soc.,* 85, 2053, 1963.
61. **Roquitte, B. C. and Wijnen, M. H. J.,** Recombination and disproportionation reactions of monochloroethyl and trichloromethyl radicals, *J. Chem. Phys.,* 38, 4, 1963.
62. **Gregory, J. E. and Wijnen, M. H. J.,** Photolysis of mixtures of diethylketone and carbon tetrachloride, *J. Chem. Phys.,* 38, 2925, 1963.
63. **Yu, W. H. S. and Wijnen, M. H. J.,** Photolysis of chloroform in the presence of ethane at 25°C, *J. Chem. Phys.,* 52, 2736, 1970.
64. **Yu, W. H. S. and Wijnen, M. H. J.,** Photolysis of carbon tetrachloride in the presence of ethyl chloride at 25°C, *J. Chem. Phys.,* 52, 4166, 1970.
65. **Matheson, I., Tedder, J., and Sidebottom, H.,** Photolysis of carbon tetrachloride in the presence of alkanes, *Int. J. Chem. Kinet.,* 14, 1033, 1982.
66. **Nguyen, T. Q. and Gaumann, T.,** Radiation chemistry of hydrocarbons: the binary system n-hexane-CCl₄ in the liquid state, *Radiat. Phys. Chem.,* 11, 183, 1978.
67. **Nguyen, T. Q., Dang, T. M., and Gaumann, T.,** Hydrogen-Deuterium isotope effect in the gamma-initiated chain reaction between carbon tetrachloride and cyclohexane in the liquid phase, *Radiat. Phys. Chem.,* 15, 223, 1980.
68. **Huyser, E. S.,** Addition and abstraction reactions of the trichloromethyl radical with olefins, *J. Org. Chem.,* 26, 3261, 1961.
69. **Skell, P. C. and Woodworth, R. C.,** The stereochemistry of radical-olefin addition reactions. Reactions of *cis-* and *trans-2-*butenes with bromotrichloromethane, *J. Am. Chem. Soc.,* 79, 4638, 1955.
70. **Alfassi, Z. B. and Feldman, L.,** The kinetics of radiation induced hydrogen abstraction by CCl₃ radicals in the liquid phase: cyclanes, *Int. J. Chem. Kinet.,* 13, 517, 1981.
71. **Tedder, J. M. and Walton, J. C.,** Free radical addition to olefins; addition of bromotrichloromethane to ethylene, *Trans. Faraday Soc.,* 60, 1769, 1964.

72. **Tedder, J. M. and Walton, J. C.,** Free radical addition to olefins: addition of trichloromethyl-radicals to fluoroethylenes, *Trans. Faraday Soc.,* 62, 1859, 1966.

73. **Tedder, J. M. and Walton, J. C.,** Free radical addition to olefins: addition of trichloromethyl radicals to propene, 2-fluoropropene and hexafluoropropene, *Trans. Faraday Soc.,* 63, 2679, 1967.

74. **Johari, D. P., Sidebottom, H. W., Tedder, J. M., and Walton, J. C.,** Free radical addition to olefins; addition of trichloromethyl radicals to chloroolefins, *J. Chem. Soc. B,* 95, 1971.

75. **Kerr, J. A. and Parsonage, M. J.,** The kinetics of the addition of trichloromethyl radicals to olefins in the gas phase; recalculation of the data, *Int. J. Chem. Kinet.,* 4, 243, 1972.

76. **Sidebottom, H. W., Tedder, J. M., and Walton, J. C.,** Photolysis of bromotrichloromethane, *Trans. Faraday Soc.,* 65, 755, 1969.

77. **Sidebottom, H. W., Tedder, J. M., and Walton, J. C.,** Arrhenius parameters for the reactions of trichloromethyl radicals, *Int. J. Chem. Kinet.,* 4, 249, 1972.

78. **Horowitz, A. and Baruch, G.,** Kinetics and mechanism of the addition of trichloromethyl radicals to chloroethylenes. The addition to cis-$C_2H_2Cl_2$, $trans$-$C_2Cl_2H_2$ and C_2Cl_3H, *Int. J. Chem. Kinet.,* 12, 883, 1980.

79. **Tedder, J. M. and Walton, J. C.,** Kinetics and mechanism of the addition of fluorine containing radicals to olefins, *Am. Chem. Soc. Symp. Ser.,* 66, 107, 1978.

80. **Kim, V., Shostenko, A. G., Zagorets, P. A., Shilin, S. A., and Malkov, A. V.,** Effect of the media on the reaction capacity of the trichloromethyl radical reacting with 1-hexene, *React. Kinet. Catal. Lett.,* 18, 159, 1981.

81. **El Soueni, A., Tedder, J. M., and Walton, J. C.,** *J. Fluorine Chem.,* 17, 51, 1981.

82. **Kooyman, E. C., and Farenhorst,** The relative reactivities of polycyclic aromatics towards trichloromethyl radicals, *Trans. Faraday Soc.,* 49, 58, 1953.

83. **Arnold, J. C., Gleicher, G. J. and Unruh, J. D.,** Substituent effects in the radical trichloromethylation of 9-x-anthracenes. An observed linear free energy relationship, *J. Am. Chem. Soc.,* 96, 787, 1974.

84. **Pryor, W. A., Fuller, D. L., and Stanley, J. P.,** Reactivity patterns of the methyl radical, *J. Am. Chem. Soc.,* 94, 1632, 1972.

85. **Weissman, M. and Benson, S. W.,** Heat of formation of the $CHCl_2$ radical. Bond dissociation energies in chloromethanes and chloroethanes, *J. Phys. Chem.,* 87, 243, 1983.

86. **Kerr, J. A. and Trotman-Dickenson, A. F.,** Strength of chemical bonds in *CRC Handbook of Chemistry and Physics,* 63rd ed., CRC Press, Boca Raton, Fla., 1982, F-205.

87. **Russel, G. A. and DeBoer, C.,** Substitutions at saturated carbon hydrogen bonds utilizing molecular bromine or bromotrichloromethane, *J. Am. Chem. Soc.,* 85, 3136, 1963.

88. **Baruch, G. and Horowitz, A.,** Liquid phase reactions of CCl_3 radicals with trimethylsilane and triethylsilane, *J. Phys. Chem.,* 84, 2535, 1980.

89. **Ormson, B. N., Perrymore, W. D., and White, M. L.,** Hydrogen abstraction from trichlorosilane by the trichloromethyl and 1,1 dichlorotrifluoroethyl radicals, *Int. J. Chem. Kinet.,* 9, 663, 1977.

90. **Huyser, E. S.,** Relative reactivities of substituted toluenes towards trichloromethyl radicals, *J. Am. Chem. Soc.,* 82, 394, 1960.

91. **Tanner, D. D., Arhart, R. J., Blackburn, E. V., Das N. C., and Wada, N.,** Polar radicals. Reversible hydrogen abstraction in the mechanism of the bromination reactions of bromotrichloromethane. Polar effects in the abstraction reactions of the trichloromethyl free radical, *J. Am. Chem. Soc.,* 96, 829, 1974.

92. **Toterow, W. D. and Gleicher, G. J.,** Steric effects in hydrogen atom abstraction, *J. Am. Chem. Soc.,* 91, 7150, 1969.

93. **Gleicher, G. J.,** Free radical hydrogen atom abstraction from substituted cumenes, *J. Org. Chem.,* 33, 332, 1968.

94. **Chang, E. P., Huang, R. L., and Lee, K. H.,** Polar influences in radical reactions. Hydrogen abstraction from nuclear-substituted ethylbenzenes, benzaldehyde dimethyl acetals and diphenylmethyl methyl ethers by the trichloromethyl radical, *J. Chem. Soc. B,* p. 878, 1969.

95. **Martin, M. M. and Gleicher, G. J.,** A Hammet study of hydrogen abstraction from substituted allylbenzenes, *J. Org. Chem.,* 28, 3266, 1963.

96. **Huang, R. L., and Lee, K. H.,** Polar influences in radical reactions. The abstraction of benzylic hydrogen atoms from substituted benzyl methyl ethers and benzaldehyde dimethyl acetals by atomic bromine, *J. Chem. Soc.,* p. 5963, 1964.

97. **Huyser, E. S., Schimke, H., and Burham, R. L.,** Competitive reactions of cycloalkanes with trichloromethanesulfonyl chloride and bromotrichloromethane, *J. Org. Chem.,* 28, 2141, 1963.

98. **Wong, R. H. W. and Gleicher, G. J.,** Internal strain in benzylic radical formation. The effect of ring size in the reaction of trichloromethyl radicals with benzocycloalkenes, *J. Org. Chem.,* 38, 1957, 1973.

99. **Unruh, J. D. and Gleicher, G. J.,** Hydrogen abstractions from arylmethanes, *J. Am. Chem. Soc.,* 93, 2008, 1971.

100. **Hautecloque, S.,** Cinetique des reactions $CCl_3 + H_2(D_2)$ et $CCl_3 + CH_4(CD_4)$. Effect isotopique, *J. Chim. Phys.,* 67, 771, 1970.

101. **Hautecloque, S.,** Reactions des radicaux CCl$_3$ avec C$_2$H$_6$ et C$_2$D$_6$. Effect isotopique, *C.R. Acad. Sci. Paris Ser. C,* 272, 2094, 1971.

102. **Hautecloque, S.,** Reactions des radicaux trichlorométhyle avec des bromomethanes et deuteriobromomethanes. Effect isotopique primaire, *J. Chim. Phys.,* 71, 13, 1974.

103. **Hautecloque, S.,** Reactions des radicaux CCl$_3$ avec des bromomethanes et deuteriobromomethanes. *C.R. Acad. Sci. Paris Ser. C,* 280, 701, 1975.

104. **Hautecloque, S. and Nguyen, T. M. N.,** Reactions des radicaux CBr$_3$ avec H$_2$ et D$_2$, *C.R. Acad. Sci. Paris Ser. C,* 280, 609, 1975.

105. **Johnston, H. S.,** *Gas Phase Reaction Rate Theory,* Ronald Press, 1966.

106. **Tedder, J. M. and Watson, R. A.,** Free radical substitution in aliphatic compounds. Reaction of trichloromethyl radicals with 1-fluorobutane and 1,1,1 trifluoropentane, *Trans. Faraday Soc.,* 64, 1304, 1968.

107. **Copp, D. E. and Tedder, J. M.,** Photolysis of bromotrichloromethane in the presence of dichloromethane, difluoromethane, chlorofluoromethane and dichlorofluoromethane, *J. Chem. Soc. Faraday Trans. 1,* 72, 1177, 1976.

108. **Kerr, J. A. and Moss, S. J.,** *CRC Handbook of Bimolecular and Termolecular Gas Reactions,* Vol. 1, CRC Press, Boca Raton, Fla. 1981, 270.

109. **Hendry, D. G., Mill, T., Piszkiewicz, L., Howard, J. A., and Eigenmann, H. K.,** A critical review of H atom transfer in the liquid phase: chlorine atom, alkyl, trichloromethyl, alkoxy and alkylperoxy radicals, *J. Phys. Chem. Ref. Data,* 3, 937, 1974.

110. **Feldman, L. and Alfassi, Z. B,** The liquid phase reaction of CCl$_3$ radicals with primary alcohols, *Int. J. Chem. Kinet.,* 14, 659, 1982.

111. **Feldman, L. and Alfassi, Z. B.,** The kinetics of radiation induced hydrogen abstraction by CCl$_3$ radicals in the liquid phase: secondary alcohols, *J. Phys. Chem.,* 85, 3060, 1981.

112. **Feldman, L. and Alfassi, Z. B.,** Radiolysis of cholesteric and cholestanic esters in CCl$_4$ solutions, *Rad. Phys. Chem.,* 15, 687, 1980.

113. **Rice, J. A., Treacy, J. J., and Sidebottom, H. W.,** Reactions of trichloromethyl radicals with organosilicon compounds. *Int. J. Chem. Kinet.,* 16, 1505, 1984.

114. **McCarthy, R. L. and MacLachlan, A.,** Observation of free-radical kinetics in the cyclohexane-oxygen system, *J. Phys. Chem.,* 35, 1625, 1961.

115. **For compilation, see Habersbergerova, A., Janovsky, I., and Kourim, P.,** Absorption spectra of intermediates formed during radiolysis and photolysis. II, *Radiat. Res. Rev.,* 4, 123, 1972.

116. **Lyons, E. H., Jr., and Dickenson, R. G.,** The photo-oxidation of liquid carbon tetrachloride, *Am. Chem. Soc.,* 57, 443, 1935.

117. **Francis, W. C. and Haszeldine, R. N.,** Oxidation of polyhalogenocompounds. I, *J. Chem. Soc.,* 2151, 1955.

118. **Marsh, D. and Heicklen, J.,** Photolysis of fluorotrichloromethane, *J. Phys. Chem.,* 69, 4410, 1965.

119. **Heicklen, J.,** Photolysis of trifluoroiodomethane in the presence of oxygen and nitric oxide, *J. Phys. Chem.,* 70, 112, 1966.

120. **Heicklen, J.,** in *Advances in Photochemistry,* Vol. 7, Wiley, New York, 1979, 57.

121. **Sanhueza, E., Hisatsune, I. C., and Heicklen, J.,** Oxidation of haloethylenes, *Chem. Rev.,* 76, 801, 1976.

122. **Mathias, F., Sanhueza, E., Hisatsune, I. C., and Heicklen, J.,** The chlorine atom sensitized oxidation and the ozonolysis of C$_2$Cl$_4$, *Can. J. Chem.,* 52, 3852, 1974.

123. **Jayanty, R. K. M., Simonaitis, R., and Heicklen, J.,** The photolysis of CCl$_4$ in the presence of O$_2$ or O$_3$ at 213.9 nm and the reaction of O(^{1}D) with CCl$_4$, *J. Photochem.,* 4, 203, 1975.

124. **Jayanty, R. K. M., Simonaitis, R., and Heicklen, J.,** The photolysis of chlorofluoromethanes in the presence of O$_2$ or O$_3$ at 213.9 nm, and their reactions with O(^{1}D), *J. Photochem.,* 4, 381, 1975.

125. **Russell, G. A.,** Deuterium-isotope effects in the autoxidation of aralkyl hydrocarbons. Mechanism of the interaction of peroxy radicals, *J. Am. Chem. Soc.,* 79, 3871, 1957.

126. **Gillespie, H. M., Carraway, J., and Donovan, R. J.,** Reaction of O(^{1}D$_2$) with halomethanes, *J. Photochem.,* 7, 29, 1977.

127. **Ohta, T. and Mizoguchi, I.,** Determination of the mechanism of the oxidation of the trichloromethyl radical by the photolysis of chloral in the presence of oxygen, *J. Chem. Kinet.,* 12, 717, 1980.

128. **Simonaitis, R. and Heicklen, J.,** The reactions of CCl$_3$O$_2$ with NO and NO$_2$ and the thermal decomposition of CCl$_3$O$_2$NO$_2$, *Chem. Phys. Lett.,* 62, 473, 1979.

129. **Cooper, R., Cumming, T. B., Gordon, S., and Mulac, W. A.,** The reactions of the halomethyl radicals CCl$_3$ and CF$_3$ with oxygen, *Radiat. Phys. Chem.,* 16, 169, 1980.

130. **Symons, M. C. R., Albano, E., Slater, T. F., and Tomasi, A.,** Radiation mechanisms. XXII. Radiolysis of tetrachloromethane, *J. Chem. Soc. Faraday Trans. 1,* 78, 2205, 1982.

131. **Basco, N., James, D. J. L., and James, F. C.,** A quantitative study of alkyl radical reactions by kinetic spectroscopy. II. Combination of methyl radical with oxygen molecule, *Int. J. Chem. Kinet.,* 4, 129, 1972.

132. **Hochanadel, C. J., Ghormley, C. A., Boyle, J. W., and Orgen, P. J.,** Absorption spectrum and rates of formation and decay of the CH_3O_2 radical, *J. Phys. Chem.,* 81, 3, 1977.

133. **Callear, A. B. and Van Den Bergh, H. E.,** Relaxation of excited methyl radicals produced in the flash photolysis of dimethyl mercury, *Int. J. Chem. Kinet.,* 8, 777, 1976.

134. **Bibler, N. E.,** Free radical scavenging at high dose rate, in the radiolysis of liquid carbon tetrachloride, *J. Phys. Chem.,* 77, 167, 1973.

135. **Henglein, A.,** in *Large Radiation Sources in Industry,* Vol. 2, I.A.E.A., Vienna, p. 142, 1960.

136. **Cumming, J. B., Cooper, R., Mulac, W. A., and Gordon, S.,** The reaction of CCl_3 radicals with NO_2, *Radiat. Phys. Chem.,* 16, 207, 1980.

137. **Millstein, R. and Rowland, F. S.,** Quantum yield for the photolysis of CF_2Cl_2 in O_2, *J. Phys. Chem.,* 79, 669, 1975.

Chapter 13

KINETICS OF CARBENE REACTIONS*

J. C. Scaiano

TABLE OF CONTENTS

---

* Issued as NRCC- 25127.

I. INTRODUCTION

Carbenes have been a subject of interest to organic chemists for over 50 years. The chemistry of these highly reactive divalent carbon species has been largely elucidated on the basis of product studies that have revealed many details of their reactivity.[1-4]

During the last decade our understanding of the chemistry of carbenes has been greatly enhanced by the user of pulsed time-resolved techniques which have allowed the determination of numerous rate constants for carbene reactions. The work by Closs and Rainbow,[5] published in 1976 is a landmark in this area. These authors examined the reactions of diphenylcarbene and some of its ring substituted derivatives using conventional flash photolysis techniques. More recently, nanosecond laser photolysis techniques and picosecond spectroscopy have extended dramatically the time domain in which carbene reactions can be studied.[6,7] This review covers kinetic aspects of carbene chemistry that have been studied using flash techniques. The majority of the kinetic data covered by *Chemical Abstracts* up to September 1985 has been included, but no attempt has been made to be exhaustive. Only solution data have been included. Rate constants based on competitive flash studies have only been included when the absolute rate constant used as a reference had been included in the same report and had been determined under the same conditions as the competitive data.

Only "free-carbenes" are included in this review. In particular carbene-metal systems are not covered.

A few carbene reactions have been repeatedly studied. Each reported value has been included as a separate entry in the tables, even in those cases when the same laboratory has reported values on several occasions that differ only slightly.

Reported rate constants frequently include a considerable degree of elaboration on the raw data. When this was the case, I have tried to stay as close as possible to the experimental raw data; this applies particularly to some rate constants for reactions of singlet carbenes when these have been derived from studies of triplet reactivity *(vide infra)*. Chart 1 shows the structures and abbreviations used for the carbenes included in this report; "X" is used to represent a generic substituent to be specified in each case.

Most of the carbenes included in Chart 1 have triplet ground states, as determined by low temperature EPR spectroscopy. Only XA, DMF*l,* and the aryl halo carbenes have low lying singlet states.

The possible involvement of two electronic states of different multiplicity (i.e. singlet and triplet) has been at the center of most reports of carbene reactivity, whether these have involved product and/or time-resolved studies. Normally the triplet and the lowest lying singlet are attributed "radical-like" and "zwitterionic" properties, respectively.

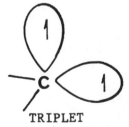

TRIPLET

SINGLET

A second singlet, isoelectronic with the triplet state, is generally believed to be of higher energy than the one shown above. Multiplicity and orbital occupancy clearly play a determining role in most carbene reactions.

CHART I

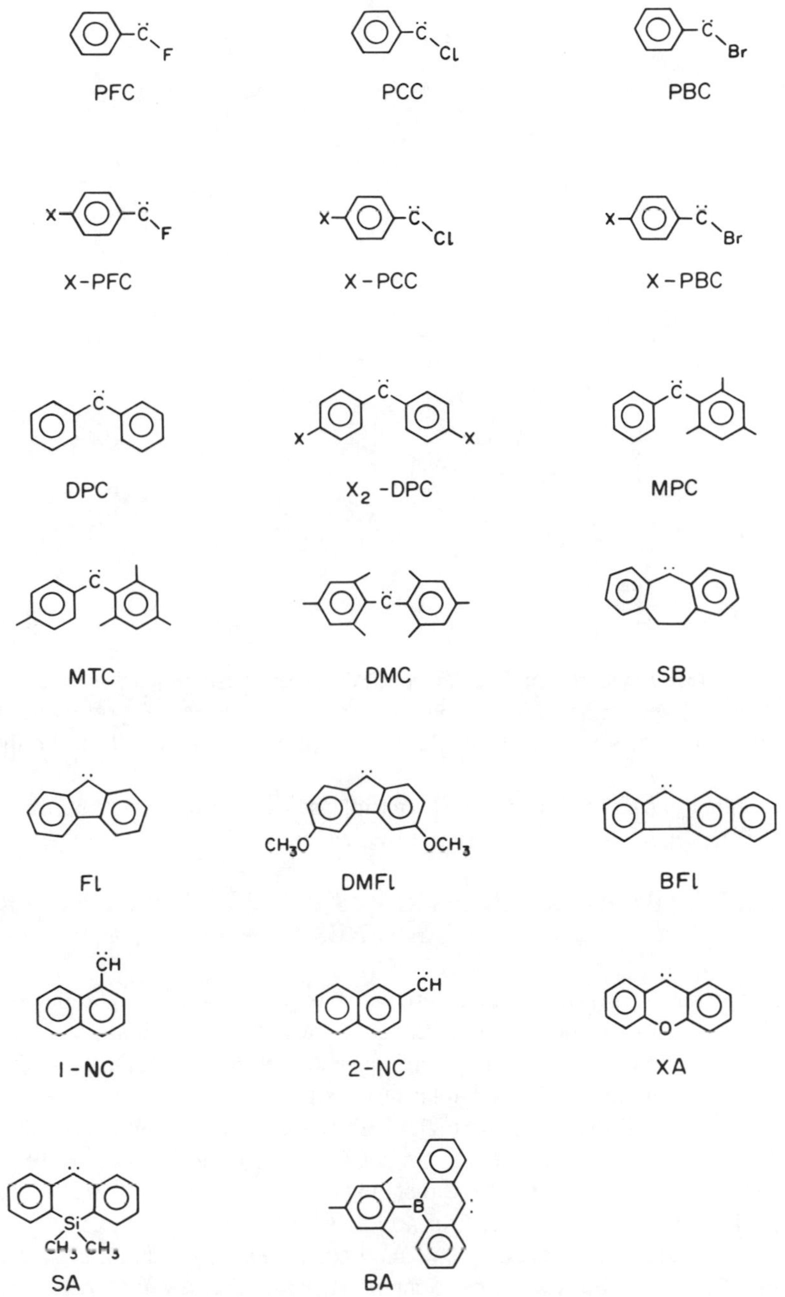

PFC

PCC

PBC

X-PFC

X-PCC

X-PBC

DPC

X₂-DPC

MPC

MTC

DMC

SB

FL

DMFL

BFL

1-NC

2-NC

XA

SA

BA

II. PREPARATION OF CARBENES

There are numerous reactions in organic chemistry which are believed to lead to, or involve the intermediacy of, carbenes. Three common sources use diazo compounds, diazirines, and ketenes as precursors; these are illustrated in that order in Scheme I.[1-4] In general, diazirines have been the precursors of choice for aryl halo carbenes[8] and diazos for diaryl or monoaryl carbenes.[6]

SCHEME I

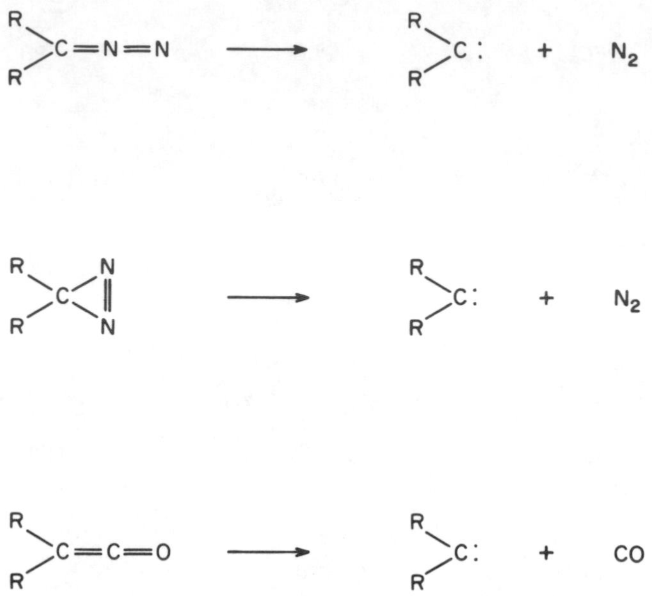

Carbene generation via thermolysis of photolysis of these precursors leads to singlet carbenes. This is a very important consideration, since even in those systems where the low lying carbene state is the triplet, the singlet-triplet equilibrium is usually fed only from the singlet.

Triplet carbenes can be generated directly in triplet sensitized reactions, but in most systems this is not a convenient approach for time-resolved studies.

III. REACTIVITY AND MECHANISMS FOR SINGLET AND TRIPLET CARBENES

Carbenes participate in numerous reactions in organic chemistry. Some of the more common reaction paths are illustrated in Scheme II, which includes both, singlet and triplet paths. In particular, in the case of the cyclopropanation reaction, path D, the singlet reaction is generally stereospecific, while the triplet process is not.

In general, the overall reactivity of singlet carbenes tends to be higher than that of triplet carbenes. If a carbene has a low lying singlet state, its reactivity is usually determined predominantly by this state, with the triplet playing a very minor, or no role at all. For example this is believed to be the case for aryl halo carbenes, such as phenylchlorocarbene (PCC), for which numerous reactions have now been studied, including substrates such as olefins, alcohols, and carboxylic acids (Scheme II, paths D and A). Experimentally these studies are straightforward: laser photolysis of the corresponding diazirine leads to the singlet carbene in a process that for all practical purposes can be regarded as instantaneous in the nanosecond time scale.[8-10] The singlet carbene is readily detectable, e.g., for PCC, λ_{max} = 308 nm in isooctane solution.[9] The carbene is long lived and insensitive to oxygen. Bimolecular rate constants can be determined directly by monitoring the change in the rate of carbene decay as a function of added quencher.[8-10]

Interpretation of the rate data for aryl halo carbenes has been far from straightforward. For example, in the case of carbene trapping by olefins, the reactions show unusual tem-

SCHEME II

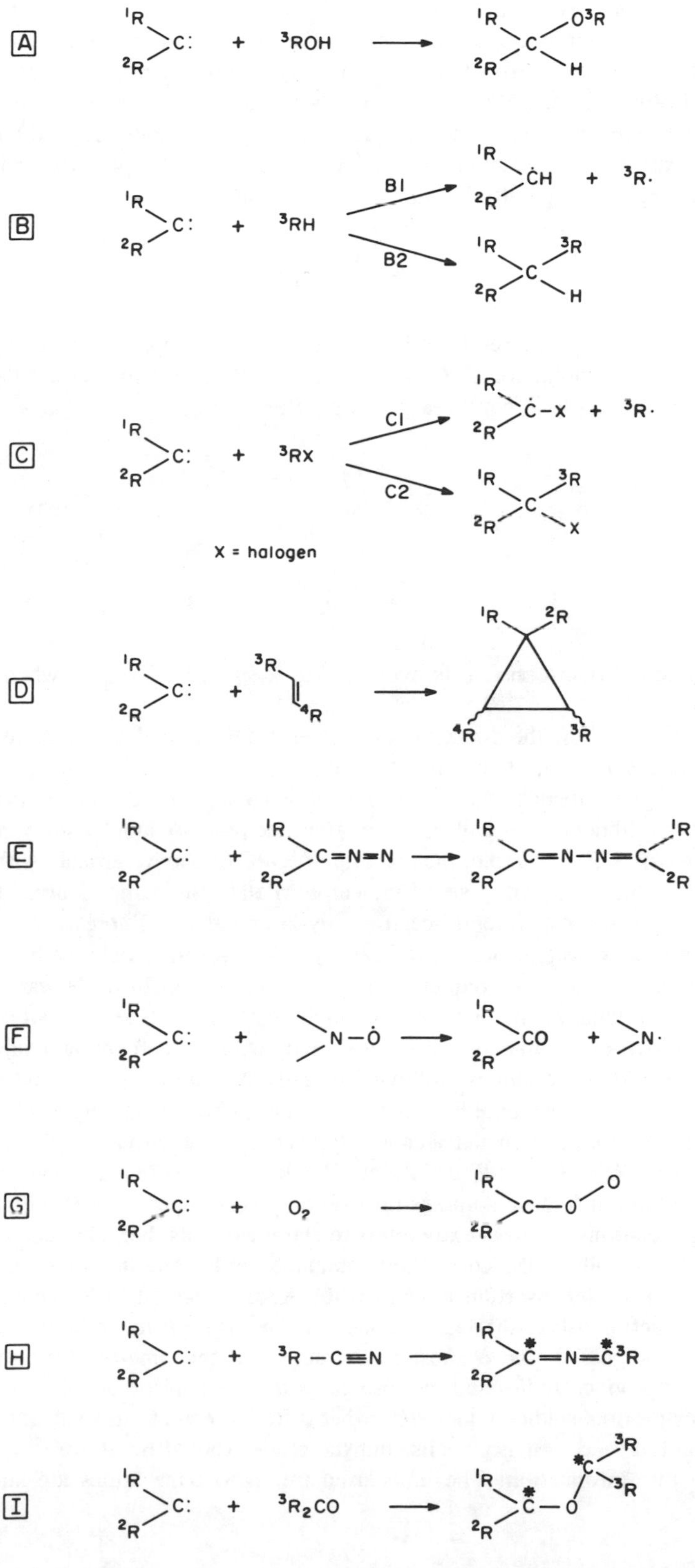

A $\quad$ $\begin{smallmatrix}{}^1R \\ {}^2R\end{smallmatrix}$C: $\quad + \quad {}^3ROH \quad \longrightarrow \quad \begin{smallmatrix}{}^1R \\ {}^2R\end{smallmatrix}C\begin{smallmatrix}O^3R \\ H\end{smallmatrix}$

B $\quad$ $\begin{smallmatrix}{}^1R \\ {}^2R\end{smallmatrix}$C: $\quad + \quad {}^3RH$

$\quad$ B1 $\longrightarrow$ $\begin{smallmatrix}{}^1R \\ {}^2R\end{smallmatrix}$ĊH $\quad + \quad {}^3R\cdot$

$\quad$ B2 $\longrightarrow$ $\begin{smallmatrix}{}^1R \\ {}^2R\end{smallmatrix}C\begin{smallmatrix}{}^3R \\ H\end{smallmatrix}$

C $\quad$ $\begin{smallmatrix}{}^1R \\ {}^2R\end{smallmatrix}$C: $\quad + \quad {}^3RX$

$\quad$ C1 $\longrightarrow$ $\begin{smallmatrix}{}^1R \\ {}^2R\end{smallmatrix}$Ċ$-$X $\quad + \quad {}^3R\cdot$

$\quad$ C2 $\longrightarrow$ $\begin{smallmatrix}{}^1R \\ {}^2R\end{smallmatrix}C\begin{smallmatrix}{}^3R \\ X\end{smallmatrix}$

X = halogen

D $\quad$ $\begin{smallmatrix}{}^1R \\ {}^2R\end{smallmatrix}$C: $\quad + \quad \begin{smallmatrix}{}^3R \\ {}^4R\end{smallmatrix}$ $\longrightarrow$ triangle: ${}^1R$ ${}^2R$ ${}^4R$ ${}^3R$

E $\quad$ $\begin{smallmatrix} \\ \end{smallmatrix}$C: $\quad + \quad \begin{smallmatrix}{}^1R \\ {}^2R\end{smallmatrix}$C=N=N $\longrightarrow$ $\begin{smallmatrix}{}^1R \\ {}^2R\end{smallmatrix}$C=N$-$N=C$\begin{smallmatrix}{}^1R \\ {}^2R\end{smallmatrix}$

F $\quad$ $\begin{smallmatrix}{}^1R \\ {}^2R\end{smallmatrix}$C: $\quad + \quad \rangle$N$-\dot{O}$ $\longrightarrow$ $\begin{smallmatrix}{}^1R \\ {}^2R\end{smallmatrix}$CO $\quad + \quad \rangle$N$\cdot$

G $\quad$ $\begin{smallmatrix}{}^1R \\ {}^2R\end{smallmatrix}$C: $\quad + \quad O_2 \quad \longrightarrow \quad \begin{smallmatrix}{}^1R \\ {}^2R\end{smallmatrix}C-O-$O

H $\quad$ $\begin{smallmatrix}{}^1R \\ {}^2R\end{smallmatrix}$C: $\quad + \quad {}^3R-C{\equiv}N \quad \longrightarrow \quad \begin{smallmatrix}{}^1R \\ {}^2R\end{smallmatrix}$C=N=C${}^3R$

I $\quad$ $\begin{smallmatrix}{}^1R \\ {}^2R\end{smallmatrix}$C: $\quad + \quad {}^3R_2CO \quad \longrightarrow \quad \begin{smallmatrix}{}^1R \\ {}^2R\end{smallmatrix}C-O-C\begin{smallmatrix}{}^3R \\ {}^3R\end{smallmatrix}$

perature dependencies, frequently leading to curved Arrhenius plots and negative activation energies; the reader is referred to the original literature for detailed discussions of these problems.[11] The reactions of carbenes with alcohols are also complicated due to the different reactivity of alcohol monomers and oligomers. For example in the case of *p*-chlorophenyl-chloro-carbene (C*l*-PCC) in isooctane at room temperature the rate constants for methanol monomer and oligomers are 2×10^7 and 4.3×10^9 $M^{-1}\text{sec}^{-1}$, respectively.[10]

When the carbene under examination has a triplet ground state (e.g., diphenylcarbene, DPC) and the carbene is generated in its singlet state, the singlet triplet equilibrium is rapidly established through singlet-to-triplet intersystem crossing

$$^1\text{DPC} \xrightarrow{k_{ST}} {}^3\text{DPC} \tag{1}$$

Reaction 1 and similar processes for other carbenes have been examined using picosecond spectroscopy. For example, for DPC $k_{ST} = 3.23 \times 10^9$ sec^{-1} in acetonitrile at 300 K.[12] Intersystem crossing leads to equilibration, e.g., Reaction 2.

$$^3\text{DPC} \underset{k_{ST}}{\overset{k_{TS}}{\rightleftharpoons}} {}^1\text{DPC} \tag{2}$$

$$K_{eq} = \frac{k_{TS}}{k_{ST}} \tag{3}$$

Note that the equilibrium constant is expressed as k_{TS}/k_{ST}, i.e., $K_{eq} < 1$ when the triplet is the ground state.

Pulsed techniques allow the direct, time-resolved study of singlet carbene reactions, even when the ground state is a triplet. These studies are carried out by direct, photochemical generation of singlet carbenes, which are then studied using picosecond spectroscopy before singlet-triplet equilibration is established. In effect the processes under study complete with intersystem crossing. For example, the reaction of ^{1}DPC (diphenylcarbene) with *tert*-butanol occurs with $k = 6.2 \times 10^9$ M^{-1} sec^{-1} in acetonitrile at room temperature;[13] the value has been determined using picosecond spectroscopy-laser induced fluorescence. It should be noted that these rate constants are "true" rate constants for the singlet carbene, rather than composite rate constants, as is frequently the case *(vide infra)* for triplet carbenes.

Tables 1 to 3 summarize the rate data for singlet carbenes. These, and all other tables in this chapter follow approximately the same format, with the carbene and substrate (if applicable) as the first two columns, followed by solvent, rate constants, and conditions of the experiment. Errors are quoted as in the original publication; only explicitly reported errors are included; i.e., statements such as "typical errors are around 10%" have not been taken into account. The "Method" and "Notes" column use abbreviations which are defined in Appendix 1. Note that these columns provide only very general information; e.g., "fer" for free energy relationships frequently refers to Hammett plots, but other correlations might be included. The details of the correlations obtained are beyond the scope of this chapter. Notes such as "ins" for insertion refer to paths A, B2, and C2 in Scheme 2 and do not exclude other reaction paths, although the one indicated is expected to be the dominant one. The reader should realize that a substantial fraction of the time-resolved studies do not include detailed product studies and in those cases the mechanistic details may be based on analogy and/or product studies under somewhat different conditions. The author has made an effort to include mechanistic details, but the reader should be aware of the limitations that apply to this information. The units used throughout the tables are summarized in Appendix 2.

Table 1

FIRST-ORDER DECAY PROCESSES OF SINGLET CARBENES FOR SYSTEMS WITH GROUND STATE TRIPLETS

Carbene	Solvent	k	T	Method	Notes	Ref.
DPC	Isooctane	$(1.052 \pm 0.095) \times 10^{10}$	300	ps,lif	ST,fer	12
	3-Methylpentane	$(1.052 \pm 0.093) \times 10^{10}$	300	ps,lif	ST,fer	12
	Diethylether	$(7.69 \pm 0.65) \times 10^{9}$	300	ps,lif	ST,fer	12
	Tetrahydrofuran	$(5.5 \pm 0.45) \times 10^{9}$	300	ps,lif	ST,fer	12
	Butyronitrile	$(3.57 \pm 0.26) \times 10^{9}$	300	ps,lif	ST,fer	12
	Acetonitrile	$(3.23 \pm 0.19) \times 10^{9}$	300	ps,lif	ST,fer	7,12
	Acetonitrile	$(2.90 \pm 0.25) \times 10^{9}$	274	ps,lif	ST,fer	12
	Acetonitrile	$(2.60 \pm 0.26) \times 10^{9}$	253	ps,lif	ST,fer	12
	Acetonitrile	$(9.1 \pm 1) \times 10^{9}$	rt	ps,lif	ST	14
SB	3-Methylpentane	$(1.66 \pm 0.06) \times 10^{10}$	298	ps,lif	ST,fer	15
	Tetrahydrofuran	$(1.54 \pm 0.05) \times 10^{10}$	298	ps,lif	ST,fer	15
	1,3-Dioxolane	$(1.33 \pm 0.06) \times 10^{10}$	298	ps,lif	ST,fer	15
	Acetonitrile	$(1.25 \pm 0.06) \times 10^{10}$	298	ps,lif	ST,fer	15
Fl	Acetonitrile	$(2.5 \pm 0.8) \times 10^{9}$	rt	ps,kas	ST	16
	Cyclohexane	$\sim 7 \times 10^{9}$	rt	ps,kas	ST	16
	Hexafluorobenzene	$(2.2 \pm 0.4) \times 10^{9}$	rt	ps,kas	ST	16
DMFl	Benzene (anh)	1.35×10^{7}	rt	lp,kas	ST	17
	Acetonitrile	3.8×10^{7}	rt	lp,kas	ST	17
	Cyclohexane	6.3×10^{7}	rt	lp,kas	ins?	17
XA	Pentane	3.0×10^{6}	rt	lp,kas		18
SA	Cyclohexane	4.4×10^{7}	rt	lp,kas	ST?	19

Table 2

INTERMOLECULAR REACTIONS OF SINGLET CARBENES (EXCEPT ARYL HALOCARBENES)

Carbene	Substrate	Solvent	k	T	Method	Notes	Ref.
DPC	CH_3OH	Diethylether	$(3 \pm 1) \times 10^{10}$	293	ps,lif	ins,fer[a]	20
	CH_3OH	Acetonitrile	$(2.1 \pm 0.5) \times 10^{10}$	rt	Comp + ps	ins	13
		Acetonitrile	$(5.0 \pm 0.8) \times 10^{9}$	293	ps,lif	ins,fer[a]	20
	i-C_3H_7OH	Acetonitrile	$(1.3 \pm 0.3) \times 10^{10}$	rt	comp + ps	ins	13
	t-Bu OH	Acetonitrile	$(6.2 \pm 1.5) \times 10^{9}$	rt	comp + ps	ins	13
	$CH_3CHOHC_3H_7$	Acetonitrile	$(2.04 \pm 0.08) \times 10^{9}$	293	ps,lif	ins,fer	20
XA	C_2H_5OH	Pentane	$(5.1 \pm 0.1) \times 10^{9}$	rt	lp,kas	ins	18
	t-Bu OH	Pentane	$(3.4 \pm 0.3) \times 10^{9}$	rt	lp,kas	ins	18
		Cyclohexane	$(1.8 \pm 0.1) \times 10^{9}$	rt	lp,kas	ins	18
	t-Bu OD	Cyclohexane	$(1.6 \pm 0.2) \times 10^{9}$	rt	lp,kas	ins	18
	$C_6H_5CH=CH_2$	Pentane	$\sim 6 \times 10^{5}$ [b]	rt	lp,kas	add?	18
	$C_6H_5C(CH_3)=CH_2$	Pentane	5.1×10^{6} [b]	rt	lp,kas	add?	18
	$(C_2H_5)_3N$	Pentane	1.4×10^{6} [b]	rt	lp,kas	?	18
	Diazo	Pentane	$\sim 3 \times 10^{7}$	rt	comp + lp	azine	18
DMFl	CH_3OH	Benzene	$(5.6 \pm 0.3) \times 10^{9}$	rt	lp,kas	ins	17
	CH_3OD	Benzene	$(4.8 \pm 0.4) \times 10^{9}$	rt	lp,kas	ins	17
	t-Bu OH	Benzene	$(4.1 \pm 0.5) \times 10^{9}$	rt	lp,kas	ins	17
	$C_6H_5CH=CH_2$	Benzene	$(4.3 \pm 0.4) \times 10^{7}$	rt	lp,kas	add	17
	m-$FC_6H_4CH=CH_2$	Benzene	$(9.8 \pm 0.8) \times 10^{7}$	rt	lp,kas	add	17
	p-$CH_3OC_6H_4CH=CH_2$	Benzene	$(1.6 \pm 0.2) \times 10^{7}$	rt	lp,kas	add	17
	$C_6H_5(CH_3)=CH_2$	Benzene	$(1.3 \pm 0.1) \times 10^{8}$	rt	lp,kas	add	17
	$(CH_3)_2C=CHCH_3$	Benzene	$(5.2 \pm 0.6) \times 10^{7}$	rt	lp,kas	add	17
SA	CH_3OH	Cyclohexane	1.5×10^{8}	rt	lp,kas	ins	19

[a]   Curved quenching plots, corrected based in $E_t(30)$.
[b]   Uncertain value, see original.

Table 3
INTERMOLECULAR REACTIONS OF ARYL HALOCARBENES

Carbene	Substrate	Solvent	k	T	Ea	log A	Method	Notes	Ref.
PFC	$(CH_3)_2C=C(CH_3)_2$	Isooctane	1.6×10^8	296			lp,kas	add	8,21
	$(CH_3)_2C=CHCH_3$	Isooctane	5.3×10^7	296			lp,kas	add	8,21
	$trans\text{-}CH_3CH=CHC_2H_5$	Isooctane	2.4×10^6	296			lp,kas	add	8,21
	$n\text{-}Bu\ CH=CH_2$	Isooctane	9.3×10^5	296			lp,kas	add	8,21
PCC	$(CH_3)_2C=C(CH_3)_2$	Isooctane	$(3.3 \pm 0.2) \times 10^8$	298	$(-1.7)^a$	$(7.28)^a$	lp,kas	add	11
		Isooctane	2.2×10^8	rt			lp,kas	add	8
		Isooctane	1.3×10^8	296			lp,kas	add	9
		Isooctane	2.8×10^8	296			lp,kas	add	8,21
		Acetonitrile	$(1.95 \pm 0.28) \times 10^8$	300			lp,kas	add	22
		Acetonitrile	1.1×10^8	rt			lp,kas	add	8
		Toluene	1.4×10^8	rt			lp,kas	add	8
	$(CH_3)_2C=CHCH_3$	Isooctane	7.7×10^7	296			lp,kas	add	9
		Isooctane	$(1.6 \times 0.1) \times 10^8$	298	-0.77	7.64	lp,kas	add	11
		Isooctane	1.3×10^8	296			lp,kas	add	8,21
	$trans\text{-}CH_3CH=CHC_2H_5$	Isooctane	3.4×10^6	296			lp,kas	add	9
		Isooctane	$(1.0 \pm 0.1) \times 10^7$	298	1.0	7.73	lp,kas	add	11
		Isooctane	5.5×10^6	296			lp,kas	add	8,21
	$n\text{-}Bu\ CH=CH_2$	Isooctane	1.3×10^6	296			lp,kas	add	9
		Isooctane	$(3.8 \pm 0.2) \times 10^6$	298	1.1	7.38	lp,kas	add	11
		Isooctane	2.2×10^6	296			lp,kas	add	8,21
	$PhCH=CH_2$	Isooctane	4.0×10^7	rt			lp,kas	add	8
	$p\text{-}ClC_6H_4CH=CH_2$	Isooctane	4.4×10^7	rt			lp,kas	add	8
	$p\text{-}CH_3C_6H_4CH=CH_2$	Isooctane	4.3×10^7	rt			lp,kas	add	8
	$p\text{-}CH_3OC_6H_4CH=CH_2$	Isooctane	1.3×10^8	rt			lp,kas	add	8
	CH_3OH (oligomer)	Isooctane	$(2.9 \pm 0.2) \times 10^9$	rt			lp,kas	ins	10
	$CH_3CHOHCHOHCH_3$	Isooctane	5.4×10^7	rt			lp,kas	ins	23
	CH_3COOH	Isooctane	$(3.1 \pm 0.6) \times 10^9$	300			lp,kas	ins	22
		Acetonitrile	$(1.78 \pm 0.36) \times 10^9$	300			lp,kas	ins	22
	CF_3COOH	Acetonitrile	$(2.40 \pm 0.12) \times 10^9$	300			lp,kas	ins	22
	$CH_3COOC_2H_5$	Acetonitrile	$<2 \times 10^6$	300			lp,kas	ins	22
$CF_3\text{-}PCC$	$(CH_3)_2C=C(CH_3)_2$	Isooctane	1.5×10^9	296			lp,kas	add,fer	8,24
	$(CH_3)_2C=CHCH_3$	Isooctane	6.8×10^8	296			lp,kas	add,fer	8,24

	Substrate	Solvent	k	T			Ref.
	trans-CH₃CH=CHC₂H₅	Isooctane	4.9×10^7	296	lp,kas	add,fer	8,24
	n-Bu CH=CH₂	Isooctane	1.8×10^7	296	lp,kas	add,fer	8,24
	PhCH=CH₂	Isooctane	3.0×10^8	rt	lp,kas	add,fer	8
	p-ClC₆H₄CH=CH₂	Isooctane	2.0×10^8	rt	lp,kas	add,fer	8
	p-CH₃C₆H₄CH=CH₂	Isooctane	4.6×10^8	rt	lp,kas	add,fer	8
	p-CH₃OC₆H₄CH=CH₂	Isooctane	6.5×10^8	rt	lp,kas	add,fer	8
Cl-PCC	(CH₃)₂C=C(CH₃)₂	Isooctane	3.3×10^8	296	lp,kas	add,fer	24
	(CH₃)₂C=CHCH₃	Isooctane	1.8×10^8	296	lp,kas	add,fer	24
	trans-CH₃CH=CHC₂H₅	Isooctane	7.5×10^6	296	lp,kas	add,fer	24
	n-Bu CH=CH₂	Isooctane	2.3×10^6	296	lp,kas	add,fer	24
	PhCH=CH₂	Isooctane	7.1×10^7	rt	lp,kas	add,fer	8
	p-ClC₆H₄CH=CH₂	Isooctane	6.0×10^7	rt	lp,kas	add,fer	8
	p-CH₃C₆H₄CH=CH₂	Isooctane	1.1×10^8	rt	lp,kas	add,fer	8
	p-CH₃OC₆H₄CH=CH₂	Isooctane	2.9×10^8	rt	lp,kas	add,fer	8
	CH₃OH (monomer)	Isooctane	2×10^7	rt	lp,kas	ins	10
	CH₃OH	Acetonitrile	6.5×10^6	rt	lp,kas	ins	10
	CH₃OH (oligomer)	Isooctane	$(4.3 \pm 0.4) \times 10^9$	rt	lp,kas	ins	10
	t-Bu OH (monomer)	Isooctane	$(2.52 \pm 0.15) \times 10^6$	rt	lp,kas	ins	10
	CH₃COOH	Isooctane	$(5.1 \pm 1.2) \times 10^9$	300	lp,kas	ins	22
		Acetonitrile	$(2.16 \pm 0.10) \times 10^9$	300	lp,kas	ins	22
CH₃-PCC	(CH₃)₂C=C(CH₃)₂	Isooctane	1.2×10^8	296	lp,kas	add,fer	24
	(CH₃)₂C=CHCH₃	Isooctane	4.9×10^7	296	lp,kas	add,fer	24
	trans-CH₃CH=CHC₂H₅	Isooctane	1.8×10^6	296	lp,kas	add,fer	24
	n-Bu CH=CH₂	Isooctane	6.2×10^5	296	lp,kas	add,fer	24
	PhCH=CH₂	Isooctane	1.8×10^7	rt	lp,kas	add,fer	8
	p-ClC₆H₄CH=CH₂	Isooctane	2.7×10^7	rt	lp,kas	add,fer	8
	p-CH₃C₆H₄CH=CH₂	Isooctane	2.2×10^7	rt	lp,kas	add,fer	8
	p-CH₃OC₆H₄CH=CH₂	Isooctane	1.3×10^8	rt	lp,kas	add,fer	8
CH₃O-PCC	(CH₃)₂C=(CH₃)₂	Isooctane	1.4×10^7	296	lp,kas	add,fer	8,24
		Acetonitrile	$(2.04 \pm 0.10) \times 10^8$	300	lp,kas	add	22
	(CH₃)₂C=CHCH₃	Isooctane	7.7×10^6	296	lp,kas	add,fer	24
	trans-CH₃CH=CHC₂H₅	Isooctane	4.4×10^5	296	lp,kas	add,fer	24
	n-Bu CH=CH₂	Isooctane	1.3×10^5	296	lp,kas	add,fer	24
	PhCH=CH₂	Isooctane	3.7×10^6	rt	lp,kas	add,fer	8
	p-ClC₆H₄CH=CH₂	Isooctane	8.8×10^6	rt	lp,kas	add,fer	8
	p-CH₃C₆H₄CH=CH₂	Isooctane	4.5×10^6	rt	lp,kas	add,fer	8
	p-CH₃OC₆H₄CH=CH₂	Isooctane	1.2×10^8	rt	lp,kas	add,fer	8

Table 3 (continued)
INTERMOLECULAR REACTIONS OF ARYL HALOCARBENES

Carbene	Substrate	Solvent	k	T	Ea	log A	Method	Notes	Ref.
PBC	$(CH_3)_2C=C(CH_3)_2$	Isooctane	3.8×10^8	296			lp,kas	add	8,21
	$(CH_3)_2C=CHCH_3$	Isooctane	1.8×10^8	296			lp,kas	add	21
	$trans\text{-}CH_3CH=CHC_2H_5$	Isooctane	1.2×10^7	296			lp,kas	add	21
	$n\text{-}Bu\ CH=CH_2$	Isooctane	4.0×10^6	296			lp,kas	add	8

Table 4
DIMERIZATION OF TRIPLET CARBENES

Carbene	Solvent	k	T	Method	Notes	Ref.
DPC	Benzene	$(5.4 \pm 1.6) \times 10^9$	298	fp,kas	dimer	5
Me$_2$DPC	Benzene	$(1.1 \pm 0.3) \times 10^9$	298	fp,kas	dimer	5
Br$_2$DPC	Benzene	$(3.5 \pm 1.1) \times 10^9$	298	fp,kas	dimer	5

Table 1 includes first order decay processes, frequently intersystem crossing, while Tables 2 and 3 give the details of intermolecular rate constants. Intermolecular reactions of aryl halo carbenes (Table 3), for which a considerable volume of information is available have been covered in a separate table.

The simplest intermolecular reaction of triplet carbenes is their self-reaction to yield the corresponding olefin. The very limited data available on this reaction have been included in Table 4. It should be noted that the rate constants for carbene self-reactions are the only rates included in this review where the absolute value of the rate constant depends directly on the determination of an extinction coefficient for the carbene.

When a triplet carbene is generated through singlet-to-triplet equilibration (i.e., feeding Reaction 2 from the right) the corresponding concentration of singlet carbene will usually be present in the system. *If* singlet-triplet equilibration is maintained, and *if* the concentration of singlet is substantially smaller than that of triplet carbene, then, the observed rate constant will be given by:

$$k_{obs} = k_T + K_{eq}k_S \tag{4}$$

where k_T and k_S are the rate constants for reaction of the triplet and singlet states, respectively with a given substrate. The value which is obtained experimentally monitoring the decay of the triplet carbene, or *any* product of reaction is always k_{obs}, which includes all modes and sites of reaction of both electronic states of the carbene with the substrate in question. We refer to k_{obs} as the "composite rate constant for the triplet carbene." Numerous studies of triplet carbenes have concentrated on the question of whether the k_T term or the $K_{eq}k_S$ term is dominant in Equation 4; in the latter case the triplet carbene behaves essentially as a reservoir for the singlet carbene.

In the case of triplet fluorenylidene (F*l*), there is good indication that its composite reactivity is dominated by the singlet term.[16,25] This is probably a reflection of the small triplet-singlet energy gap. It should be noted that many of the rate constants for triplet fluorenylidene were originally attributed to the singlet state, but later reinterpreted.[26,27] Only the currently accepted interpretation of the data has been included here. This reinterpretation applies specifically to References 28 to 32 (Table 5). By contrast, in the case of DPC its reactivity seems to be controlled by the k_T term, i.e., k_{obs} usually corresponds to mainly triplet reactivity. For example, in the case of hydrogen donors DPC shows typical radical-like behavior.[36] Even in the case of the reaction with methanol (path A in Scheme 2) there is good evidence that k_{obs} is dominated by the triplet term,[37] in spite of the singlet nature of the products and of many reports that have suggested that the triplet acts exclusively as a reservoir for the singlet in this particular reaction.[5,7,14,38] Table 6 summarizes the data on the composite reactivity of triplet diphenylcarbene.

Table 7 gives the kinetic data for triplet carbenes other than F*l* and DPC.

Many literature reports have included lifetimes for triplet carbenes under a variety of experimental conditions. These lifetimes usually reflect reactive decay and are closely linked to the data included in Tables 5 to 7. Table 8 provides a summary of selected carbene triplet lifetimes.

Many of the reactions included in Tables 5 to 8 involve an addition of the carbene to the

Table 5
COMPOSITE REACTIVITY OF TRIPLET FLUORENYLIDENE

Substrate	Solvent	k	T	Ea	logA	Method	Notes	Ref.
CH_3OH	Hexafluorobenzene	$(6.30 \pm 1.23) \times 10^8$	300			lp,kas	ins	25
CH_3OD	Hexafluorobenzene	$(4.59 \pm 0.85) \times 10^8$	300			lp,kas	ins	25
CD_3OH	Hexafluorobenzene	$(7.48 \pm 1.56) \times 10^8$	300			lp,kas	ins	25
CD_3OD	Hexafluorobenzene	$(5.25 \pm 0.69) \times 10^8$	300			lp,kas	ins	25
CH_3OH	Acetonitrile	8.95×10^8	298			comp + lp	ins	28
	Acetonitrile-d_3	8.4×10^8	300			comp + lp	ins	25
CH_3OD	Acetonitrile-d_3	6.1×10^8	300			comp + lp	ins	25
CH_3OH	Freon-113	8.16×10^8	rt			lp,kas	ins	32
	Acetonitrile	8.6×10^8	rt			comp + lp	ins	16
	Hexafluorobenzene	1.2×10^8	rt			lp,kas	ins	16
	Acetonitrile	8.6×10^8	rt			comp + lp	ins	33
CH_3OD	Acetonitrile	4.7×10^8	rt			comp + lp	ins	33
C_2H_5OH	Spiropentane	$(2.0 \times 0.4) \times 10^8$	rt			Comp + lp	ins	27
	Acetonitrile	7.3×10^8	rt			lp,kas	ins	16
	Hexafluorobenzene	2.4×10^8	rt			comp + lp	ins	16
$(CH_3)_2CHOH$	Acetonitrile	5.2×10^8	rt			comp + lp	ins	16
	Hexafluorobenzene	9.8×10^7	rt			comp + lp	ins	16
$(CH_3)_3COH$	Acetonitrile	1.6×10^8	rt			comp + lp	ins	33
$(CH_3)_3COD$	Acetonitrile	4.2×10^7	rt			comp + lp	ins	33
$(CH_3)_3COH$	Acetonitrile-d_3	1.51×10^8	298			comp + lp	ins	28
CH_3COOH	Acetonitrile	5.6×10^8	300			comp + lp	ins	25
$(CH_3)_3CNH_2$	Acetonitrile	6.3×10^7	rt			comp + lp	ins + H-abs	33
$(CH_3)_3CND_2$	Acetonitrile	7.0×10^7	rt			comp + lp	ins + D-abs	33
$(CH_3)_3CSH$	Acetonitrile	1.1×10^9	rt			comp + lp	ins	33
$(CH_3)_3CSD$	Acetonitrile	1.2×10^9	rt			comp + lp	ins	33
$CH_3(CH_2)_3SH$	Acetonitrile	1.4×10^9	rt			comp + lp	ins	33
O_2	Freon-113	$(1.4 \pm 0.2) \times 10^9$	300			lp,kas	co	34
CH_3COCH_3	Acetonitrile-d_3	$(1.0 \pm 0.5) \times 10^7$	300			lp,kas	yld + H-abs	35
CD_3COCD_3	Acetonitrile-d_3	$(0.99 \pm 0.01) \times 10^7$	300			lp,kas	yld + H-abs	35
cyclo-C_6H_{12}	Acetonitrile-d_3	8.3×10^7	303			lp,kas	ins + H-abs	29

Substrate	Solvent	k	T					Ref.
Diethylether	Freon-113	$(1.3 \pm 0.2) \times 10^8$	300			lp,kas	ins + H-abs?	34
1,4-Cyclohexadiene	Freon-113	$(1.1 \pm 0.2) \times 10^9$	300			lp,kas	ins + H-abs?	34
Isooctane	Freon-113	$(9 \pm 1) \times 10^6$	300			lp,kas	ins + H-abs?	34
Toluene	Acetonitrile-d_3	6.4×10^8	300			comp + lp	?	25
PhCH=CH$_2$	Acetonitrile	9.4×10^8	rt			comp + lp	add	16,28
	Hexafluorobenzene	4.6×10^8	rt			comp + lp	add	31
cis-PhCH=CHCH$_3$	Freon-113	2.24×10^8	rt	0.16 ± 0.4	$\Delta S^{\ddagger} = -17.6 \pm 1.5$	comp + lp	add	16,32
	Acetonitrile	1.94×10^8	rt			comp + lp	add	16,32
trans-PhCH=CHCH$_3$	Acetonitrile	2.86×10^8	rt			comp + lp	add	32
PhC(CH$_3$)=CH$_2$	Freon-113	1.4×10^8	300	-0.31 ± 0.2	$\Delta S^{\ddagger} = -18.6 \pm 0.8$	comp + lp	add	31
						comp + lp	add	34
(CH$_3$)$_2$C=C(CH$_3$)$_2$	Acetonitrile-d_3	$(5.1 \pm 0.6) \times 10^8$	300	1.93	10.23	lp,kas	add	25
cis-2-Pentene	Acetonitrile	1.29×10^8	rt	$-0.20 = 0.25$	$\Delta S^{\ddagger} = -22.2 \pm 0.94$	comp + lp	add + ins	16,32
	Acetonitrile					comp + lp	add + ins	31
Trans-2-Pentene	Acetonitrile	1.43×10^8	rt	0.19 ± 0.21	$\Delta S^{\ddagger} = -20.1 \pm 0.80$	comp + lp	add + ins	16,32
	Acetonitrile					comp + lp	add + ins	31
2-Methyl-2-butene	Acetonitrile	3.51×10^8	298	-0.94 ± 0.3	$\Delta S^{\ddagger} = -22.5 \pm 1.01$	comp + lp	add	28,16
	Acetonitrile					comp + lp	add	31
cis-4-Methyl-2-pentene	Hexafluorobenzene	$(2.1 \pm 0.5) \times 10^8$	300			lp,kas	add + H-abs	25,26
2-Methyl-2-pentene	Hexafluorobenzene	$(7.7 \pm 1.7) \times 10^8$	300			lp,kas	add + H-abs	25
2,4-Hexadiene	Acetonitrile	7.8×10^8	rt			comp + lp	add	16
2,5-Dimethyl-2,4-hexadiene	Acetonitrile	6.7×10^8	rt			comp +	add	16
	Freon-113	6.7×10^8	300			lp,kas	add	34
trans-1,3-Pentadiene	Freon-113	1.0×10^9	rt			comp + lp	add	16,28
Cyclohexene	Acetonitrile	3.2×10^8	rt			comp + lp	add	16,28
β-Pinene	Freon-113	$(1.2 \pm 0.4) \times 10^9$	300			lp,kas	add	34
	Acetonitrile-d_3	$(1.5 \pm 0.3) \times 10^9$	300	0.29	9.37	lp,kas	add	25
Diethylfumarate	Acetonitrile	1.5×10^9	rt			comp + lp	add	16
Diethylmaleate	Acetonitrile	1.27×10^8	298			comp + lp	add	28
trans-PhCH=CHCC$_2$CH$_3$	Acetonitrile	4.38×10^8	298			comp + lp	add	28
CH$_2$=CHCO$_2$CH$_3$	Acetonitrile	3.85×10^8	298			comp + lp	add	28

Table 5 (continued)
COMPOSITE REACTIVITY OF TRIPLET FLUORENYLIDENE

Substrate	Solvent	k	T	Ea	logA	Method	Notes	Ref.
Methylmethacrylate	Acetonitrile	6.44×10^8	298			comp + lp	add	28
	Acetonitrile			-1.2 ± 0.5	$\Delta S\ddagger = -22.3 \pm 1.6$	comp + lp	add	31
cis-CHCl=CHCl	Acetonitrile	1.7×10^7	rt			comp + lp	add	16
trans-CHCl=CHCl	Acetonitrile	2.8×10^7	rt			comp + lp	add	16

Table 6
COMPOSITE REACTIVITY OF TRIPLET DIPHENYLCARBENE

Substrate	Solvent	k	T	Ea	log A	Method	Notes	Ref.
DPC	Benzene	5.4×10^9	298			fp,kas	dimer	5
diazo	Benzene	$\sim 3 \times 10^7$	rt			lp,dielectric	azine	39
O_2	Benzene	1.0×10^9	298			comp + fp	co	5
	Benzene	$\sim 3.5 \times 10^9$	rt			lp,dielectric	co	39
	Acetonitrile	5×10^9	300			lp,kas	co	40
CH_3OH	Benzene	$(6.8 \pm 2.1) \times 10^6$	298			fp,kas	ins	5
	Benzene	1.97×10^7	293	1.17 ± 0.36	8.17 ± 0.26	lp,kas	ins	37
	Isooctane	6.43×10^6	293	3.61 ± 0.36	9.51 ± 0.27	lp,kas	ins	37
	Chlorobenzene	2.75×10^7	293	2.48 ± 0.44	9.30 ± 0.34	lp,kas	ins	37
	Acetonitrile	2.39×10^7	293	1.66 ± 0.20	8.62 ± 0.16	lp,kas	ins	37
	Acetonitrile	2.4×10^7	rt			lp,kas	ins	7
$cyclo\text{-}C_6H_{11}OH$	Isooctane	$(6.3 \pm 0.3) \times 10^6$	rt			lp,kas	ins	41
	Chlorobenzene	$(2.2 \pm 0.1) \times 10^7$	rt			lp,kas	ins	41
	Acetonitrile	$(1.3 \pm 0.1) \times 10^7$	rt			lp,kas	ins	41
$CH_3CHOHCH_3$	Acetonitrile	1.5×10^7	rt			lp,kas	ins	7
$t\text{-}Bu\ OH$	Acetonitrile	6.1×10^6	rt			lp,kas	ins	7
$CH_3CH_2C(CH_3)_2OH$	Acetonitrile	6.6×10^6	rt			lp,kas	ins	7
H_2O	Acetonitrile	2.1×10^7	rt			lp,kas	ins	7
$(C_2H_5)_3N$	Cyclohexane	$(3.4 \pm 1.4) \times 10^5$	300			lp,kas	H-abs	36
$n\text{-}Bu\ NH_2$	Acetonitrile	$(1.8 \pm 0.4) \times 10^6$	300			lp,kas	ins	42
$n\text{-}Bu\ ND_2$	Acetonitrile	$(1.3 \pm 0.2) \times 10^6$	298			lp,kas	ins	42
$t\text{-}Bu\ NH_2$	Acetonitrile	$(1.4 \pm 0.3) \times 10^5$	298			lp,kas	ins	42
$t\text{-}Bu\ ND_2$	Acetonitrile	$(1.1 \pm 0.3) \times 10^5$	298			lp,kas	ins	42
$CH_2=CH\text{-}C(CH_3)=CH_2$	Acetonitrile	$(3.5 \pm 0.5) \times 10^5$	rt			lp,kas	add	14
	Acetonitrile	$(1.36 \pm 0.15) \times 10^6$	rt			lp,kas	add	7
$CH_2=CHCH=CH_2$	Benzene	6.4×10^5	298			fp,kas	add	5
$C_6H_5CH=CH_2$	Benzene	3.8×10^5	298			comp + fp	add	5
$(C_6H_5)_2C=CH_2$	Cyclohexane	$(1.4 \pm 0.3) \times 10^6$	300			lp,kas	add	36
	Benzene	4.8×10^5	298			comp + fp	add	5
CCl_4	Cyclohexane	$(3.6 \pm 0.4) \times 10^5$	300			lp,kas	Cl-abs	36
$HCCl_3$	Cyclohexane	$(8.9 \pm 0.9) \times 10^5$	300			lp,kas	H-abs?	36
Tetrahydrofuran	Cyclohexane	$(2.0 \pm 0.3) \times 10^5$	300			lp,kas	H-abs?	36

Table 6 (continued)
COMPOSITE REACTIVITY OF TRIPLET DIPHENYLCARBENE

Substrate	Solvent	k	T	Ea	log A	Method	Notes	Ref.
Cyclohexene	Benzene	$(1.4 \pm 0.2) \times 10^5$	300			lp,kas	H-abs?	36
Cyclooctene	Cyclohexane	$(2.8 \pm 0.1) \times 10^5$	300			lp,kas	H-abs + add?	36
CH$_2$=CHC$_6$H$_5$	Cyclohexane	$(3.3 \pm 0.3) \times 10^5$	300			lp,kas	H-abs + add?	36
(CH$_3$)$_2$C=C(CH$_3$)$_2$	Cyclohexane	$(1.0 \pm 0.1) \times 10^5$	300			lp,kas	H-abs + add?	36
1,4-Cyclohexadiene	Cyclohexane	$(2.1 \pm 0.3) \times 10^5$	300			lp,kas	add + H-abs?	36
1,3-Cyclohexadiene	Cyclohexane	$(1.0 \pm 0.1) \times 10^7$	300			lp,kas	H-abs	36
1,5-Cyclooctadiene	Cyclohexane	$(2.6 \pm 0.2) \times 10^6$	300			lp,kas	H-abs + add?	36
1,3-Cyclooctadiene	Cyclohexane	$(2.6 \pm 1.2) \times 10^6$	300			lp,kas	H-abs + add?	36
1,3-Octadiene	Cyclohexane	$(2.8 \pm 0.1) \times 10^5$	300			lp,kas	H-abs? + add?	36
	Cyclohexane	$(1.5 \pm 0.1) \times 10^6$	300			lp,kas	add + H-abs?	36
TEMPO	Isooctane	$(7.3 \pm 0.1) \times 10^8$	rt			lp,kas	O-abs	41
	Acetonitrile	$(2.7 \pm 0.1) \times 10^8$	rt			lp,kas	O-abs	41
4-HO-TEMPO	Chlorobenzene	$(1.7 \pm 0.1) \times 10^8$	rt			lp,kas	O-abs + ins	41
	Acetonitrile	$(3.0 \pm 0.2) \times 10^8$	rt			lp,kas	O-abs + ins	41
4-NH$_2$-TEMPO	Isooctane	$(4.8 \pm 0.3) \times 10^8$	rt			lp,kas	O-abs	41
	Acetonitrile	$(2.4 \pm 0.1) \times 10^8$	rt			lp,kas	O-abs	41

Table 7
COMPOSITE REACTIVITY OF VARIOUS TRIPLET CARBENES (EXCEPT F*l* AND DPC)

Carbene	Substrate	Solvent	k	T	Ea	log A	Method	Notes	Ref.
BF*l*	CH_3OH	Acetonitrile	3.6×10^9	rt			lp,kas	ins	43
		Benzene	6.3×10^9	rt			lp,kas	ins	43
	$PhCH{=}CH_2$	Acetonitrile	2.8×10^8	rt			lp,kas	add	43
SB	CH_3OH	Toluene	$(2.5 \pm 0.2) \times 10^6$	298	5.0 ± 1.0	9.1 ± 0.8	lp,kas	ins	44
		Cyclohexane	$(1.9 \pm 0.2) \times 10^6$	298			lp,kas	ins	44
	1,4-Cyclohexadiene	Cyclohexane	$(3.0 \pm 0.2) \times 10^6$	298			lp,kas	H-abs	44
	$PhCH{=}CH_2$	Benzene	$(9 \pm 2) \times 10^4$	298			lp,kas	add	44
	Tetrahydrofuran	Benzene	$(8.8 \pm 0.8) \times 10^4$	298			lp,kas	H-abs?	44
SA	O_2	Cyclohexane	2×10^9	rt			lp,kas	co	45
	CH_3OH	Cyclohexane	6.9×10^6	rt			lp,kas	ins	45
	2-Methyl-2-butene	Cyclohexane	4.2×10^6	rt			lp,kas	add?	45
	2-Methyltetrahydrofuran	Cyclohexane	2.8×10^7	rt			lp,kas	H-abs?	45
BA	Cyclohexane	Benzene	$(8.1 \pm 2.1) \times 10^5$	rt			comp + lp	H-abs,(H/D~7)	46
	$PhCH{=}CH_2$	Benzene	$(1.2 \pm 0.2) \times 10^7$	rt			lp,kas	add	46
	$CH_3CHO{\vdots}CH_3$	Benzene	$(3.9 \pm 0.9) \times 10^6$	rt			comp + lp	ins,H-abs(H/D~5)	46
1NC	CH_3CN	Benzene	$(4.6 \pm 0.2) \times 10^5$	300			lp,kas	yld	47
2NC	Cyclohexane	Isooctane	$(1.48 \pm 0.04) \times 10^6$	297			lp,kas	H-abs	48
	$PhCH{=}CH_2$	Isooctane	$(4.33 \pm 0.1) \times 10^7$	297			lp,kas	add	48
	CH_3OH	Isooctane	$(7.25 \pm 0.5) \times 10^6$	297			lp,kas	ins	48
	CCl_4	Isooctane	$(3.35 \pm 0.07) \times 10^6$	297			lp,kas	Cl-abs?	48
	CH_3CN	Isooctane	$(5.28 \pm 0.1) \times 10^5$	297			lp,kas	yld	48
DMC	O_2	Cyclopentane	1.9×10^8	298			lp,kas	co	49

solvent or substrate to produce an ylide. Typical examples are the reactions of fluorenylidene with nitriles, ketones, and oxygen, paths H, I, and G in Scheme 2, respectively. Many of these ylides have played an integral role in the study of carbene kinetics and some of their reactions will be covered in a separate section.

IV. EXCITED TRIPLET CARBENES

In the last few years a number of reports have been concerned with the lifetime and quenching of excited triplet carbenes, specifically DPC and Me$_2$DPC. The lifetime data have been included in Table 9 and the data available on intermolecular reactions in Table 10. In most cases the details of the products and their yields have not been firmly established.

V. CARBENE-DERIVED YLIDES

While this Section deals formally with "carbene-derived" ylides, some of the sources included do not necessarily involve a carbene. For example, carbonyl oxides (also referred as Criegee intermediates) can be prepared by either reaction of the carbene with oxygen (path G in Scheme 2) or by reaction of singlet-oxygen with the parent diazo compound. For example, in the case of DPC

$$^3Ph_2C\colon + \, ^3O_2 \rightarrow Ph_2COO \tag{5}$$

$$^1O_2 + Ph_2CN_2 \rightarrow Ph_2COO + N_2 \tag{6}$$

Intermediates of the same type are also produced in the ozonolysis of olefins.[55]

The criteria followed in this section has been to include those ylides that can be produced in carbene reactions regardless of whether the actual source employed involves a carbene. The coverage is not necessarily complete. It should also be noted that in the case of fluorenylidene much of the ylide data were originally reported as carbene data but were later reinterpreted;[26,27] the currently accepted interpretation has been included in this chapter. Table 11 gives some representative lifetime data, and Table 12 summarizes the intermolecular reactivities.

Table 8

SELECTED LIFETIMES FOR TRIPLET CARBENES

Carbene	Solvent	τ,ns	T	Ea	log A	Method	Notes	Ref.
DPC	Cyclohexane	1750	300	2.5 ± 0.4	7.5 ± 0.3	lp,kas	H-abs,(H/D=2.6)	36
	Toluene	2100	300	3.2 ± 0.7	8.0 ± 0.5	lp,kas	H-abs,(H/D=6.5)	36
	Cyclopentane	450	300	2.9 ± 0.3	7.9 ± 0.2	lp,kas	H-abs	36
	Isooctane	1700	298			lp,kas	H-abs	23
	Tetrahydrofuran	430	300			lp,kas	H-abs	36
	Cyclohexene	250	300	3.0 ± 0.3	8.7 ± 0.2	lp,kas	H-abs	36
	Methylcyclohexane	1400	300			lp,kas	H-abs	36
	2-Methyltetrahydrofuran	260	300			lp,kas	H-abs	36
	Triethylamine	400	300			lp,kas	H-abs?	36
MPC	Isooctane	2×10^4	298			lp,kas	H-abs + dimer?	23
MTC	Isooctane	9×10^4	298			lp,kas	H-abs + dimer?	23
DMC	Isooctane	2×10^5	298			lp,kas	dimer	23
	Benzene	2×10^5	298			lp,kas	dimer	50
Fl	Acetonitrile	25 ± 3	rt		$\Delta S\ddagger = -17.0 \pm 2.0$	lp,kas	yld	16,28
	Acetonitrile	27	300	1.99 ± 0.05		lp,kas	yld	31
	Acetonitrile	38.7	300	2.85 ± 0.09	9.64 ± 0.07	lp,kas	yld	25,29
	Acetonitrile-d_3		300	2.47 ± 0.92	9.23 ± 0.25	lp,kas	yld	25,29
	Cyclohexane	2.0 ± 0.9	rt			ps,kas	ins + H-abs	27
	Spiropentane	42 ± 9	rt			lp,kas	ins + H-abs	27
	Hexafluorobenzene	100 ± 6	rt			lp,kas	add?	16
	Hexafluorobenzene	96	300			lp,kas	add?	25
	Benzene	16	300			lp,kas	add?	34
	Freon-113	200	300			lp,kas	Cl-abs + ins?	34
	Freon-113	290	300			lp,kas	Cl-abs + ins?	25
	Perfluorotoluene	98	300			lp,kas	add?	25
	Chlorobenzene	23	300			lp,kas	add?	25
	Di-tert-butylperoxide	29	293			lp,kas	?	25
	Di-tert-butylperoxide	94:	239			lp,kas	?	25
	Methylene chloride	31	245			lp,kas	?	25
	Isopentane	55	191			lp,kas	ins + H-abs	25
BFl	Acetonitrile	90	rt			lp,kas	yld	43
	Cyclohexane	0.38	rt			ps,kas	ins + H-abs	43

Table 8 (continued)
SELECTED LIFETIMES FOR TRIPLET CARBENES

2NC	n-Pentane	180 ± 10	297			lp,kas	H-abs + ?	48
	Isooctane	300 ± 36	297			lp,kas	H-abs + ?	48
	Freon-113	330 ± 43	297			lp,kas	?	48
	Benzene	61 ± 7	297			lp,kas	?	48
SB	Toluene	4500	298	1.8	6.8	lp,kas	H-abs,(H/D = 3.0)	44
	Triethylamine	1200	298			lp,kas	H-abs + ?	44
	Tetrahydrofuran	1000	298			lp,kas	H-abs	44
	Cyclohexane	2000	298	2.7	7.4	lp,kas	H-abs,(H/D = 6.7)	44
	Pyridine	2800	298			lp,kas	?	44
SA	Cyclohexane	866 ± 54	rt			lp,kas	H-abs	19

Table 9
LIFETIME FOR EXCITED TRIPLET CARBENES

Carbene	Solvent	τ,ns	T	Method	Ref.
DPC	Acetonitrile	3.8 ± 0.3	293	ps,fluo	13
	Acetonitrile	9.1 ± 1.7	296	lp,kas	51
	Diisopropylether	3.8 ± 0.3	296	lp,kas	51
	Cyclohexane	4.2 ± 0.3	296	lp,kas	51
	Isooctane	4.3 ± 0.3	296	lp,kas	51
Me$_2$DPC	Acetonitrile	5.6	300	lp,kas + fluo	52
	Benzene	7.2	300	lp,kas + fluo	52
	Isooctane	8.0	300	lp,kas + fluo	52

Table 10

INTERMOLECULAR REACTIVITY OF EXCITED TRIPLET CARBENES

Carbene	Substrate	Solvent	k	T	Ea	log A	Method	Notes	Ref.
DPC	DABCO	Acetonitrile	$(2.6 \pm 0.2) \times 10^{10}$	293			ps,fluo	ct,fer	53
	$(C_2H_5)_3N$	Acetonitrile	$(5.4 \pm 0.4) \times 10^9$	293			ps,fluo	ct,fer	53
		Acetonitrile	3.3×10^9	296	-1.4 ± 0.2	8.66 ± 0.15	lp,kas	ct,fer	51
	ABCO	Acetonitrile	$(2.7 \pm 0.3) \times 10^9$	293			ps,fluo	ct,fer	53
	$(C_2H_5)_2NH$	Acetonitrile	$(9.8 \pm 1.2) \times 10^8$	293			ps,fluo	ct,fer	53
		Acetonitrile	9×10^8	296			lp,kas	ct,fer	51
	$(CH_3)_2NH$	Acetonitrile	$(9.2 \pm 0.6) \times 10^8$	293			ps,fluo	ct,fer	53
	i-Pr$_2$NH	Acetonitrile	5×10^8	296			lp,kas	ct,fer	51
	s-Bu NH$_2$	Acetonitrile	4×10^7	296	1.2 ± 0.2	8.62 ± 0.13	lp,kas	ct,fer	51
	n-Bu NH$_2$	Acetonitrile	$(5.3 \pm 0.6) \times 10^7$	293			ps,fluo	ct,fer	53
	t-Bu NH$_2$	Acetonitrile	$(2.7 \pm 0.4) \times 10^7$	293			ps,fluo	ct,fer	53
	n-Bu$_2$S	Acetonitrile	$(3.2 \pm 0.6) \times 10^9$	293			ps,fluo	ct,fer	53
	isoprene	Acetonitrile	$(2.1 \pm 0.3) \times 10^9$	293			ps,fluo	?	54
	CH$_3$OH	Acetonitrile	$(3.1 \pm 0.4) \times 10^8$	293			ps,fluo	ins?	54
Me$_2$DPC	CCl$_4$	Isooctane	1.1×10^9	300			lp,kas	Cl-abs	52

Table 11
REPRESENTATIVE YLIDE LIFETIMES

Carbene	Ylide source	Solvent	$\tau, \mu s$	T	Ea	log A	Method	Ref.
DPC	O_2	Cyclohexane	28	rt			lp,dielectric	39
Fl	O_2	CH_2Cl_2	$t_{1/2} = 9$	300			lp,kas	34
		Freon-113	$t_{1/2} = 370$	300			lp,kas	34
		Cyclohexane	$t_{1/2} > 500$	300			lp,kas	34
		Acetone-d_6	$t_{1/2} = 26$	300			lp,kas	34
		Benzene	$t_{1/2} > 800$	300			lp,kas	34
		Chlorobenzene	$t_{1/2} > 500$	300			lp,kas	34
		CCl_4	$t_{1/2} = 35$	300			lp,kas	34
		Acetonitrile	$t_{1/2} = 35$	300			lp,kas	34
	Acetone	Acetone	3.4	300			lp,kas	34
	Acetone-d_6	Acetone-d_6	4.3	300	(10.96 ± 0.12)	(13.26 ± 0.11)	lp,kas	35
	2-Butanone	2-Butanone	1.2	300			lp,kas	35
	Cyclobutanone	Cyclobutanone	3.6	300			lp,kas	35
	Cyclopentanone	Cyclopentanone	6.7	300			lp,kas	35
	Cyclohexanone	Cyclohexanone	1.9	300			lp,kas	35
	Cycloheptanone	Cycloheptanone	5.1	300			lp,kas	35

Table 12
INTERMOLECULAR REACTIVITY OF SOME YLIDES

Carbene	Ylide source	Substrate	Solvent	k	T	Ea	log A	Method	Notes	Ref.
DPC	O_2	Diazo	Acetonitrile	4×10^5	300			lp,kas	add	40
FI	O_2	acetaldehyde	Freon-113	$(1.21 \pm 0.04) \times 10^9$	300			lp,kas	add	34
		Phenylacetaldehyde	Freon-113	$(5.0 \pm 1.0) \times 10^8$	300			lp,kas	add	34
		n-Octanal	Freon-113	$(4.5 \pm 0.5) \times 10^7$	300			lp,kas	add	34
		5-Norbornene-2-carboxaldehyde	Freon-113	$\sim 1.5 \times 10^7$	300			lp,kas	add	34
	Acetone	Diethylfumarate	Acetonitrile-d_3	$(4.47 \pm 0.30) \times 10^7$	300			lp,kas	add	35
		Fumaronitrile	Acetonitrile-d_3	$(1.45 \pm 0.02) \times 10^7$	300			lp,kas	add	35
		Oxygen	Acetonitrile-d_3	$(7.24 \pm 0.43) \times 10^7$	300			lp,kas	?	35
	2-Butanone	Diethylfumarate	Acetonitrile-d_3	$(3.14 \pm 0.30) \times 10^7$	300			lp,kas	add	35
	Cyclobutanone	Diethylfumarate	Acetonitrile-d_3	$(3.30 \pm 0.30) \times 10^8$	300			lp,kas	add	35
	Cyclopentanone	Diethylfumarate	Acetonitrile-d_3	$(3.75 \pm 0.24) \times 10^7$	300			lp,kas	add	35
	Cyclohexanone	Diethylfumarate	Acetonitrile-d_3	$(2.06 \pm 0.02) \times 10^7$	300			lp,kas	add	35
	Cycloheptanone	Diethylfumarate	Acetonitrile-d_3	$(4.40 \pm 0.40) \times 10^6$	300			lp,kas	add	35
	Acetonitrile	Diethylfumarate	Acetonitrile	9.17×10^7	298			lp,kas	add	16,28
		Diethylfumarate	Acetonitrile	1.0×10^8	rt			lp,kas	ncs,add	16
		Diethylfumarate	Acetonitrile	1.6×10^8	293	-2.04	6.66	lp,kas	add	25
		Dimethylfumarate	Acetonitrile	1.7×10^8	293	-1.65	6.94	lp,kas	add	25
		Diethylmaleate	Acetonitrile	6.63×10^6	298			lp,kas	add	28
		Diethylmaleate	Acetonitrile	1.9×10^6	293	0.0	6.28	lp,kas	add	25
		Maleic anhydride	Acetonitrile	9.81×10^8	298			lp,kas	add	16,28
		Maleic anhydride	Acetonitrile	9.8×10^8	293			lp,kas	add	25
		Dimethylmaleic anhydride	Acetonitrile	6.2×10^5	rt			lp,kas	add	16
		Dimethylmaleic anhydride	Acetonitrile	5.3×10^5	rt			lp,kas	add	16
FI	Acetonitrile	Diethylketomalonate	Acetonitrile	6.3×10^8	rt			lp,kas	ncs,add	16
		Diethylketomalonate	Acetonitrile	4.5×10^8	rt			lp,kas	add	16
		2,3-Dimethyl-2-butene	Acetonitrile	1.3×10^5	293			lp,kas	ncs,add	16
		Fumaronitrile	Acetonitrile	4.7×10^7	rt			lp,kas	add	25
		Fumaronitrile	Acetonitrile	5.7×10^7	293	-0.37	7.47	lp,kas	add	25,26
		Acrylonitrile	Acetonitrile	7.1×10^5	rt			lp,kas	add	16,28
		Methyl acrylate	Acetonitrile	2.0×10^6	rt			lp,kas	add	16,28
		trans-NCCH=CHCN	Acetonitrile	4.70×10^7	298			lp,kas	add	28
		trans-Phenyldicyanoethylene	Acetonitrile	2.21×10^7	298			lp,kas	add	28

INC	Trimethylacetonitrile acetonitrile	Fumaronitrile	1.5×10^7	293	lp,kas	add	25,26
		Acrylonitrile	$(5.24 \pm 0.21) \times 10^6$	300	lp,kas	add	56
		Acrylonitrile	$(4.77 \pm 0.34) \times 10^6$	300	lp,kas	ncs, add	56
		Diethylfumarate	$(5.9 \pm 0.2) \times 10^8$	300	lp,kas	add	47
		2,3-Dimethyl-2-butene	$(1.1 \pm 0.2) \times 10^5$	300	lp,kas	add	47
		2,5-Dimethyl-2,4-hexadiene	$(3.5 \pm 0.5) \times 10^4$	300	lp,kas	add	47
		cis-Dichloroethylene	$(2.3 \pm 0.2) \times 10^4$	300	lp,kas	add	47
		trans-Dichloroethylene	$(3.8 \pm 0.4) \times 10^4$	300	lp,kas	add	47
		tert-Butanol	$(5.5 \pm 3.2) \times 10^3$	300	lp,kas	add	47

APPENDIX 1

SUMMARY OF ABBREVIATIONS

ABCE : Azabicyclooctane.

abs : Abstration, for example, H-abs, C*l*-abs, and O-abs for hydrogen, chlorine, and oxygen abstraction, respectively.

add : Addition, for example path D in Scheme 2.

anh : Anhydrous.

azine : Path E in Scheme 2.

co : Carbonyl oxide, for example DPC leads to benzophenone oxide, Ph_2COO.

comp : Competitive, frequently involving a Stern-Volmer type of approach.

ct : Charge transfer.

diazo : When used alone the term is an abbreviation for the diazo compound corresponding to the carbene being discussed; e.g., from DPC, Ph_2CN_2.

dielectric : Dielectric loss experiments leading to the measurement of dipole moments.

dimer : Dimerization reaction (self-reaction).

fer : Free energy relationships, including correlations of the Hammett type, as well as ET(30), ionization potential, etc.

fluo : Fluorescence.

fp : Conventional flash photolysis.

H/D : Hydrogen-deuterium isotope effect. See original publication to determine if it involves only primary, or primary plus secondary isotope effects.

ins : Insertion, paths A, B2, and C2 in Scheme 2.

kas : Kinetic absorption spectroscopy.

lif : Laser induced fluorescence.

lp : Laser photolysis (with nanosecond resolution).

ncs : Noncarbene source, such as Reaction 6 in the case of carbonyl oxides, or ring opening in the case of nitrile and carbonyl ylides.

ps : Picosecond spectroscopy.

rt : Room temperature.

ST : Singlet-triplet intersystem crossing.

TEMPO : 2,2,6,6-tetramethylpiperidine-*N*-oxide.

yld : Ylide, such as the product of reaction paths H and I in Scheme 2.

APPENDIX 2

UNITS AND DEFINITIONS

Activation energy : (Ea) units of kcal/mol.

Entropy of activation : ($\Delta S\ddagger$) units of gibbs/mol; listed in the same column as log A and reported as in the original publication.

Half-lives : ($t_{1/2}$) units as specified. Used instead of τ when the decay has been reported to follow complex kinetics.

Lifetimes : (τ) units as specified. Refers to the reciprocal of the rate constant for first-order decay.

logA : With A in units of sec^{-1} or $M^{-1} sec^{-1}$ for first- and second-order processes, respectively.

Rate constants : (k), same units as A (see log A).
Temperature : (T), Kelvin.

REFERENCES

1. **Kirmse, W.**, *Carbenes*, 2nd ed., Academic Press, New York, 1971.
2. **Moss, R. A. and Jones, M., Jr.**, *Carbenes*, Wiley, New York, 1973.
3. **Jones, M., Jr. and Moss, R. A.**, *Carbene Chemistry*, Wiley, New York, 1975.
4. **Wentrup, C.**, *Reactive Molecules*, Wiley, New York, 1984.
5. **Closs, G. L. and Rabinow, B. E.**, Kinetic studies on diarylcarbenes, *J. Am. Chem. Soc.*, 98, 8190, 1976.
6. **Griller, D., Nazran, A. S., and Scaiano, J. C.**, Flash photolysis studies of carbenes and their reaction kinetics, *Acc. Chem. Res.*, 17, 283, 1984.
7. **Eisenthal, K. B., Turro, N. J., Sitzmann, E. V., Gould, I. R., Hefferon, G., Langan, J., and Cha, Y.**, Singlet-triplet interconvesion of diphenylmethylene. Energetics, dynamics and reactivities of different spin states, *Tetrahedron*, 41, 1543, 1985.
8. **Gould, I. R., Turro, N. J., Butcher, J., Jr., Doubleday, C., Hacker, N. P., Lehr, G. F., Moss, R. A., Cox, D. P., Guo, W., Munjal, R. C., Perez, L. A., and Fedorynski, M.**, Time-resolved flash spectroscopic investigations of the reactions of singlet aryl halo carbenes, *Tetrahedron*, 41, 1587, 1985.
9. **Turro, N. J., Butcher, J. A., Jr., Moss, R. A., Guo, W., Munjal, R. C., and Fedorynski, M.**, Absolute rate constants for additions of phenylchlorocarbene to alkenes, *J. Am. Chem. Soc.*, 102, 7576, 1980.
10. **Griller, D., Liu, M. T. H., and Scaiano, J. C.**, Hydrogen bonding in alcohols: its effect on the carbene insertion reaction, *J. Am. Chem. Soc.*, 104, 5549, 1982.
11. **Turro, N. J., Lehr, G. F., Butcher, J. A., Jr., Moss, R. A., and Guo, W.**, Temperature dependence of the cycloaddition of phenylchlorocarbene to alkenes. Observation of "negative activation energies", *J. Am. Chem. Soc.*, 104, 1754, 1982.
12. **Sitzmann, E. V., Langan, J., and Eisenthal, K. B.**, Intermolecular effects on intersystem crossing studied on the picosecond time scale: the solvent polarity effect on the rate of singlet to triplet intersystem crossing in diphenylcarbene, *J. Am. Chem. Soc.*, 106, 1868, 1984.
13. **Sitzmann, E. V., Wang, Y., and Eisenthal, K. B.**, Picosecond laser studies on the reaction of excited triplet diphenylcarbene with alcohols, *J. Phys. Chem.*, 87, 2283, 1983.
14. **Eisenthal, K. B., Turro, N. J., Aikawa, M., Butcher, J. A., Jr., DuPuy, C., Hefferon, G., Hetherington, W., Korenowski, G. M., and McAuliffe, M. J.**, Dynamics and energetics of the singlet-triplet interconversion of diphenylcarbene, *J. Am. Chem. Soc.*, 102, 6563, 1980.
15. **Langan, J. G., Sitzmann, E. V., and Eisenthal, K. B.**, Picosecond laser studies on the effect of structure and environment on intersystem crossing in aromatic carbenes, *Chem. Phys. Lett.*, 110, 521, 1984.
16. **Grasse, P. B., Brauer, B.-E., Zupancic, J. J., Kaufmann, K. J., and Schuster, G. B.**, Chemical and physical properties of fluorenylidene: equilibration of singlet and triplet carbenes, *J. Am. Chem. Soc.*, 105, 6833, 1983.
17. **Chuang, C., Lapin, S. C., Schrock, A. K., and Schuster, G. B.**, 3,6-Dimethoxy-fluorenylidene: a ground state singlet carbene, *J. Am. Chem. Soc.*, 107, 4238, 1985.
18. **Lapin, S. C. and Schuster, G. B.**, Chemical and physical properties of 9-xanthylidene; a ground state singlet aromatic carbene, *J. Am. Chem. Soc.*, 107, 4243, 1985.
19. **Sugawara, T., Iwamura, H., Hayashi, H., Sekiguchi, A., Ando, W., and Liu, M. T. H.**, Time resolved and low-temperature absorption spectroscopic studies on 10-silaanthracen-9(10H)-ylidene, *Chem. Lett.*, p. 1257, 1983.
20. **Sitzmann, E. V., Langan, J. G., and Eisenthal, K. B.**, Picosecond laser studies of the effects of reactants on intramolecular energy relaxation of diphenylcarbene: reaction of diphenylcarbene with alcohols, *Chem. Phys. Lett.*, 112, 111, 1984.
21. **Cox, D. P., Gould, I. R., Hacker, N. P., Moss, R. A., and Turro, N. J.**, Absolute rate constants for the additions of halophenylcarbenes to alkenes: a reactivity-selectivity relation, *Tetrahedron Lett.*, 24, 5313, 1983.
22. **Griller, D., Liu, M. T. H., Montgomery, C. R., Scaiano, J. C., and Wong, P. C.**, Absolute rate constants for the reactions of some arylchlorocarbenes with acetic acid, *J. Org. Chem.*, 48, 1359, 1983.
23. **Griller, D., Nazran, A. S., and Scaiano, J. C.**, Flash photolysis studies of carbenes and their impact on the Skell-Woodworth rules, *Tetrahedron*, 41, 1525, 1985.

24. **Moss, R. A., Perez, L. A., Turro, N. J., Gould, I. R., and Hacker, N. P.,** Hammett analysis of absolute carbene addition rate constants, *Tetrahedron Lett.,* 24, 685, 1983.

25. **Griller, D., Hadel, L., Nazran, A. S., Platz, M. S., Wong, P. C., Savino, T. G., and Scaiano, J. C.,** Fluorenylidene: kinetics and mechanisms, *J. Am. Chem. Soc.,* 106, 2227, 1984.

26. **Griller, D., Montgomery, C. R., Scaiano, J. C., Platz, M. S., and Hadel, L.,** A critical examination of transient assignments in the laser flash photolysis of 9-diazofluorene, *J. Am. Chem. Soc.,* 104, 6813, 1982.

27. **Brauer, B.-E., Grasse, P. B., Kaufmann, K. J., and Schuster, G. B.,** Irradiation of diazofluorene on a picosecond time scale and at very low temperature: a reassignment of transient structures, *J. Am. Chem. Soc.,* 104, 6814, 1982.

28. **Zupancic, J. J. and Schuster, G. B.,** Chemistry of fluorenylidene. Direct observation of, and kinetic measurements on, a singlet and a triplet carbene at room temperature, *J. Am. Chem. Soc.,* 102, 5958, 1980.

29. **Wong, P. C., Griller, D., and Scaiano, J. C.,** Radical-like reactions of singlet fluorenylidene. Hydrogen and halogen abstractions, *J. Am. Chem. Soc.,* 103, 5934, 1981.

30. **Wong, P. C., Griller, D., and Scaiano, J. C.,** Unusual temperature dependence of the reactions of triplet fluorenylidene with olefins, *Chem. Phys. Lett.,* 83, 69, 1981.

31. **Zupancic, J. J. and Schuster, G. B.,** Direct determination of the temperature dependence of the reactions of a singlet carbene. Intersystem crossing and the cyclopropanation of olefins by fluorenylidene in acetonitrile solution, *J. Am. Chem. Soc.,* 103, 944, 1981.

32. **Zupancic, J. J., Grasse, P. B., and Schuster, G. B.,** Nonstereoselective cyclopropanation of olefins by singlet fluorenylidene, *J. Am. Chem. Soc.,* 103, 2423, 1981.

33. **Zupancic, J. J., Grasse, P. B., Lapin, S. C., and Schuster, G. B.,** The reactions of fluorenylidene with heteroatomic nucleophiles, *Tetrahedron,* 41, 1471, 1985.

34. **Casal, H. L., Tanner, M., Werstiuk, N. H., and Scaiano, J. C.,** Fluorenone oxide: transient spectroscopy and kinetics of its formation and reactions, *J. Am. Chem. Soc.,* 107, 4616, 1985.

35. **Wong, P. C., Griller, D., and Scaiano, J. C.,** A kinetic study of the reactions of carbonyl ylides formed by the addition of fluorenylidene to ketones, *J. Am. Chem. Soc.,* 104, 6631, 1982.

36. **Hadel, L. M., Platz, M. S., and Scaiano, J. C.,** Study of hydrogen atom abstraction reactions of triplet diphenylcarbene in solution, *J. Am. Chem. Soc.,* 106, 283, 1984.

37. **Griller, D., Nazran, A. S., and Scaiano, J. C.,** Reaction of diphenylcarbene with methanol, *J. Am. Chem. Soc.,* 106, 198, 1984.

38. **Bethell, D., Stevens, G., and Tickle, P.,** The reaction of diphenylmethylene with isopropyl alcohol and oxygen: the question of reversibility of singlet-triplet interconversion of carbenes, *Chem. Commun.,* p. 792, 1970.

39. **Fessenden, R. W. and Scaiano, J. C.,** The dipole moment of carbonyl-oxides, *Chem. Phys. Lett.,* 117, 103, 1985.

40. **Werstiuk, N. H., Casal, H. L., and Scaiano, J. C.,** Reaction of diphenylcarbene with oxygen: a laser flash photolysis study, *Can. J. Chem.,* 62, 2391, 1984.

41. **Casal, H. L., Werstiuk, N. H., and Scaiano, J. C.,** The reaction of diphenycarbene with nitroxides, *J. Org. Chem.,* 49, 5214, 1984.

42. **Nazran, A. S. and Griller, D.,** Reaction of diphenylcarbene with amines and the question of triplet-singlet equilibration, *J. Am. Chem. Soc.,* 107, 4613, 1985.

43. **Grasse, P. B., Zupancic, J. J., Lapin, S. C., Hendrich, M. P., and Schuster, G. B.,** Chemical and physical properties of 2,3-benzofluorenylidene. Closing the gap between singlet and triplet carbenes, *J. Org. Chem.,* 50, 2352, 1985.

44. **Hadel, L. M., Platz, M. S., Wright, B. B., and Scaiano, J. C.,** A laser flash photolysis study of dibenzocycloheptadienylidene, *Chem. Phys. Lett.,* 105, 539, 1984.

45. **Sugawara, T., Iwamura, H., Hayashi, H., Sekiguchi, A., Ando, W., and Liu, M. T. H.,** Time resolved absorption spectroscopic detection of 10,10-dimethyl-10-silaanthracen-9 (10H)-one oxide, *Chem. Lett.,* p. 1261, 1983.

46. **Lapin, L. C., Brauer, B.-E., and Schuster, G. B.,** 9-Mesityl-9,10-dihydro-9-boraanthrylidene: a probe of structure and reactivity for aromatic carbenes, *J. Am. Chem. Soc.,* 106, 2092, 1984.

47. **Hadel, L. M., Platz, M. S., and Scaiano, J. C.,** Laser flash photolysis studies of 1-naphthyldiazomethane. Formation of nitrile ylides, *Chem. Phys. Lett.,* 97, 446, 1983.

48. **Horn, K. A. and Chateauneuf, J. E.,** Spectroscopic characterization of 2-naphthylcarbene, *Tetrahedron,* 41, 1465, 1985.

49. **Nazran, A. S. and Griller, D.,** Singlet and triplet dimesitylcarbene, *J. Am. Chem. Soc.,* 106, 543, 1984.

50. **Nazran, A. S. and Griller, D.,** Dimesitylcarbene: the distinct chemistries of its singlet and triplet states, *J. Chem. Soc. Chem. Commun.,* p. 850, 1983.

51. **Horn, K. A. and Allison, B. D.,** The absorption spectrum and reactivity of excited triplet state diphenylcarbene, *Chem. Phys. Lett.,* 116, 114, 1985.

52. **Johnston, L. J. and Scaiano, J. C.,** Absorption, fluorescence, lifetime and reactivity of excited di(*p*-methylphenyl)methylene in solution, *Chem. Phys. Lett.,* 116, 109, 1985.
53. **Sitzmann, E. V., Langan, J., and Eisenthal, K. B.,** Picosecond laser studies of the charge-transfer reaction of excited triplet diphenylcarbene with electron donors, *Chem. Phys. Lett.,* 102, 446, 1983.
54. **Wang, Y., Sitzmann, E. V., Novak, F., Dupuy, C., and Eisenthal, K. B.,** Reactions of excited triplet diphenylcarbene studied with picosecond lasers, *J. Am. Chem. Soc.,* 104, 3238, 1982.
55. **Criegee,** Mechanism of ozonolysis, *Angew. Chem. Int. Ed. Engl.,* 14, 745, 1975.
56. **Barcus, R. L., Wright, B. B., Platz, M. S., and Scaiano, J. C.,** Chemical, kinetic and spectroscopic evidence for the reaction of 1-naphthylcarbene with acetonitrile to form a ylid, *Tetrahedron Lett.,* 24, 3955, 1983.

Chapter 14

KINETICS AND MECHANISMS OF METHYLIDYNE RADICAL REACTIONS

W. A. Sanders and M. C. Lin

TABLE OF CONTENTS

I. INTRODUCTION

The CH radical has been proposed as a key intermediate in a wide variety of complex reaction systems. Examples include chemiionization in hydrocarbon flames,[1] NO production,[2-4] and chemiluminescence.[5] It has also been implicated in the chemistry of planetary atmospheres.[6,7]

Chemiluminescence originating from the $A^2\Delta \to X^2\Pi$ transition of CH at 430 nm has been observed in a number of hydrocarbon and hydrocarbon-doped systems. It has also been found to accompany the low-temperature reactions of O atoms with hydrocarbons, especially in discharge-flow studies involving unsaturated compounds. However, the mechanisms responsible for the formation of CH radicals are still unclear, especially in cases involving excited states.

The reaction $CH + N_2$ has been of interest recently because of its role in the formation of so-called "prompt NO" in hydrocarbon-air flames. This process is clearly different from that which explains the formation of NO in the postcombustion region. Fenimore[2] has proposed that $CH + N_2$, followed by oxidation of the products, is one of the possibilities, as will be discussed in detail below.

The suspected presence of CH radicals in the atmospheres of Titan[6] and Jupiter[7] is thought to be a result of photodissociation of CH_4 by VUV radiation from the sun. This speculation provided the motivation for studies of reactions of CH with a number of small alkanes and unsaturated hydrocarbons which are likely products of VUV photolysis of CH_4.

The reactions of CH with O, O_2, and NO were believed to be responsible for the observed strong CO stimulated emissions at 5 μm in the flash photolysis ($\lambda \geq 165$ nm) of $CHBr_3$ mixtures containing SO_2,[8,9] O_2,[10] and NO[11] with appropriate diluents. In the $CHBr_3$-O_2 system, 10-μm CO_2 laser emission was also detected. It was assumed to result from the direct and very exothermic reaction $CH + O_2 \to CO_2 + H$. The direct formation of CO in the $CH + NO$ reaction could only occur via a four-centered activated complex generating CO and NH simultaneously. Production of $NH(A^3\Pi)$ from this process has indeed been detected recently by its chemiluminescence.[5]

The CH radical is of fundamental theoretical interest, since it is the simplest representative of the family of monovalent carbon radicals known as carbynes. It presents a special challenge to theoreticians, because it is both simple enough to be amenable to accurate *a priori* calculations and chemically unique. A major challenge is to explain the unusually high reactivity of CH compared with other members of the family CH_x (x = 0, 1, 2, 3), which is illustrated dramatically by the data reproduced in Table 1. Consequently, measurements of accurate rate constants for reactions of CH with simple molecules or radicals provide important data for testing contemporary models of structure, energetics, and reaction rates.

The very important work on the generation of potential energy surfaces for reactions of CH with H_2 by Brooks and Schaefer[12] provides an illustrative example of the physical insights which can be derived when accurate *ab initio* calculations are possible. They carried out multiconfiguration SCF calculations, using large basis sets which were verified by accurately predicting the structural and energetic parameters of the reactants and products. The insertion reaction $^2\Pi$ $CH + ^1\Sigma_g^+ H_2 \to ^2A_2''CH_3$ was investigated in two limiting geometries, illustrated in Figures 1 and 2. The least-motion insertion path, in which the CH radical is perpendicular to the H_2 molecule, was found to have an energy barrier of about 75 kcal/mol (see Figure 1). For the nonleast motion, or parallel approach, on the other hand, there was little or no energy barrier (see Figure 2). The dramatic differences between these two pathways can be seen very clearly in the potential energy surfaces reported by Brooks and Schaeffer,[12] also reproduced in these two figures. This unusual combination of high reactivity and marked geometric selectivity makes CH an interesting species indeed.

In a recent review of the gas phase chemistry of carbynes, James et al.[13] discussed the

Table 1
ROOM TEMPERATURE REACTIVITIES OF CH$_x$ (x = 0, 1, 2, 3) RADICALS IN THE GROUND ELECTRONIC STATE[a]

Reactants	C	CH	CH$_2$	CH$_3$
H$_2$	—[b]	$\sim$1 $\times$ 10^{-10} [d]	<5 $\times$ 10^{-14} [f]	2 $\times$ 10^{-20} [i]
O$_2$	3 $\times$ 10^{-11} [c]	5 $\times$ 10^{-11} [e]	2 $\times$ 10^{-12} [g]	1.1 $\times$ 10^{-12} [j]
N$_2$	—[b]	$\sim$2 $\times$ 10^{-11} [d]	$\sim$10 − 15[h]	—[b]
CH$_4$	<3 $\times$ 10^{-15} [c]	1 $\times$ 10^{-10} [e]	<5 $\times$ 10^{-14} [f]	4 $\times$ 10^{-23} [i]

[a] Bimolecular rate constants in cm^3/molecule/sec.

[b] Not measured or not available in the second-order region.

[c] Husain, D. and Kirsch, L. J., Reactions of atomic C(2^{3}P$_J$) by kinetic absorption spectroscopy in the vacuum ultraviolet, *Trans. Faraday Soc.,* 67, 2025, 1971.

[d] Extrapolated high-pressure second order rate constant, see text.

[e] See Table 3.

[f] Braun, W., Bass, A. M., and Pilling, M., Flash photolysis of ketene and diazomethane: the production and reaction kinetics of triplet and singlet methylene, *J. Chem. Phys.,* 52, 5131, 1970.

[g] Laufer, A. H. and Bass, A. M., Rate constants for reactions of methylene with carbon monoxide, oxygen, nitric oxide and acetylene, *J. Phys. Chem.,* 78, 1344, 1974.

[h] Computed from the rate constant expression estimated by Benson and O'Neal (Reference 47).

[i] Computed from the rate constant expressions compiled by Kerr, J. A. and Parsonage, M. J., *Evaluated Kinetic Data on Gas Phase Hydrogen Transfer Reactions of Methyl Radicals,* Butterworths, London, 1976.

[j] Pilling, M. J. and Smith, M. J. C., A laser flash photolysis study of the reaction CH$_3$ + O$_2$ → CH$_3$O$_2$ at 298 K, *J. Phys. Chem.,* 89, 4713, 1985.

relationship between structure and reactivity of CH. The ground-state electronic configuration of CH is $(1\sigma^2)(2\sigma^2)(3\sigma)^2(1\pi)^1$, giving rise to the observed $^2\Pi$ spectroscopic state. Of the three doubly-occupied σ orbitals, one corresponds to the $(1s)^2$ core of the C atom, one is bonding, and the third is essentially nonbonding. Since the H atom has no low-lying atomic p orbitals to overlap with those of carbon, the π orbitals (one empty and one singly-occupied) are highly localized on the carbon atom. The resulting electron deficiency of the carbon atom helps explain the unusually high reactivity of CH in comparison with other CX species. When X is a halogen atom, for example, its p orbitals can overlap both of the πp orbitals on carbon and relieve at least a portion of the electron deficiency. This rationale is consistent with the experimental data, which indicate that the rates of insertion of CCl and CBr into C–H bonds are very slow or unobservable.[14-16]

The lowest excited state of CH is $^4\Sigma^-$, which is derived from the electron configuration $(1\sigma)^2(2\sigma)^2(3\sigma)^1(1\pi)^2$. Although conventional spectroscopic studies of this state have not been reported, it has been observed by laser photoelectron spectroscopy[17] and found to lie 17 kcal/mol above the ground state. The experimental observation that the excited state is apparently less reactive than the ground state may be explainable on the basis of spin and orbital symmetry conservation rules.[13]

Typical steady-state concentrations of CH radicals in hydrocarbon combustion systems are very low and there is no known conventional method for the reliable detection of the radical in its (X$^2\Pi$) electronic ground state. Although fluorescence from electronically excited CH was well known as a contributor to the luminescence of hydrocarbon flames, it was not possible to exploit this knowledge for quantitative detection prior to the advent of tunable dye lasers. For this reason, most of the definitive work on CH kinetics has been published in the past 6 to 7 years. Prior to the more recent investigations using lasers, in fact, there were only two such studies — both at room temperature. Braun et al.[18] measured the rates

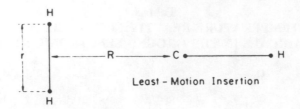

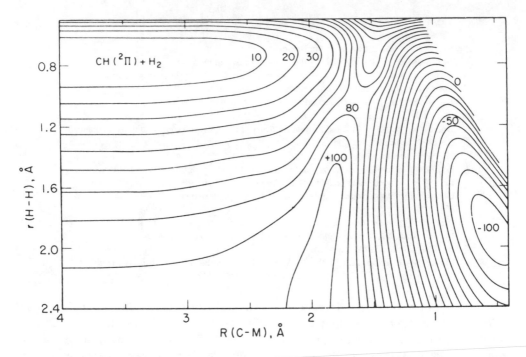

FIGURE 1. Potential energy surface for the least-motion insertion reaction CH(X²II) + H₂ → CH₃ with the geometrical approach shown at the top. (From Brooks, B. R. and Schaefer, H. F., *J. Chem. Phys.*, 67, 5146, 1977. With permission.)

of reactions of CH with H_2, N_2, and CH_4 by the vacuum UV flash photolysis/UV absorption technique, monitoring the disappearance of CH via the $C^2\Sigma^+ \leftarrow X^2\Pi$ transition at 314 nm. Bosnali and Perner[19] used a combination of electron beam initiation and UV absorption (at 314 nm) to determine the rates of CH reactions with a number of reaction partners. No studies of temperature or pressure effects were included in this early work.

In this review we will discuss the reactions of CH radicals with H_2, O_2, N_2, O, N, water, ammonia, formaldehyde, oxides of carbon and nitrogen, and with a number of saturated and unsaturated hydrocarbons. These substrates have been selected because of their importance in combustion and planetary atmospheric chemistry.

II. EXPERIMENTAL

A. Sources of CH Radicals

In order to perform reliable measurements of rate constants of the reactions of CH, it is necessary to find a clean and convenient method for generating the radicals in the reaction system. Unlike the cases of CH_2 and CH_3, there is no known method for generating CH directly. Because of its very high reactivity (See Table 1), it cannot be produced successfully by standard flash photolysis techniques. For example, flash photolysis of $CHBr_3$ in the UV

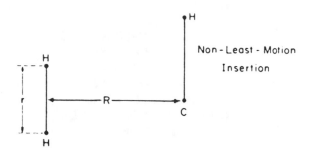

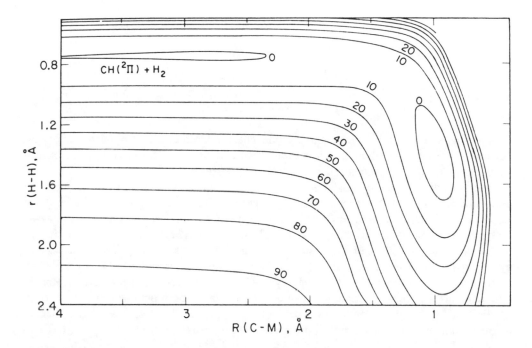

FIGURE 2. Potential energy surface for the nonleast-motion insertion reaction $CH(X^2\Pi) + H_2 \rightarrow CH_3$ with the geometrical approach shown at the top. (From Brooks, B. R. and Schaefer, H. F., *J. Chem. Phys.*, 67, 5146, 1977. With permission.)

region has been shown[20] to generate high concentrations of CBr radicals which could be detected directly by plate photometry. However, it was impossible to confirm the presence of CH radicals in this system by the same technique because of its high reactivity, which decreased the apparent concentration rapidly. On the other hand, CO stimulated emissions and mass spectrometric product analysis had clearly pointed to CH as the key active species in these laser systems.[8-11]

Direct confirmation of the presence of CH in the UV-photolyzed CHBr$_3$ system was achieved by laser-induced fluorescence (LIF), using tunable dye lasers. A combination of pulsed UV photodissociation and LIF has proven to be the best technique for generating and detecting CH in kinetic studies.[4,21-25] The use of UV laser dissociation sources has the added advantage that it makes it possible to investigate the effects of pressure on the reaction kinetics over a broad range of relevant pressures.

Using an ArF laser (193 nm) as the dissociation source, CH radicals have been generated by photolysis of numerous molecules, including C_2H_2, $CHBr_3$, CH_2Br_2, CH_3C_2H, C_2H_4, and C_6H_6. Of all the sources investigated, CHBr$_3$ was found to be the most efficient (See Table 2).[26] In later experiments, the ArF laser photolysis source was replaced by the quadrupled

Table 2
RELATIVE EFFICIENCIES OF CH RADICAL PRODUCTION FROM MULTIPHOTON DISSOCIATION OF DIFFERENT COMPOUNDS AT 193 nm[a]

Compound	Relative efficiency
$CHBr_3$	1.0
C_2H_2	0.033
C_2H_4	~0.0
CH_3C_2H	0.49
C_6H_6	~0.0

[a]   Data from Reference 26.

output of a Nd:YAG laser operating at 266 nm. This produced much better results, apparently because the longer wavelength is less destructive and generates fewer competing radicals, particularly from the reactants of interest. The mechanism of CH formation from $CHBr_3$ photodissociation, either by flash photolysis or by pulsed UV laser photolysis, is not clearly understood. It is probably produced by consecutive absorption of photons, leading to the stepwise breaking of the weaker C–Br bonds in $CHBr_3$. At high laser power levels, however, direct elimination of Br_2 from superexcited $CHBr_3$** or $CHBr_2$** cannot be ruled out.[5]

In recent studies with laser diagnostics, CH radicals have also been generated successfully by infrared multiphoton dissociation (IRMPD), using high power pulsed CO_2 lasers.[27-30] The radical sources included CH_3OH, C_3NH_2, cyclo-C_3H_6, and several other compounds. Reliable kinetic data for several reactions at room temperature have been obtained by the use of the IRMPD-LIF technique. A major drawback of this approach, however, is the fact that collisional relaxation limits the range of pressures that can be used in studies of pressure effects.

B. CH Diagnostics

The most common and successful method for kinetic measurements of the concentration of CH radical in its ($X^2\Pi$) electronic ground state has proven to be the detection of laser-induced fluorescence associated with the transition $A^2\Delta \leftarrow X^2\Pi$ at 430 nm.[27,31] The possible alternative LIF mechanisms, $B^2\Sigma^- \leftarrow X^2\Pi$ ($\lambda_{oo} = 388.9$ nm) and $C^2\Sigma^+ \leftarrow X^2\Pi$ ($\lambda_{oo} = 314.4$ nm), might be equally useful in principle, but they have been avoided because of concern over predissociation of the B and C states.[32]

Figure 3 shows a typical set of LIF spectra obtained from the A ← X excitation during the multiphoton dissociation of $CHBr_3$ at 266 nm, with and without added D_2. The rapid growth of new lines when D_2 is added can be attributed to LIF of the CD radical. This example clearly demonstrates the sharp lines of the spectra and provides experimental evidence for the exchange reaction CH + $D_2 \to$ CD + HD, as will be discussed later. For kinetic measurements, the probing laser is usually tuned to strong transitions such as $R_{1(ef,fe)}$ or $R_{2(ef,fe)}$. Tests with different rotational lines have yielded essentially identical kinetic results. To ensure complete thermalization of the rotational energy level populations, the kinetic measurements are normally carried out in the presence of relatively high mole fractions of inert gases such as Ar or He.

C. Apparatus for Kinetic Measurements

Details of experimental procedures vary slightly from one laboratory to another. However,

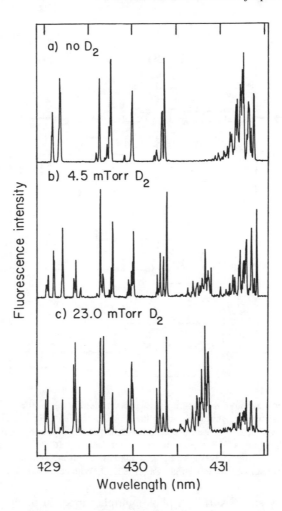

FIGURE 3.   Laser-induced fluorescence excitation spectra of CH and CD radicals obtained at 20 μsec after the photolysis of CHBr$_3$ (∼1 mtorr) at 266 nm with and without added D$_2$ as indicated. (From Berman, M. R. and Lin, M. C., *J. Chem. Phys.*, 81, 5743, 1984. With permission.)

the setup used in the later work of Lin and co-workers is representative and will be described here as a typical example. A schematic diagram of the experiment is shown in Figure 4, which is reproduced from Reference 22. The photolyzing and probing laser beams propagated anticollinearly through the cell. The photolyzing pulses were provided by a Quanta-Ray Nd^{3+}:YAG laser operating on the fourth harmonic at 266 nm. Typical energies were 3 to 5 mJ/pulse at a repetition rate of 10 Hz. The laser beam, which had a doughnut-shaped cross section with a diameter of 0.8 cm, was focused into the center of the cell by a BaF$_2$ lens with a focal length of 50 cm. The probing dye laser was a Molectron DL-14 using coumarin 102, and it was pumped by a second Nd^{3+}:YAG laser at the same repetition rate (10 Hz). The dye laser output, which was normally 50 to 100 μJ/pulse, was focused into the center of the reaction cell by a crown glass lens with a focal length of 50 cm. For detection of CH by LIF, the dye laser was operated at a fixed wavelength of 429.8 nm, which corresponds to the R$_{1(cd,dc)}$ (2) rotational line of the (A^{2}Δ ← X^{2}Π), (0-0) vibronic transition.

LIF from the confocal region in the center of the reaction cell was detected at right angles

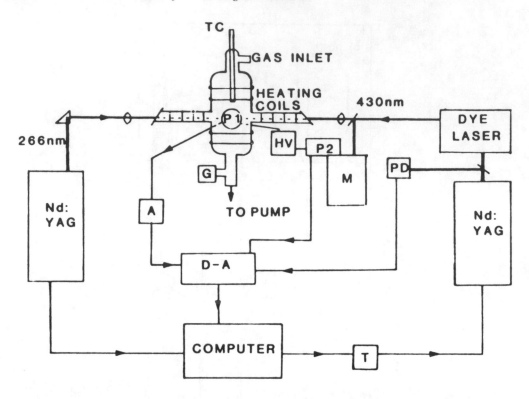

FIGURE 4.   A schematic diagram of the experimental apparatus used in the kinetic measurements of CH radical reactions at different temperatures and pressures using the photodissociation/LIF probe technique. A = amplifier, D-A = data acquisition system, G = pressure gauge, HV = high voltage power supply, M = monochromator, P1 and P2 = photomultiplier tubes, PD = photodiode, T = trigger, and TC = retractable thermocouple, (From Berman, M. R., Fleming, J. W., Harvey, A. B., and Lin, M. C., *Nineteenth Symposium (International) on Combustion,* The Combustion Institute, Pittsburgh, Pa., 1982, 73. With permission.)

to the laser beams by an RCA 1P28A photomultiplier tube. In order to achieve both high collection efficiency and discrimination against scattered light and other fluorescence sources, the LIF was collected through f/3 glass optics, filtered to pass radiation in the range 390 to 500 nm (Schott BG-12, KV-388, KV-389), and spatially defined by a 1.0 × 20-mm slit placed over the aperture of the photomultiplier tube. The amplified (Tektronix AM 502) output of the photomultiplier was sampled 200 nsec after the dye laser pulse, digitized, and relayed to a computer (HP 9825A) for storage and subsequent processing. Prior to each experiment, the following signals were sampled and stored: electronic noise (both lasers blocked), scattered light from the dye laser (photolysis laser blocked), and scattered light and visible emission due to the photolysis laser alone. These samples were used by the data analysis program to correct the LIF measurements for the various background sources. A sample of the dye laser output was deflected by a beam-splitter, passed through a Spex 1401 monochromator, and measured by a second photodetector. For each laser pulse, the LIF signal was divided by this sample in order to normalize the data to the laser intensity. Because of the normal shot-to-shot variations of the dye laser intensity, this procedure improves the signal-to-noise ratio substantially.

The kinetic data were obtained by using a programmable pulse generator to vary the time interval between the photolyzing and probing laser pulses. The pulse delay was controlled by the computer and was normally incremented by 1 μsec between each pair of data points, each of which was obtained by averaging over 5 laser pulses. Data were collected for a total time sufficient to ensure that the kinetics could be established, which normally meant two

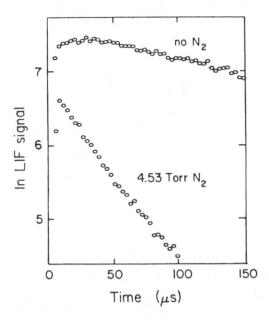

FIGURE 5.   Typical $A^2\Delta$-$X^2\Pi$ LIF profiles of CH radicals vs. time between the photolysis (266 nm) and probing (429.8 nm) pulses for the CH + N_2 reaction. P_{CHBr_3} = 0.2 mtorr, P_{total} = 100 torr Ar. (From Berman, M. R. and Lin, M. C., *J. Phys. Chem.*, 87, 3933, 1983. With permission.)

or three lifetimes. In a typical kinetic run, the time delay commenced at -5 μsec and was incremented to a maximum value of 300 μsec.

For experiments carried out at elevated temperatures, provision was made for resistive heating of the reaction cell. The temperature in the reaction zone was measured by a retractable thermocouple before and after the experiment. A similar cell with a jacket through which coolant could be flowed was used for studying reactions below room temperature.

D. Data Analysis Procedures

Details of the data analysis procedure may vary slightly for different reaction systems, but the case CH + N_2 is typical[4] and will be discussed here as an example. The LIF intensity is the primary measure of CH radical concentration and was plotted as a function of time, as shown in Figure 5. The concentration of CH radicals decreased with time in the absence of substrate due to diffusion out of the detection zone. In the presence of N_2, the LIF intensity decayed exponentially after an initial relaxation period. The observed signal was least-squares fitted to an equation of the form

$$I = Ae^{-k_1 t} + B \tag{1}$$

where k_1 is the pseudo-first-order rate constant for loss of CH due to all loss processes, i.e., reaction plus diffusion out of the probing beam. The decay equation was fitted over a period of time corresponding to at least two lifetimes.

The observed pseudo-first-order rate constant was found to increase linearly with the N_2 concentration, and could be fitted to an equation of the form

$$k_1 = k_0 + k_2[N_2] \tag{2}$$

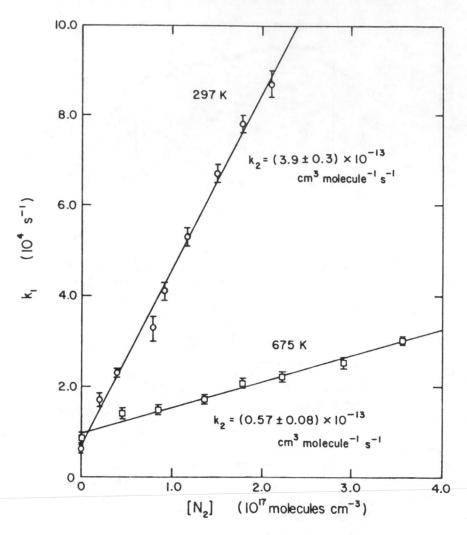

FIGURE 6.   Plots of the pseudo-first-order decay rate constants vs. N_2 concentrations at different temperatures. The total pressure was maintained constant at 100 torr with Ar dilution. (From Berman, M. R. and Lin, M. C., *J. Phys. Chem.*, 87, 3933, 1983. With permission.)

where k_o is the rate constant for loss of CH in the absence of N_2 and k_2 is the second-order rate constant for the reaction $CH + N_2$. Figure 6 illustrates this linear relation for the lowest and highest temperatures investigated. The values of k_2 shown in the figure were obtained from the slopes of the fitted lines.

The data plotted in Figure 6 were obtained at a constant total pressure of 100 torr. As shown in Figure 7, however, the second-order rate constant exhibits a significant dependence on the total pressure. This indicates the importance of a third-body collisional stabilization process. In principle, this effect could be used to determine the relative efficiencies of different collision partners. If Ar is the diluent gas, for example, the observed pseudo-first-order rate constants can be fitted to the equation

$$k_1 = k_0 + k_3^{sc}P\{\beta_{c,Ar}X_{Ar} + \beta_{c,N_2}X_{N_2}\}[N_2] \qquad (3)$$

where k_3^{sc} is the third-order rate constant in the so-called strong-collision limit, where every collision is assumed to lead to deactivation, P is the total pressure, $\beta_{c,i}$ is the collision

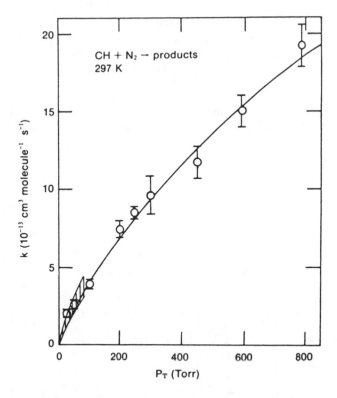

FIGURE 7. Second order rate constant for CH + N_2 at 297 K as a function of total pressure (Ar). The shaded area contains the data of Reference 29. The solid curve is the result of a TST-RRKM calculation (Model II of Reference 4).

efficiency of the ith collision partner, and the X_i are the mole fractions. This equation can be rearranged to give

$$(k_1 - k_0)/[N_2] = k_3^{sc}P[\beta_{c,Ar} + (\beta_{c,N_2} - \beta_{c,Ar})X_{N_2}] \tag{4}$$

which makes it possible to determine the relative collision efficiencies of Ar and N_2. A plot of $(k_1-k_0)/[N_2]$ vs. X_{N_2} showed that the efficiencies are essentially equal.[4] The common collision efficiency β_c can then be obtained from the equation

$$(k_1 - k_0)/[N_2]^2 = k_3^{sc}\beta_c([Ar]/[N_2] + 1) \tag{5}$$

If [M] is the total concentration of collision partners, then $k_3^{sc} \beta_c = k_2/[M]$.

III. RESULTS AND DISCUSSION

Since most of the earlier work on CH radical kinetics was limited to room temperature measurements, the discussion of results will focus primarily on the more recent experiments in which rate constants were measured as functions of temperature. Table 3 summarizes the Arrhenius parameters of reactions which have been studied to date. Room temperature rate constants are included for comparison with earlier work. The individual systems are discussed in detail below.

Table 3
SUMMARY OF ARRHENIUS PARAMETERS [$k = A \exp(-E_a/RT)$] AND ROOM TEMPERATURE RATE CONSTANTS FOR CH RADICAL REACTIONS

Reactant	Temperature range (K)	$A \times 10^{11}$ (cm³/molecule/sec)	E_a (cal/mol)	$k_{RT} \times 10^{11}$ (cm³/molecule/sec)	Ref.
H_2	150—300[a]	0.237 ± 0.043	-1041 ± 85	1.40 ± 0.10	33
	400—658	$36 + 12; 36 - 3$	3900 ± 1400	—	33
	372—675	23.8 ± 3.1	3500 ± 130	—	34
	—	—	—	2.6 ± 0.5	31
	—	—	—	1.74 ± 0.20	19
	—	—	—	0.10	18
CH_4	167—652	5.0 ± 0.5	-398 ± 62	10.2 ± 0.4	25
	—	—	—	3.34 ± 0.08	19
	—	—	—	0.025	18
C_2H_6	162—650	18 ± 2	-263 ± 59	27 ± 2	25
C_3H_8	—[b]	—	—	37 ± 6	25
	—	—	—	$14 + 2; 14 - 3$	19
C_4H_{10}	257—653	44 ± 8	-55 ± 120	48 ± 5	25
	—	—	—	13 ± 1	19
C_2H_4	160—652	22.3 ± 2.7	-173 ± 35	42 ± 3	24
	—	—	—	25.5 ± 1.4	26
	—	—	—	11.5 ± 1	19
C_2H_2	171—657	34.9 ± 4.2	-61 ± 36	42 ± 2	24
	—	—	—	14.0 ± 4.5	26
	—	—	—	$7.5 + 1.5; 7.5 - 0.7$	19
C_6H_6	297—674	43 ± 3	0	43 ± 3	24
	—	—	—	7.9 ± 3.2	26
C_7H_8	(c)	—	—	50 ± 4	24
H_2O	297—669	0.949 ± 0.005	-751 ± 41	3.23 ± 0.13	40
	—	—	—	$4.5 + 0.8; 4.5 - 0.3$	19
NH_3	(c)	—	—	$9.8 + 0.8\ 9.8 - 1.5$	19
CH_2O	297—670	15.7 ± 1.4	-515 ± 60	37.2 ± 3.0	40
N_2	297—675[a]	0.0017 ± 0.003	-1950 ± 130	0.039 ± 0.003	4
	—	—	—	0.0366 ± 0.0027[a]	29
	—	—	—	0.093 ± 0.01[a]	23
	—	—	—	$0.10 + 0.02; 0.10 - 0.05$[d]	19
	—	—	—	0.0071[e]	18
	—	—	—	0.0071[f]	30
	—	—	—	0.0076 ± 0.0005[f]	29
O_2	297—676	5.4 ± 1.0	100 ± 100	5.1 ± 0.5	22
	—	—	—	3.3 ± 0.4	27
	—	—	—	$\leqq 4.0$	19
	—	—	—	0.21	30
CO	297—669[a]	0.046 ± 0.01	-1700 ± 200	0.83 ± 0.04	22
	—	—	—	0.48	19
CO_2	297—670	0.57 ± 0.09	690 ± 110	0.18 ± 0.01	22
N_2O	—[c]	—	—	7.8 ± 1.4	29
NO	297—676	19 ± 3	-100 ± 100	19 ± 3	22
	—	—	—	20 ± 3	29
NO_2	—[c]	—	—	16.7 ± 1.1	29
O	—[c]	—	—	9.5 ± 1.4	28
N	—[c]	—	—	2.1 ± 0.5	28

Table 3 (continued)
SUMMARY OF ARRHENIUS PARAMETERS [k = A exp (−Eₐ/RT)] AND
ROOM TEMPERATURE RATE CONSTANTS FOR CH RADICAL
REACTIONS

a   The rate constant is pressure-dependent in this low-temperature region. These values are for a total pressure of 100 torr (Ar).

b   Values estimated by interpolation.

c   Measurements have been reported at room temperature only.

d   Total pressure not reported, but apparently in the range 15 to 20 torr.

e   Pressure range 1 to 40 torr.

f   Total pressure 10 torr.

A. Reactions of CH and CD with H_2 and D_2

As mentioned above, the reaction of CH with H_2 is of considerable interest to experimentalists and theoreticians alike because it is a prototype for an entire family of carbyne reactions and because it is simple enough to be accessible to accurate *ab initio* calculations. Prior to the development of the current laser photolysis/LIF techniques, however, experimental studies had only been carried out at room temperature.[18,19,23] These experiments were performed under different conditions, and the rate constants obtained varied by a factor of 20. Recently Berman and Lin[33] used the two-laser technique to obtain kinetic data for the reaction CH + H_2 and its isotopic analogs over rather wide ranges of temperature and pressure.

For the reaction of CH with H_2 there are two possible product channels, shown in Equations 6a and 6b,

$$CH(^2\Pi) + H_2 \rightarrow CH_2(^3B_1) + H \quad \Delta H^\circ = +2.5 \text{ kcal/mol} \tag{6a}$$

$$\xrightarrow{M} CH_3 \qquad \Delta H^\circ = -105.3 \text{ kcal/mol} \tag{6b}$$

The abstraction channel (Equation 6a) is endothermic by only 2.5 kcal/mol, which suggests that it might compete with insertion at experimentally accessible temperatures. Room temperature rate constants were measured at total pressures ranging from 25 to 600 torr and at a constant total pressure of 100 torr in the temperature range 159 to 658 K in an attempt to obtain information on the reaction mechanism. Additional insights were provided by studies of the isotopic variants CD + D_2, CH + D_2, and CD + H_2. Each of these is discussed briefly below.

1. CH + H_2

At 297 K the rate constant increased monotonically with pressure over the range 25 to 600 torr, with a total increase of approximately a factor of 5. The data could be fitted reasonably well by a model based on transition state (RRKM) theory, but it was not possible to extract a reliable value for the zero-pressure rate constant. The experimentally measured and calculated rate constants are similar to those for the CH + N_2 reaction shown in Figure 7, with the exception that the rate constants for CH + H_2 are larger by about a factor of 30.

Figure 8 shows the temperature dependence of the rate constant at 100 torr, which could not be represented by a single Arrhenius function. In the temperature range 150 to 300 K the reaction exhibited a negative temperature dependence which could be described quite

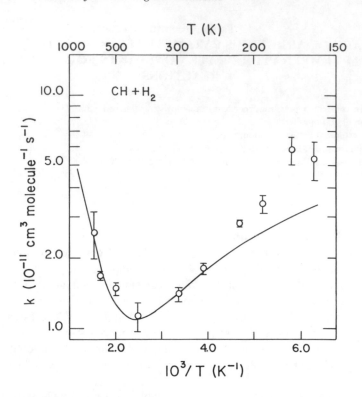

FIGURE 8.   An Arrhenius plot of the second order rate constant for the CH + H$_2$ reaction at 100 torr total pressure (Ar). The solid curve is the result of the TST-RRKM calculation.[33]

well by an Arrhenius expression with parameters A = (2.37 ± 0.43) × 10^{-12} cm^3/molecule/sec and E$_a$ = −1040 ± 85 cal/mol. At temperatures above 400 K, on the other hand, the rate constants were observed to increase with increasing temperature. If the extrapolated values predicted by the low-temperature Arrhenius expression were subtracted, the high-temperature data could be fitted reasonably well by an Arrhenius function with parameters A = (3.6 $^{+11.5}_{-2.7}$) × 10^{-10} cm^3/molecule/sec and E$_a$ = 3900 ± 1400 cal/mol. An alternative extrapolation procedure using the expression k = BT^{-n} gave very similar results. The rate constants in the high temperature regime (T > 400 K) have been measured more recently by Zabarnick, Fleming, and Lin[34] in greater detail. Their results are presented in Figure 9, together with those obtained by Berman and Lin.[33] A least-squares analysis of these newer data yielded A = (2.38 ± 0.31) × 10^{-10} cm^3/molecule/sec and E$_a$ = 3504 ± 131 kcal/mol, which are in close agreement with the earlier results.

2. CD + D$_2$

The behavior of this system was qualitatively similar to that of CH + H$_2$, but the low-temperature rate constants were higher by about 50%.[33] In the temperature range 175 to 403 K a linear least squares fit to the Arrhenius function gave A = (3.5 ± 0.58) × 10^{-12} cm^3/molecule/sec and E$_a$ = −1030 ± 90 cal/mol. The opening of the high-temperature channel appeared to occur at a somewhat higher temperature than for CH + H$_2$, and the high-temperature rate constants were substantially smaller. A fit of the limited data above 500 K gave the approximate values A = 5 × 10^{-11} cm^3/molecule/sec and E$_a$ = 3000 cal/mol.

3. CH + D$_2$ and CD + H$_2$

Berman and Lin[33] measured rate constants for the two mixed-isotope reactions at 297 K

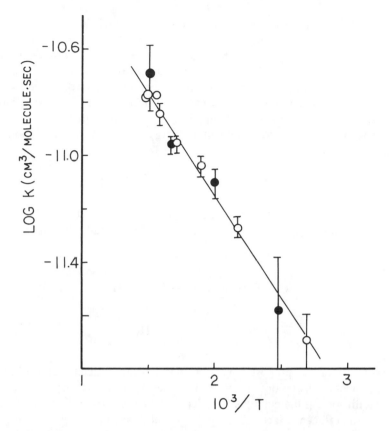

FIGURE 9. An Arrhenius plot of the rate constant for $CH + H_2 \rightarrow CH_2 + H$ in the high temperature regime. Open circles — data from Reference 34; filled circles — data from Reference 33.

and 100 torr. The rate constant corresponding to loss of CH in the $CH + D_2$ system was $(1.14 \pm 0.07) \times 10^{-10}$ cm³/molecule/sec, which is essentially the same as the rate constant for loss of CD in the $CD + H_2$ system, $(1.17 \pm 0.12) \times 10^{-10}$ cm³/molecule/sec. In each of these systems the loss rates were 5 to 8 times larger than those of the single-isotope variants.

The enhanced loss rates were attributed to fast isotope exchange reactions, exemplified by

$$CH + D_2 \rightleftarrows CHD_2^{\ddagger} \rightleftarrows CD + HD \qquad (7)$$

with an analogous process occurring in the $CD + H_2$ system. This process is not observable in the single-isotope systems, of course, because it results in no net change of the CH/CD concentration. Evidence supporting the contribution of Reaction 7 was obtained by monitoring the time evolution of LIF signals from both CH and CD. This was made possible by the existence of nonoverlapping spectral lines, as shown in Figure 3. The concentration of CD was found to rise initially at a rate equal to that of the CH decay. After all of the CH was converted to CD, the CD signal decayed at a rate which agreed very closely with that observed in the $CD + D_2$ system. Analogous results were obtained from $CD + H_2$, thus lending further support to the hypothesis.

4. Theoretical Interpretation

The observed temperature and pressure dependence of the $CH + H_2$ reaction suggests a competition between two alternate reaction channels, one of which dominates at temperatures below 300 K while the other becomes rate determining above 400 K. In the low-temperature region the observed pressure dependence suggests that the dominant process is collisional stabilization of the intermediate $CH_3^{\dagger}$. The small negative activation energy is consistent with previous observations on similar reaction systems known to involve the formation of long-lived complexes.[4,22,25] The observed isotopic exchange, which should be expected to proceed through the same complex, lends further support to this mechanism.

Above 400 K there is a positive activation energy of 3.5 ± 0.1 kcal/mol, which is consistent with the reaction

$$CH + H_2 \rightarrow CH_2 + H \quad \Delta H_{298}^{\circ} = 2.5 \pm 0.1 \text{ kcal/mol} \tag{8}$$

At higher temperatures the positive temperature dependence of this reaction causes it to predominate over the collisional stabilization process. The overall mechanism can then be written as

$$CH + H_2 \underset{b}{\overset{a}{\rightleftarrows}} CH_3^{\dagger} \overset{c}{\rightarrow} CH_2 + H$$

$$\underset{[M]}{\overset{d}{\rightarrow}} CH_3 \tag{9}$$

As a further test of the proposed reaction mechanism, Berman and Lin[33] performed transition state theory calculations, using RRKM theory[35] to model the competitive dissociation and stabilization of the complex $CH_3^{\dagger}$. The calculations made use of the temperature-dependent collision efficiency factor β_c derived by Troe,[36] and the geometry of the transition state was taken from the potential energy surface of Brooks and Schaefer.[12] The results of the modeling calculations, reproduced in Figure 8, show that the experimental temperature dependence is reproduced quite well except at the lowest temperatures, where the rate constants are underestimated. This may be due to inadequacies of the model for collisional deactivation at low temperatures, such as collision diameter and collision efficiency, to errors in the assumed parameters of the transition state for the addition step (a), or perhaps to a combination of both.

On the basis of spin and orbital symmetry conservation rules and the results of their numerical calculations, Brooks and Schaefer[12] suggested that the reactivity of ground state ($^2\Pi$) CH should be similar to that of the $^1A_1$ excited state of CH_2. The actual observed rate constants depend upon the accessible product channels. In the case of CH + H_2, the dissociation of the $CH_3^{\dagger}$ complex to ground state ($^3B_1$) CH_2 and H atoms is endothermic by 2.5 kcal/mol at room temperature. At low temperatures, therefore, collisional stabilization is the dominant process and the observed rate constant is pressure dependent. For $CH_2(^1A_1)$ + H_2, on the other hand, the dissociation of the $CH_4^{\dagger}$ adduct to CH_3 + H is exothermic by 5 kcal/mol and collisional stabilization is not competitive. This reaction may, in fact, be more nearly comparable to the CH + CH_4 system, for which the dissociation of $C_2H_5^{\dagger}$ to form C_2H_4 + H is also exothermic. The observed rate constants and pressure dependence tend to bear this out (see the discussions of reactions of CH with hydrocarbons below).

B. Reactions of CH with Alkanes

Earlier work on the rates of reactions of CH with hydrocarbons at room temperature[18,19,23] produced results which varied widely. For example, the spread of reported values for the rate constant of the reaction CH + CH_4 was a factor of 40. Berman and Lin[25] used the laser photolysis/LIF technique to measure the absolute rate constants for CH reacting with CH_4, C_2H_6, and n-C_4H_{10} over a wide range of temperatures.

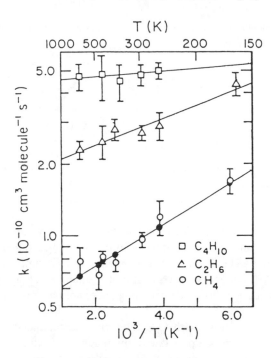

FIGURE 10. Arrhenius plots of the bimolecular rate constants for the reactions of CH with CH_4, C_2H_6, and C_6H_{10}. Open circles — experimental data of Reference 25; filled circles — the result of a TST-RRKM calculation for CH + CH_4. (From Berman, M. R. and Lin, M. C., *Chem. Phys.*, 82, 435, 1983. With permission.)

The second-order rate constants for the reactions of $CH(X^2\Pi)$ with these three small alkanes obtained in the temperature studies at a fixed total pressure of 100 torr are shown in Figure 10. All of these reactions exhibited negative temperature dependences in the range of temperatures investigated and could be fitted quite well by Arrhenius functions. The Arrhenius parameters are also given in Table 3.

For the reaction CH + CH_4, the pressure dependence of the rate constant was also investigated. Measurements were made at pressures in the range 25 to 200 torr at a constant temperature of 297 K. Within experimental error, there was no observed pressure dependence, as was predicted by the result of a TST-RRKM calculation.[25]

All of the measured rate constants were of the order of the gas kinetic collision rate. The results were in good agreement with the most recent prior work,[23] with improved confidence limits. In all three cases the values obtained are substantially larger than those of the two earlier studies.[18,19] In both cases the CH radicals were generated by dissociation of CH_4, and significant depletion of CH_4 during the course of the initiation reaction might have accounted for the smaller measured rate constants.

The room temperature rate constants were found to depend linearly on the number of C–H bonds in the reactant RH. By performing a least-squares fit of the measured rate constants, Berman and Lin estimated the value of the rate constant for the reaction of CH with C_3H_8 to be k = $(3.7 \pm 0.6) \times 10^{-10}$ cm³/molecule/sec. This procedure could not be extrapolated to yield an estimated rate constant for CH + H_2, since the latter reaction is known to be pressure dependent (see above).

For the reaction CH + CH_4, the dominant product channel is expected to be C_2H_4 + H, which is exothermic by 37 kcal/mol. The alternative channel CH_2 + CH_3 is endothermic by 3 kcal/mol and hence is not competitive at normal temperatures. Thus the reaction mechanism can be written as

$$CH + CH_4 \underset{b}{\overset{a}{\rightleftarrows}} C_2H_5^\dagger \overset{a}{\rightarrow} C_2H_4 + H$$

$$\underset{[M]}{\overset{c}{\rightarrow}} C_2H_5 \qquad (10)$$

Again the observed negative temperature dependence is consistent with the assumption of a long-lived intermediate complex. The fact that the room temperature rate constant is independent of pressure shows that collisional stabilization of the adduct is not the dominant process.

A further test of the mechanism (Equation 10) was provided by transition state calculations. The observed rate constants could be reproduced by assuming a rather loose transition state with no energy barrier. The RRKM theory predicted a lifetime of 4×10^{-13} sec for dissociation of the $C_2H_5^\dagger$ adduct (with 96 kcal/mol of internal energy) to $C_2H_4 + H$, which precludes the importance of collisional quenching under normal pressure conditions as alluded to above.

The products of CH reactions with C_2H_6 and other larger members of the homologous series are expected to be very complex due to the much larger numbers of reaction channels rendered accessible by the vast amount of available internal energy.[25]

C. Reactions of CH with Unsaturated Hydrocarbons

Berman et al.,[24] measured bimolecular rate constants for the reactions of CH with C_2H_2, C_2H_4, C_6H_6, and $C_6H_5CH_3$. The rate constants for the C_2H_2 and C_2H_4 reactions were found to have small negative temperature coefficients (-121 and -344 cal/mol, respectively). The rate constant for the reaction of CH with C_6H_6 was found to be independent of temperature. The Arrhenius parameters and room temperature values of these rate constants, including that of $C_6H_5CH_3$ determined at 297 K, are summarized in Table 3. Also included in this table are the room temperature rate constants for the C_2H_2 and C_2H_4 reactions reported by Bosnali and Perner,[19] 7.5×10^{-11} and 1.1×10^{-10} cm³/molecule/sec, respectively. These values are lower than those of Berman et al.[24] by about a factor of 4. It is interesting to note that the room temperature rate constants for the above four reactions are approximately the same (within 10%), in spite of the large differences in both the size and structure of the reactants.

The reaction of CH with unsaturated hydrocarbons may take place via addition to the unsaturated C–C π-bond, as well as by insertion into C–H bonds in the manner of the CH + alkane reactions discussed above. In the case of CH + C_2H_4, for example, the two possibilities are

$$CH + C_2H_4 \rightarrow \begin{array}{c} \cdot CH \\ \diagup \quad \diagdown \\ CH_2 - CH_2^\dagger \end{array} \quad \text{(addition)} \qquad (11a)$$

$$\rightarrow \cdot CH_2CH{=}CH_2^\dagger \quad \text{(insertion)} \qquad (11b)$$

These processes are very exothermic and the adducts possess as much as 89 and 114 kcal/mol, respectively, of internal energy. Both are expected to break down into C_3H_4 (allene and/or methylacetylene) plus H atoms, with an overall exothermicity of about 47 kcal/mol. On the other hand, the abstraction reaction which produces $CH_2 + C_2H_3$ is endothermic by

6.5 kcal/mol and is therefore not expected to be competitive with either addition or insertion.

The mechanism of the CH + C_2H_4 reaction via the two paths depicted above has been studied recently by Strausz and co-workers,[37] employing *ab initio* molecular orbital theory. They computed potential energy surfaces with configuration interaction, but utilizing the molecular geometry optimized at the SCF level. Analysis of these surfaces predicts that the energy barrier for the addition process is zero if an asymmetric nonleast-motion reaction path is followed. On the other hand, a barrier as high as 15 kcal/mol is suggested for the insertion process. The former prediction is consistent with the small negative temperature coefficient (-344 cal/mol) observed by Berman et al.,[24] while the computed insertion barrier is considerably higher than the small negative temperature coefficients observed for the CH + alkane reactions discussed in the preceding section. Strausz and co-workers[37] also attempted to calculate the pre-exponential factor for the addition reaction by transition state theory, based on their optimized activated complex geometry. However, their predicted A-factor was more than two orders of magnitude smaller than the experimental value of 2.2 $\times$ 10^{-10} cm³/molecule/sec quoted in Table 3.

The mechanism for CH + C_2H_2 is probably similar, with the reaction occurring exothermically via either addition or insertion:

$$CH + C_2H_2 \rightarrow \quad \begin{matrix} \overset{\displaystyle CH^\dagger}{\diagup \diagdown} \\ CH - CH \end{matrix} \text{ (addition)} \tag{12a}$$

$$\rightarrow \cdot CH_2 - C \equiv CH^\dagger \text{ (insertion)} \tag{12b}$$

The stability of the C_3H_3 radical ring adduct is not known. Because of the expected large amount of internal energy available from the addition process (the combined enthalpy of the reactants is 196 kcal/mol), it probably converts readily into the more stable linear form, $CH_2C\equiv CH$, which is also the direct C–H insertion product as indicated above. On the basis of the known heat of formation of the $CH_2C\equiv CH$ (propargyl) radical, $\Delta H^\circ_f = 86.5$ kcal/mol,[38] the nascent radical is expected to carry as much as 110 kcal/mol of internal energy. This large amount of excess energy is probably sufficient for the following unimolecular decomposition to occur spontaneously,

$$\cdot CH_2C\equiv CH^\dagger \rightarrow {:}CH-C\equiv CH + H \tag{13}$$

prior to undergoing significant collisional deactivation. This may account for the facility and high efficacy of this reaction in comparison with the reactions of CH with the other unsaturated hydrocarbons discussed below (also see Table 3). The bond dissociation energy $D(C_2HCH-H)$ is not expected to be greater than 100 kcal/mol in view of the possible resonance stabilization,

$$:CH-C\equiv CH \leftrightarrow \cdot CH=C=CH\cdot \leftrightarrow CH\equiv C-CH: \tag{14}$$

If such is the case, then the CH + C_2H_2 reaction can occur without any pressure effect under normal experimental conditions. The direct abstraction (or similarly the alternate decomposition channel from $CH_2C_2H^\dagger$) to produce CH_2 + C_2H is endothermic by as much as 30 kcal/mol, according to the latest reported heat of formation of C_2H (134 $\pm$ 2 kcal/mol.)[39] It is therefore not expected to be important, even under combustion conditions.

The reactions of CH with C_6H_6 and $C_6H_5CH_3$ can produce a variety of fragmentation products exothermically, including those derived from ring-opening processes following the initial addition or insertion. In this respect they are similar to the reaction CH + C_2H_4, and therefore the rates are independent of both temperature and pressure.

Fleming et al.[26] measured the rate of the reaction CH + C_3H_4 (methylacetylene) at room temperature, using an ArF laser to generate CH from $CHBr_3$. Although the value they obtained [(4.4 ± 0.9) × 10^{-10} cm^3/molecule/sec] is close to that for CH + C_2H_2 reported by Berman et al.,[24] it is 3 times as large as their own result for CH + C_2H_2.

D. Reactions of CH with H_2O, NH_3, and CH_2O

These reactions of CH, particularly with H_2O and CH_2O, are very important to hydrocarbon combustion chemistry. There are no reported studies of the temperature dependence of these reactions. However, room temperature measurements for the reactions with H_2O and NH_3 were made some time ago by Bosnali and Perner,[19] who produced CH radicals by pulse radiolysis of CH_4 in the presence of the reactant gas. The time dependence of concentration during the course of the reaction was monitored by kinetic absorption spectroscopy. Their results for these reactions are listed in Table 3.

The temperature dependence of the reactions of CH with H_2O and CH_2O has been measured recently by Zabarnick, Fleming, and Lin[40] using the two-laser pump-probe method. Their results, summarized in Table 3, indicate that both processes have slight negative activation energies, reflecting the dominance of the insertion mechanism. The CH + H_2O reaction could, in principle, only occur by insertion, since the abstraction channel is endothermic by 17 kcal/mol. Possible reaction channels for these three processes are as follows:

$$CH + H_2O \rightarrow CH_2OH^\dagger \quad \rightarrow CH_2O + H \quad \Delta H° = -60 \text{ kcal/mol}$$

$$\rightarrow CH_2 + OH \quad \Delta H° = +17 \text{ kcal/mol}$$

$$CH + NH_3 \rightarrow CH_2NH_2^\dagger \quad \rightarrow CH_2NH + H \quad \Delta H° \simeq -53 \text{ kcal/mol}$$

$$\rightarrow CH_2 + NH_2 \quad \Delta H° = +7 \text{ kcal/mol}$$

$$CH + CH_2O \rightarrow CH_2CHO^\dagger \rightarrow CH_3 + CO \quad \Delta H° = -105 \text{ kcal/mol}$$

$$\rightarrow CH_2 + CHO \quad \Delta H° = -4 \text{ kcal/mol}$$

The direct abstraction channel is open only to CH + CH_2O, as seen above. Its importance at low temperatures is questionable because of a possible energy barrier.

E. Reactions of CH with N_2 and CO

The reaction of CH with N_2 is of considerable importance to high temperature combustion of hydrocarbons, particularly with regard to its possible role as a non-Zeldovich NO source. It was first proposed by Fenimore,[2] together with the $C_2 + N_2 \rightarrow 2CN$ reaction, as a possible precursor for the NO formed at the fronts of hydrocarbon-air flames, where the Zeldovich processes $O + N_2 \rightarrow NO + N$ and $N + O_2 \rightarrow NO + O$ are not important. More recently, the results of kinetic modeling of data obtained from flame studies[3,41] favor either CH + $N_2 \rightarrow$ HCN + N or $CH_2 + N_2 \rightarrow$ HCN + NH as the more likely sources for "prompt" NO formation. Blauwens et al.[3] concluded that the observed NO generation rate in the flame-fronts of several hydrocarbon flames studied could be well described by either $R_{NO} = 1.3 \times 10^{-12}$ [CH][N_2]exp($-11,000/RT$) or $R_{NO} = 4.6 \times 10^{-12}$ [CH_2][N_2]exp($-22,500/RT$) molecules/cm^3/sec. On the other hand, Miyauchi et al.[41] favored the $CH_2 + N_2$ process on the grounds of the spin-conservation rule, and they arrived at the rate expression $R_{NO} = 3.7 \times 10^{-12}$ [CH_2][N_2]exp($-25,020/RT$) molecules/cm^3/sec. On the basis of the detailed kinetic data obtained by Berman and Lin[4] for CH + N_2 and other data relevant to $CH_2 + N_2$, we can demonstrate convincingly with the aid of transition-state (RRKM) theory calculations that the CH + N_2 reaction is kinetically a much more likely process for NO precursor formation. A similar conclusion was also reached earlier by Benson[42] from thermochemical arguments.

Berman and Lin[4] studied the reaction $CH + N_2$ at total pressures between 25 and 787 torr at a constant temperature of 297 K and at temperatures from 297 to 675 K at a constant total pressure of 100 torr. The second-order rate constant at 297 K was found to be pressure-dependent, exhibiting a pronounced increase with increasing pressure in the range investigated (see Figure 7). The slight curvature of the pressure dependence could be reproduced quite satisfactorily by a transition-state theory model. A plot of the measured second-order rate constant vs. the mole fraction of N_2 at a constant total pressure of 25 torr had zero slope within experimental error, which led to the conclusion that the collision efficiencies of N_2 and the Ar diluent are equal. The third-order rate constant at 297 K was plotted as a function of total pressure and extrapolated to yield a value for the low-pressure limiting rate constant of $k^\circ = (2.8 \pm 0.5) \times 10^{-31}$ cm^6/molecule^2/sec.

Wagal et al.[29] also investigated the reaction $CH + N_2$ at room temperature and found the second-order rate constant to be pressure-dependent. Their data also suggested a slight nonlinearity, although their measurements were limited to the pressure range 2 to 75 torr (see Figure 7). They estimated a value for the low-pressure limiting rate constant of $k^\circ = (2.6 \pm 0.3) \times 10^{-31}$ cm^6/molecule^2/sec, in good agreement with the value of Berman and Lin. Using a fitting function based on the Lindemann model for a collision-stabilized recombination mechanism, they extrapolated their data to yield an estimate of the room temperature rate constant at a total pressure of 100 torr. The resulting value, $k_2 = (3.66 \pm 0.27) \times 10^{-13}$ cm^3/molecule/sec, agrees closely with the value obtained by Berman and Lin. However, their predicted value of $k^\infty = (6.3 \pm 1.3) \times 10^{-13}$ cm^3/molecule/sec is a factor of 30 smaller than that estimated by Berman and Lin on the basis of the RRKM model. In fact, Berman and Lin measured rate constants greater than the estimated high-pressure limit of Wagal et al. at pressures as low as 200 torr. The RRKM calculations suggest that total pressures of the order of 10^4 atm are required to reach the high-pressure limit.

An Arrhenius plot of the second-order rate constant for $CH + N_2$ at a total pressure of 100 torr[4] showed significant curvature at both the high and low ends of the 297 to 675 K temperature range, as shown in Figure 11. The rate constant value quoted in Table 3 was obtained from a weighted linear least squares fit of the experimental data. Also shown in Figure 11 are the data of Blauwens et al.[3] for $CH + N_2$ derived from their flame studies. As noted above, their data indicated a positive, instead of negative, activation energy of more than 11 kcal/mol. These two sets of contrasting data, as revealed by the results presented in Figure 11, could only be reconciled by a nonconventional transition-state (RRKM) calculation based on a mechanism similar to that discussed above for the $CH + H_2$ reaction:[4]

$$CH + N_2 \rightleftarrows CHN_2^\ddagger \rightarrow HCN + N \qquad (15a)$$

$$\xrightarrow{M} CHN_2 \qquad (15b)$$

The key difference between the present system and the $CH + H_2$ reaction lies in the formation of the N-atom product, which involves a spin change from doublet for both the reactant and the adduct to quartet for the product. This nonconservation of the total spin has led to some reluctance in accepting the probability that this reaction can occur. On the other hand, there is ample experimental and theoretical evidence that the spin-conservation rule does not hold rigorously for complex-forming bimolecular processes such as $O(^1D) + N_2/CO \rightarrow O(^3P) + N_2/CO$.[43,44] The formation of long-lived intermediates (N_2O and CO_2 for these two reactions) allows very effective electronic quenching and $E \rightarrow V$ energy transfer.[45]

The schematic energy diagram for the $CH + N_2$ system is shown in Figure 12. The various energies given in the figure, $E_a = 0.5$, $E_a' = 3.3$, and $E_o \simeq 57$ kcal/mol,[4] were

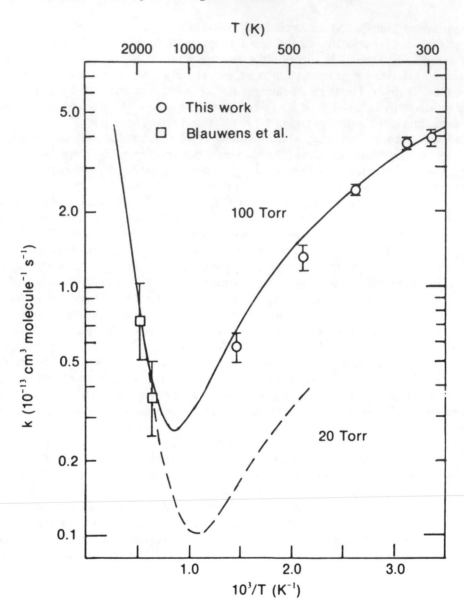

FIGURE 11.   Arrhenius plots of the experimental and calculated second-order rate constants for the CH + N$_2$ reaction at different pressures. Open circles — data of Berman and Lin (Reference 4); squares — data of Blauwens, et al. (Reference 3). The curves are the calculated values based on the TST-RRKM theory for different pressures as indicated (see Reference 4).

obtained by fitting the observed temperature (297 to 1820 K) and pressure (25 to 787 torr) data. The computed V-shaped, non-Arrhenius temperature dependence of the second-order rate constant for the first time accounted for the grossly contrasting data. The calculated results also reveal that the CH + N$_2$ reaction becomes pressure-independent above 1500 K, where the endothermic process forming the NO precursors HCN + N dominates the reaction.

To establish the relative importance of CH$_2$ + N$_2$ and CH + N$_2$, we have carried out a similar calculation[46] for the former process using the known unimolecular decomposition rate constant for CH$_2$N$_2$ → CH$_2$ + N$_2$ and other relevant thermochemical data.[47] The results of this calculation show that the reaction is very slow at low temperatures and the second-

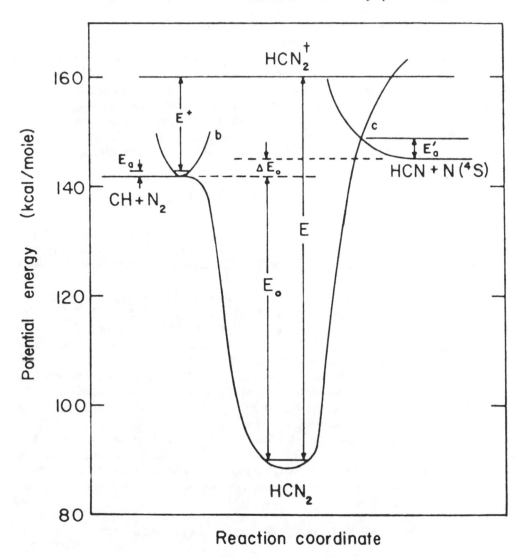

FIGURE 12.   A schematic energy diagram for the CH + N$_2$ $\rightleftarrows$ CHN$_2$ $\rightleftarrows$ HCN + N system. E$_0$ = 57, E$_a$ = 0.5, and E$_a'$ = 3.3 kcal/mol were found to provide a reasonable fit of the experimental data over the entire temperature range (the details of the TST-RRKM formulation and calculations can be found in Reference 4.)

order rate constant for the formation of HCN + NH in the temperature range 1000 to 2500 K is k = 8 × 10^{-12} exp (−36000/RT) cm^3/molecule/sec. The importance of this process for prompt NO production is therefore questionable.

Berman et al.[22] determined the room temperature rate constant of the reaction CH + CO as a function of pressure in the range 50 to 640 torr. Like the N$_2$ analog, this reaction is strongly pressure-dependent and appears to proceed by the same mechanism. No estimate of the high-pressure limiting rate constant was reported.

The Arrhenius plot for the reaction of CH with CO in the temperature range 297 to 676 K at a total pressure of 100 torr was linear, with a negative temperature dependence. The negative activation energy is of the same order of magnitude as that observed for N$_2$, which lends further support to the conclusion that these reactions proceed by the same stabilized-adduct mechanism. The substantial difference between the magnitudes of the room temperature rate constants can be rationalized readily by the much larger well depth for the formation of the HC$_2$O adduct, which facilitates the stabilization process.

F. Reactions of CH with O_2, NO, NO_2, N_2O, and CO_2

The reactions of CH with these oxidants are important to combustion chemistry and chemical laser pumping processes.[8-11] With the possible exception of CH + CO_2, these reactions are highly exothermic and a variety of product channels are accessible, including electronically excited diatomic products such as $CN(A^2\Pi)$, $OH(A^2\Sigma^+)$, and $NH(A^3\Pi)$.

Messing et al.[27] measured the rate of the reaction CH + O_2 at 298 K, obtaining the value $k_2 = (3.3 \pm 0.4) \times 10^{-11}$ cm^3/molecule/sec. The reaction rate was determined both by following the LIF signal of the CH radical and by monitoring the chemiluminescence from $OH(A^2\Sigma^+)$. From the close agreement of the two methods, they concluded that one of the primary reactions was

$$CH(X^2\Pi) + O_2(X^3\Sigma_g^-) \rightarrow CO(X^1\Sigma^+) + OH(A^2\Sigma^+) \tag{16}$$

The temperature dependence of this reaction was investigated by Berman et al., who found no significant variation in the range 297 to 676 K. The value they obtained at room temperature, 5.1×10^{-11} cm^3/molecule/sec (see Table 3), is much higher than that of Messing et al.

Wagal et al.[29] determined rate constants for the reactions of CH with the three oxides of nitrogen, obtaining the following values at 300 K: $k_{NO} = (2.0 \pm 0.3) \times 10^{-10}$, $k_{NO_2} = (1.67 \pm 0.11) \times 10^{-10}$, and $k_{N_2O} = (7.8 \pm 1.4) \times 10^{-11}$, all in units of cm^3/molecule/sec. Berman et al.[22] studied the reaction CH + NO and found the rate constant to be independent of temperature in the range 297 to 676 K. Their rate constant value (included in Table 3) is in excellent agreement with that quoted above. No studies of the temperature dependence of the reactions of CH with N_2O and NO_2 have been reported to date.

The reaction CH + CO_2 was investigated by Berman et al.[22] in the temperature range 297 to 676 K. The reaction was found to have a positive temperature coefficient, with an activation energy of about 700 cal/mol. It is interesting to note that the rate constant for this reaction at room temperature, $(1.8 \pm 0.1) \times 10^{-12}$ cm^3/molecule/sec, is smaller than that of its isoelectronic counterpart CH + N_2O by more than a factor of 40. These two isoelectronic reactions may occur by the following accessible channels, which involve either addition or direct abstraction:

$$CH + CO_2 \rightarrow HCO + CO \quad \Delta H° = -67 \text{ kcal/mol}$$

$$\rightarrow H + 2CO \quad \Delta H° = -49 \text{ kcal/mol}$$

$$CH + N_2O \rightarrow HCN + NO \quad \Delta H° = -108 \text{ kcal/mol}$$

$$\rightarrow HCO + N_2 \quad \Delta H° = -153 \text{ kcal/mol}$$

$$\rightarrow H + CO + N_2 \quad \Delta H° = -135 \text{ kcal/mol}$$

G. Reactions of CH with O and N Atoms

The reactions of CH with O and N atoms, as should be expected, are highly exothermic and can have numerous accessible reaction channels. Some of these channels are listed below, together with their associated exothermicities.[28]

$$CH(^2\Pi) + O(^3P) \rightarrow H + CO(d^3\Delta) \quad \Delta H° = -2.54 \text{ kcal/mol}$$

$$\rightarrow HCO^+(^1A') \quad \Delta H° = -11.5 \text{ kcal/mol}$$

$$\rightarrow H + CO(a'^3\Sigma^+) \quad \Delta H° = -17.8 \text{ kcal/mol}$$

$$\rightarrow OH(^2\Pi) + C(^3P) \quad \Delta H° = -21.4 \text{ kcal/mol}$$

$$\rightarrow H + CO(a'^3\Pi) \quad \Delta H^\circ = -37.4 \text{ kcal/mol}$$

$$\rightarrow H + CO(X^1\Sigma^+) \quad \Delta H^\circ = -176 \text{ kcal/mol}$$

$$\rightarrow HCO(X^2A') \quad \Delta H^\circ = -189 \text{ kcal/mol}$$

$$CH(^2\Pi) + N(^4S) \rightarrow C(^3P) + NH(^3\Sigma^-) \quad \Delta H^\circ = 0.0 \text{ kcal/mol}$$

$$\rightarrow H(^2S) + CN(B^2\Sigma^+) \quad \Delta H^\circ = -25.1 \text{ kcal/mol}$$

$$\rightarrow H(^2S) + CN(A^2\Pi) \quad \Delta H^\circ = -72.9 \text{ kcal/mol}$$

$$\rightarrow H(^2S) + CN(X^2\Sigma^+) \quad \Delta H^\circ = -99.4 \text{ kcal/mol}$$

$$\rightarrow HCN(\tilde{X}^1\Sigma^+) \quad \Delta H^\circ = -221.4 \text{ kcal/mol}$$

As indicated, the exothermicity of CH + O is high enough to form the CHO^+ chemi-ion.[1] This particular reaction channel has been an object of interest for more than 20 years because of its possible role in soot formation.[48] The chemi-ionization channel, however, is only a minor contributor with a rate constant of 2.4×10^{-14} cm^3/molecule/sec at 295 K,[49] which is about 2.5×10^{-4} times the overall rate constant measured by Messing et al.[28] It should be mentioned that chemi-ionization is not possible in the CH + N reaction because of the high ionization potential of HCN (313.6 kcal/mol).

Messing et al.[28] measured the rate constants for the reactions of CH with O and N atoms at 298 K in a discharge flow system, using both chemiluminescence and LIF to monitor the time-dependent concentrations of CH and products. The concentrations of O and N atoms were measured by conventional titration methods using NO_2 and NO, respectively. Emissions from various excited products, such as C_2, CN, NO, CO, etc., were detected, but they could not be verified with certainty. The overall rate constants for CH reactions with O and N atoms were determined to be $(9.5 \pm 1.4) \times 10^{-11}$ and $(2.1 \pm 0.5) \times 10^{-11}$ cm^3/molecule/sec, respectively.

H. Comparison of the Reactivity of CH(X$^2\Pi$) with CCl(X$^2\Pi$) and CBr(X$^2\Pi$)

The reactions of these two carbyne analogs of CH have been studied extensively. Tyerman[50,51] investigated the reactions of CCl(X$^2\Pi$), while James and co-workers[14-16,52] studied both CCl(X$^2\Pi$) and CBr(X$^2\Pi$). Rate constants for reactants common to the CH studies are shown in Table 4 for comparison.

In all cases for which comparative data exist, it is seen that the reactions of ground state CCl and CBr are very much slower than the CH analogs. These observations are consistent with the argument discussed above. That is, overlap of a halogen p-orbital with the unoccupied pπ-orbital of the carbon atom is expected to reduce the electron deficiency of the carbon and hence decrease the reactivity of the halocarbyne radical relative to CH.

I. Comparison of the Reactivity of CH(X$^2\Pi$) with CH(A$^2\Delta$) and CH$_2$(ā^{1}A$_1$)

The lowest excited state of CH is $a^4\Sigma^-$, which has been detected by laser photoelectron spectroscopy[17] and found to lie 17 kcal/mol above the ground state. As discussed by James et al.,[13] it can be argued on the basis of spin and orbital symmetry conservation rules that parallels can be drawn between states of CH and CH$_2$. According to this line of reasoning, the X$^2\Pi$ ground state of CH should be similar in reactivity to the ā^{1}A$_1$ excited state of CH$_2$, whereas the behavior of CH(a$^4\Sigma^-$) should more nearly resemble that of the ground state CH$_2$ ($\tilde{X}^3$B$_1$).

The *ab initio* calculations of Brooks and Schaefer[12] provide some support for the state correlation argument in the specific case of CH + H$_2$, as discussed above. They found a very low insertion barrier for CH(X$^2\Pi$), analogous to CH$_2$(ā^{1}A$_1$), and an abstraction barrier

Table 4
COMPARISON OF ROOM TEMPERATURE RATE CONSTANTS FOR THE REACTIONS OF CH(X²Π), CCl(X²Π), AND CBr(X²Π) WITH SELECTED SUBSTRATES

$k \times 10^{11} (cm^3/molecule/sec)$

Substrate	CBr	Ref.	CCl	Ref.	CH	Ref.
H_2	$<5 \times 10^{-4}$	60	$<7 \times 10^{-4}$	62	1.4 ± 0.10^a	33
CH_4	$<5 \times 10^{-4}$	60			10.2 ± 0.4	25
C_3H_8			$<7 \times 10^{-4}$	12	37 ± 6	25
C_2H_4	0.076 ± 0.01	60	0.055 ± 0.007	51	42 ± 3	24
C_6H_6	0.037 ± 0.003	61			43 ± 3	24
NH_3			<0.002	62	$9.8 + 0.8; 9.8 - 1.5$	19
N_2	$<7 \times 10^{-4}$	60	$<2.5 \times 10^{-4}$	51	0.0039 ± 0.0003^a	4
O_2	0.22 ± 0.15	60	0.42 ± 0.005	51	5.1 ± 0.5	22
NO	2.2 ± 0.3	60			19 ± 3	22

[a]   Pressure-dependent value at 100 torr.

for CH(a⁴Σ⁻) somewhat lower than that of CH₂(X̃³B₁). More recently, Gosavi et al.[37] performed *ab initio* calculations for the system CH(X̃²Π) + C₂H₄(X̃¹A₁). They found no activation energy for asymmetric addition, but predicted a barrier of about 15 kcal/mol for insertion (which is, however, inconsistent with the absence of a reaction barrier for the CH + alkane reactions as pointed out above).

The predictions are indeed borne out by the observation that CH undergoes insertion and concerted stereospecific cycloadditions with hydrocarbons. By analogy, then, CH(a⁴Σ⁻) should be expected to abstract hydrogen atoms from hydrocarbons or add nonstereospecifically. Unfortunately, no experimental data are available to test the latter prediction.

By contrast, the second excited state of the CH radical, A²Δ, is quite reactive. Its total (reaction + deactivation) rate constants[53] at room temperature parallel quite closely those of CH₂(ã¹A₁).[54] The data are summarized in Table 5 for comparison with the reaction rate constants of CH(X²Π), using a few selected reaction partners as examples. It is interesting to note that, with the sole exception of CO (which is undoubtedly due to a pressure effect), CH(X²Π) rate constants are typically about a factor of 2 to 5 larger than those of CH(A²Δ) and CH₂(ã¹A₁). Rate constants for the quenching of CH₂(ã¹A₁) by rare gas atoms have been measured[54,55] and found to lie in the range from 3×10^{-12} to 1.6×10^{-11} cm³/molecule/ sec. This high quenching efficiency may suggest that large fractions of the observed rate constants for its interactions with hydrocarbons (which are about 50% of the CH(X²Π) values) may be attributable to physical quenching rather than to the insertion reaction. On the other hand, analysis of insertion product yields[56,57] for some of the reactions of CH₂(ã¹A₁) with alkanes indicates that quenching rates are less than 10% of total removal rates.[54,58,59] It would be theoretically very important to elucidate the mechanisms of these reactions, particularly with regard to quenching vs. reaction probabilities.

IV. CONCLUDING REMARKS

In this chapter we have reviewed rather completely what is presently known about perhaps the most reactive of all free radicals, methylidyne (CH). Its kinetics and possible mechanisms associated with reactions of interest to combustion and planetary atmospheric chemistry have been discussed in some detail. We have placed greatest emphasis on the results of direct measurements using the two-laser photoinitiation/laser-induced fluorescence method. With

Table 5
COMPARISON OF ROOM TEMPERATURE RATE CONSTANTS FOR THE REACTIONS/ DEACTIVATIONS OF CH(X²Π), CH(A²Δ), AND CH₂(ã¹A₁) WITH SELECTED SUBSTRATES

	$k \times 10^{11}$(cm³/molecule/sec)		
Substrate	CH(X²Π)[a]	CH(A²Δ)[b]	CH₂(ã¹A₁)[c]
H_2	1.4 ± 0.3[d]	0.90 ± 0.08	11 ± 0.5
O_2	5.1 ± 0.5	1.6 ± 0.1	7.4 ± 5
NO	20 ± 2	10.4 ± 0.4	16 ± 2
CO	2.6 ± 0.2[d]	5.2 ± 0.3	4.9 ± 0.4
N_2	2 ± 1[e]	—	1.1 ± 0.1
N_2O	7.8 ± 1.4	0.46 ± 0.1	—
CH_4	$10.2 + 0.4$	2.0 ± 0.1	7.4 ± 0.4
C_2H_6	27 ± 2	$11 \pm .5$	19 ± 2
C_4H_{10}	48 ± 5	24 ± 2	—
C_2H_4	42 ± 3	19 ± 1	15 ± 6
C_2H_2	42 ± 2	19 ± 1	—

[a] From Table 3.
[b] From Reference 53.
[c] From Reference 54.
[d] Pressure dependent-value at 100 torr.
[e] Extrapolated high-pressure value.

the aid of *ab initio* multiconfiguration SCF calculations by Brooks and Schaefer,[12] measurements of temperature and pressure dependence of rate constants, and transition-state theory (RRKM) interpretations performed at the Naval Research Laboratory, it has been possible to improve considerably our understanding of the reactions of CH with substrates such as H_2, N_2, CO, and simple alkanes. The reactions of CH with H_2 and N_2 both exhibit strongly curved, V-shaped Arrhenius plots which could not be comprehended readily without the aid of the TST-RRKM calculations.[4,33]

In addition to the examples just cited, other reactions involving unsaturated hydrocarbons, oxides of nitrogen, H_2O, CO_2, CH_2O, NH_3, and the two atomic species O and N have been discussed. For the majority of these reactions, accurate measurements of the temperature dependence are yet to be reported. As has been demonstrated for several reactions studied at NRL, information on temperature and pressure effects is very useful for elucidating mechanisms. Measurements of the pressure dependence are particularly important in cases where there are high exit energy barriers.

We have perhaps only begun to understand the behavior of this unique free radical. Numerous questions remain unanswered. For example, how does CH react with sulfur-containing compounds and what is its reactivity toward carbon-halogen bonds? Since CH is so reactive, does it differentiate among primary, secondary, and tertiary C–H bonds, and how do the rate constants scale as the number of C–H bonds increases? Further work is currently underway at NRL to address these questions.

ACKNOWLEDGMENTS

The authors wish to express their gratitude for the partial support of this work by the U.S. Department of Energy under contract DE-FG05-85ER13373 (WAS) and by the NASA Planetary Atmosphere Program (MCL).

REFERENCES

1. **Kistiakowsky, G. B. and Michael, J. V.,** Mechanism of chemi-ionization in hydrocarbon oxidations, *J. Chem. Phys.,* 40, 1447, 1964.
2. **Fenimore, C. P.,** Formation of nitric oxide in premixed hydrocarbon Flames, in *Thirteenth Symposium (International) on Combustion,* The Combustion Institute, Pittsburgh, Pa., 1971, 373.
3. **Blauwens, J., Smets, B., and Peeters, J.,** Mechanism of "prompt" nitrogen oxide (NO) formation in hydrocarbon flames, in *Sixteenth Symposium (International) on Combustion,* The Combustion Institute, Pittsburgh, Pa., 1977, 1055.
4. **Berman, M. R. and Lin, M. C.,** Kinetics and mechanism of the CH + N$_2$ reaction. Temperature- and pressure-dependence studies and transition-state-theory analysis, *J. Phys. Chem.,* 87, 3933, 1983.
5. **Lichtin, D. A., Berman, M. R., and Lin, M. C.,** Imidogen (NH)(A$^3\Pi \rightarrow$ X$^3\Sigma^-$) chemiluminescence from the methylidyne (CH) (X$^2\pi$) + nitric oxide reaction, *Chem. Phys. Lett.,* 108, 18, 1984.
6. **Strobel, D. F.,** Chemistry and evolution of Titan's atmosphere, *Plant. Space Sci.,* 30, 839, 1982.
7. **Prinn, R. G. and Owen, T.,** Chemistry and spectroscopy of the jovian atmosphere, in *Jupiter,* Gehrels, T., Ed., University of Arizona, Tucson, 1976, 319.
8. **Lin, M. C.,** Chemical lasers produced from O(^{3}P) atom reactions. I. Observation of CO and HF laser emissions from several O atom reactions, *Int. J. Chem. Kinet.,* 6, 173, 1973.
9. **Lin, M. C.,** Chemical lasers produced from O(^{3}P) atom reactions. III. 5-μm CO laser emission from the O + CH reaction, *Int. J. Chem. Kinet.,* 6, 1, 1974.
10. **Lin, M. C.,** Chemical CO and CO$_2$ lasers produced from the CH + O$_2$ reaction, *J. Chem. Phys.,* 61, 1835, 1974.
11. **Lin, M. C.,** The mechanism of CO laser emission for the CH + NO reaction, *J. Phys. Chem.,* 77, 2726, 1973.
12. **Brooks, B. R. and Schaefer, H. F.,** Reactions of carbynes. Potential energy surfaces for the doublet and quartet methylidyne (CH) reactions with molecular hydrogen, *J. Chem. Phys.,* 67, 5146, 1977.
13. **James, F. C., Choi, H. K. J., Ruzsicska, B., and Strausz, O. P.,** The gas phase chemistry of carbynes, in *Frontiers of Free Radical Chemistry,* Pryor, W. A., Ed., Academic Press, New York, 1980, 139.
14. **McDaniel, R. S., Dickson, R., James, F. C., Strausz, O. P., and Bell, T. N.,** Rate parameters for the reactions of the bromomethylidyne radical, *Chem. Phys. Lett.,* 43, 130, 1976.
15. **James, F. C., Ruzsicska, B., McDaniel, R. S., Dickson, R., Strausz, O. P., and Bell, T. N.,** Rate constants for the reaction of the bromomethyne radical with alkynes, *Chem. Phys. Lett.,* 45, 449, 1977.
16. **James, F. C., Choi, J., Strausz, O. P., and Bell, T. N.,** Rate constants for the reaction of the chloromethyne radical with alkynes, *Chem. Phys. Lett.,* 53, 206, 1978.
17. **Kasdan, A., Herbst, E., and Lineberger, W. C.,** Laser photoelectron spectroscopy of methylidyne ($-$) ion, *Chem. Phys. Lett.,* 31, 78, 1975.
18. **Braun, W., McNesby, J. R., and Bass, A. M.,** Flash photolysis of methane in the vacuum ultraviolet. II. Absolute rate constants for reactions of CH with methane, hydrogen, and nitrogen, *J. Chem. Phys.,* 46, 2071, 1967.
19. **Bosnali, M. W. and Perner, D. Z.,** Reactions of pulse radiolytically generated methylidyne radical with methane and other substances, *Naturforscher,* 26a, 1768, 1971.
20. **Simons, J. P. and Yarwood, A. J.,** Decomposition of hot radicals. I. Production of CCl and CBr from halogen-substituted methyl radicals, *Trans. Faraday Soc.,* 57, 2167, 1961; Decomposition of hot radicals. II. Mechanisms of excitation and decomposition, *Trans. Faraday Soc.,* 59, 90, 1963.
21. **Lin, M. C.,** in *Summary Report on the Workshop on High Temperature Chemical Kinetics: Applications to Combustion Research,* NBS Special Publication No. 531, 1978, 56.
22. **Berman, M. R., Fleming, J. W., Harvey, A. B., and Lin, M. C.,** Temperature dependence of CH radical reactions with oxygen, nitric oxide, carbon monoxide, and carbon dioxide, in *Nineteenth Symposium (International) on Combustion,* The Combustion Institute, Pittsburgh, Pa., 1982, 73.
23. **Butler, J. E., Fleming, J. W., Goss, L. P., and Lin, M. C.,** Kinetics of CH radical reactions with selected molecules at room temperature, *Chem. Phys.,* 56, 355, 1981.
24. **Berman, M. R., Fleming, J. W., Harvey, A. B., and Lin, M. C.,** Temperature dependence of the reactions of CH radicals with unsaturated hydrocarbons, *Chem. Phys.,* 73, 27, 1982.
25. **Berman, M. R. and Lin, M. C.,** Kinetics and mechanisms of the reactions of CH with CH$_4$, C$_2$H$_6$ and n-C$_4$H$_{10}$, *Chem. Phys.,* 82, 435, 1983.
26. **Fleming, J. W., Fujimoto, G. T., Lin, M. C., and Harvey, A. B.,** Applications of multiphoton dissociation and laser induced fluorescence to combustion: reactions of CH radicals with unsaturated hydrocarbons, *Proc. Int. Conf. Lasers 1979,* 1980, 246.
27. **Messing, I., Sadowski, C. M., and Filseth, S. V.,** Absolute rate constant for the reaction of methylidyne radical with molecular oxygen, *Chem. Phys. Lett.,* 66, 95, 1979.
28. **Messing, I., Filseth, S. V., Sadowski, C. M., and Carrington, T.,** Absolute rate constants for the reactions of CH with O and N atoms, *J. Chem. Phys.,* 74, 3874, 1981.

29. **Wagal, S. S., Carrington, T., Filseth, S. V., and Sadowski, C. M.,** Absolute rate constants for the reactions of $CH(X^2\pi)$ with NO, N_2O, NO_2, and N_2 at room temperature, *Chem. Phys.,* 69, 61, 1982.

30. **Duncanson, J. A., Jr. and Guillory, W. A.,** The state-selective reactions of vibrationally excited $CH(X^2\Pi)$ with oxygen and nitrogen, *J. Chem. Phys.,* 78, 4958, 1983.

31. **Butler, J. E., Goss, L. P., Lin, M. C., and Hudgens, J. W.,** Production, detection, and reactions of the methylidyne radical, *Chem. Phys. Lett.,* 63, 104, 1979.

32. **Huber, K. P. and Herzberg, G.,** *Molecular Spectra and Molecular Structure.* Vol. 4, Van Nostrand Reinhold, New York, 1979.

33. **Berman, M. R. and Lin, M. C.,** Kinetics and mechanisms of the reactions of CH and CD with H_2 and D_2, *J. Chem. Phys.,* 81, 5743, 1984.

34. **Zabarnick, S. S., Fleming, J. W., and Lin, M. C.,** Rate constants for the CH + H_2 reaction between 372 and 675 K, to be published, 1987.

35. **Robinson, P. J. and Holbrook, K. A.,** *Unimolecular Reactions,* Wiley, New York, 1972.

36. **Troe, J.,** Theory of thermal unimolecular reactions at low pressures. I. Solutions of the master equation, *J. Chem. Phys.,* 66, 4745, 1977.

37. **Gosavi, R. K., Safarik, I., and Strausz, O. P.,** Molecular orbital studies of carbyne reactions: addition and insertion reaction paths for the $CH(\tilde{X}^2\Pi)$ + $C_2H_4(\tilde{X}^1A_1)$ reaction, *Can. J. Chem.,* 63, 1689, 1985.

38. **Kerr, J. A. and Trotman-Dickenson, A. F.,** in *Handbook of Chemistry and Physics,* Weast, R. C., Ed., CRC Press, Boca Raton, Fla., 1980, F240.

39. **Wodtke, A. M. and Lee, Y. T.,** Photodissociation of acetylene at 193.3 nm, *J. Phys. Chem.,* 85, 4744, 1985.

40. **Zabarnick, S. S., Fleming, J. W., and Lin, M. C.,** Temperature dependence of the reaction of CH with H_2O and CH_2O, to be published, 1987.

41. **Miyauchi, T., Mori, Y., and Imamura, A.,** A study of nitric oxide formation in fuel-rich hydrocarbon flames: role of cyanide species, H, OH and O, in *Sixteenth Symposium (International) on Combustion,* The Combustion Institute, Pittsburgh, Pa., 1977, 1073.

42. **Benson, S. W.,** Comments, in *Sixteenth Symposium (International) on Combustion,* The Combustion Institute, Pittsburgh, Pa., 1977, 1062.

43. **Zahr, G. E., Preston, R. K., and Miller, W. H.,** Theoretical treatment of quenching in $O(^1D)$ + N_2 collisions, *J. Chem. Phys.,* 62, 1127, 1975.

44. **Tully, J. C.,** Collision complex model for spin forbidden reactions: Quenching of $O(^1D)$ by N_2, *J. Chem. Phys.,* 61, 61, 1974; Reactions of $O(^1D)$ with atmospheric molecules, *J. Chem. Phys.,* 62, 1893, 1975.

45. **Shortridge, R. G. and Lin, M. C.,** The dynamics of the $O(^1D_2)$ + $CO(X^1\Sigma^+,$ v = 0) reaction, *J. Chem. Phys.,* 64, 4076, 1976.

46. **Sanders, W. A., Lin, C. Y., and Lin, M. C.,** On the importance of the reaction CH_2 + $N_2 \to$ HCN + NH as a precursor for prompt NO formation, *Combust. Sci. Tech.,* 51, 103, 1987.

47. **Benson, S. W. and O'Neal, H. E.,** *Kinetic Data on Gas Phase Unimolecular Reactions,* National Bureau of Standards, Washington, D.C., 21, 481, 1970.

48. **Calcote, H. F.,** Mechanisms of soot nucleation in flames — a critical review, *Combust. Flame,* 42, 215, 1981.

49. **Vinckier, C.,** Determination of the rate constant of the reaction CH + $O \to CHO^+$ + e^-, *J. Phys. Chem.,* 83, 1234, 1979.

50. **Tyerman, W. J. R.,** Flash photolysis of haloethylenes. Formation of CCl from 1,1-dichloroethylene, *Trans. Faraday Soc.,* 65, 2948, 1969.

51. **Tyerman, W. J. R.,** Rate constants for reaction of the chloromethylydine radical with olefins, alkynes, chloroalkanes, hydrogen, and oxygen, *J. Chem. Soc. A,* 2483, 1969.

52. **James, F. C., Choi, H. K. J., Strausz, O. P., and Bell, T. N.,** Rate constants for the reactions of chloromethylidyne radical [CCl $(X^2\Pi)$] with a series of silanes, *Chem. Phys. Lett.,* 68, 131, 1979.

53. **Nokes, C. J. and Donovan, R. J.,** Time-resolved kinetic studies of electronically excited CH radicals. II. Quenching efficiencies for $CH(a^2\Delta)$, *Chem. Phys.,* 90, 167, 1984.

54. **Langford, A. O., Petek, H., and Moore, C. B.,** Collisional removal of $CH_2(^1A_1)$: absolute rate constants for atomic and molecular collisional partners at 295 K, *J. Chem. Phys.,* 78, 6650, 1983.

55. **Ashfold, M. N. R., Fullstone, M. A., Hancock, G., and Ketley, G. W.,** Singlet methylene kinetics: direct measurements of removal rates of $\tilde{a}^1A_1$ and $\tilde{b}^1B_1$ methylene and methylene-D_2, *Chem. Phys.,* 55, 245, 1981.

56. **Halberstadt, M. L. and Crump, J.,** Insertion of methylene into the carbon-hydrogen bonds of the C_1 to C_4 alkanes, *J. Photochem.,* 1, 295, 1972/73.

57. **Hase, W. L. and Simons, J. W.,** Excitation energies of chemically activated isobutane and neopentane and the correlation of their decomposition rates with radical recombination rates, *J. Chem. Phys.,* 54, 1277, 1971.

58. **Eder, T. W. and Carr, R. W. Jr.,** Collision induced singlet-triplet intersystem crossing of methylene and methylene-d_2, *J. Chem. Phys.,* 53, 2258, 1970.

59. **Zabransky, V. and Carr, R. W., Jr.,** Photodissociation of ketene at 313 nm, *J. Chem. Phys.,* 79, 1618, 1975.
60. **McDaniel, R. S., Dickson, R., James, F. C., Strausz, O. P., and Bell, T. N.,** Rate parameters for the reactions of the bromomethylidyne radical, *Chem. Phys. Lett.,* 43, 130, 1976.
61. **Ruzsicska, B. P.,** Data quoted in Reference 13.
62. **Choi, H. K. J.,** Data quoted in Reference 13.

Chapter 15

CHEMICAL KINETICS OF REACTIONS OF SULFUR-CONTAINING ORGANORADICALS

Osamu Ito and Minoru Matsuda

TABLE OF CONTENTS

I. INTRODUCTION

This chapter deals with both sulfur-centered radicals and carbon-centered radicals containing sulfur atoms. These radicals are important intermediates in various reactions; i.e., the thermal and photochemical reactions are of great synthetic utility. Gas-phase reactions are also important in relation to the air pollution and to fundamental kinetic studies. The radiation reactions are related to the radioprotective actions of the sulfur compounds.

Fundamental roles of the sulfur-containing radicals in the reactions are revealed in reviews appeared until 1978.[1-5] These reaction mechanisms have been mainly deduced from a pile of data which were accumulated by the analysis of the reaction products. Recently, fast kinetic spectroscopic methods which can observe the radical intermediates have been applied to sulfur-radical chemistry. From the time-dependences of the radical-concentrations, the reaction rates can be evaluated in the form of the absolute values instead of the relative rates. From the absolute rate parameters, the comparisons among the reactivities of a wide variety of radicals become possible. The new kinetic methods frequently discovered some "hidden" reactions and reversible processes which make the mechanistic studies of the sulfur-radical reactions difficult. This review deals with mainly such kinetic studies.

In this review we include the alkyl- and phenyl-substituted, sulfur-containing radicals; for some reactions, the different reactivities between both radicals are prominent. The oxidation of the sulfide groups yielding the sulfinyl and sulfonyl moieties results in considerably strong perturbation on the electronic structures of the sulfur atoms. Although some properties remain similar, many kinds of the reactivities become considerably different from those of the thiyl radicals.

II. CARBON-CENTERED RADICALS CONTAINING SULFUR ATOMS

A. α-Position

1. Decomposition of Azo Compounds

Kinetics of decomposition of azo compounds is quite simple, since the reaction rate constants can be measured by the nitrogen gas evolution (Reaction 1). A reliable set of data for the rate constants and free energies of activation was reported by Ohno et al.[6]

$$X-C(CH_3)_2-N=N-C(CH_3)_2-X \xrightarrow{k_d} 2\ X-\dot{C}(CH_3)_2\ +\ N_2 \tag{1}$$

Typical stabilization energies, which are calculated from the difference between the free energies of activation for the unsubstituted and substituted systems, are collected in Table 1.[7] Since the difference in the stabilization energies between $-OCH_3$ and $-CH_3$ is small, the resonance participation from the oxygen atom is not conspicuous. Sulfides show very low activation energies, indicating the prominent participation of the sulfur atom to the stabilization of the radical center at the α-position. This can be attributable to the electron-sharing conjugation (**1b**) and/or electron-transfer resonance structure (**1c**).[8]

$$-\ddot{\underset{..}{S}}-\dot{C}< \leftrightarrow -\dot{\underset{..}{S}}=C< \leftrightarrow \overset{+}{-}\underset{..}{S}-\ddot{\underset{..}{C}}\overset{..}{\lessgtr}$$

1a   **1b**   **1c**

The effect of the sulfur-atom is less than that of directly bonded phenyl π-system. It is noteworthy that the difference between the alkyl and phenyl group attached to the sulfur or oxygen atoms looks to be small, although in the canonical structures such as **1b** and **1c**, the

Table 1
**RELATIVE DISSOCIATION RATES (k_d IN
REACTION 1) AND STABILIZATION ENERGIES (SE
IN kcal/mol) FOR X–$\overset{\centerdot}{C}$(CH$_3$)$_2^{6,7}$**

X	k_d	SE	X	k_d	SE
–CH$_3$	(1.0)	4.6	–Ph	4.1×10^6	15.9
–OCH$_3$	10.4	6.4	–OPh	50	7.6
–SCH$_3$	2.3×10^4	12.1	–SPh	6.6×10^3	11.2

Note: SE = 0 for X = H; k_d at 100°C in tetraline.

odd electron on the sulfur or oxygen atom looks to delocalize onto the adjacent phenyl π-system.

On the basis of the MO calculations,[7] it is shown that the participation of 3d orbitals of the sulfur atom in the bonding is negligibly small. Consequently, the stabilization can be attributed to the interaction of the singly occupied radical orbital (p-orbital of the carbon atom) with the doubly occupied nonbonding molecular orbital of the adjacent sulfur atom. This is three-electron interaction in two orbitals, and it leads to net one-electron stabilization. The large stabilizing effect of the sulfur atom is caused by the greater proximity of these interacting energy levels, because the ionization potential of sulfur atom (12.50 eV) is similar to the carbon atom (11.42 eV) rather than that of oxygen (17.28 eV).[7] It may be interesting to compare these results with sulfoxides and sulfones. Such data, however, have not been reported.

2. Hydrogen-Atom Abstraction

The relative rates for the radical abstraction from the hydrogen attached to the carbon atom adjacent to the sulfur atoms also give useful information about the α-substituent effect of the sulfur atoms. For example, the phenyl radical generated from the radical sources such as phenylazotriphenylmethane [PhN=NC(Ph)$_3$] reacts with RH (k_H).[9]

$$Ph^{\centerdot} + RH \xrightarrow{k_H} PhH + R^{\centerdot} \tag{2}$$

In Table 2, the relative rate constants reported by Russell et al. for the phenyl radical are listed;[9] the reactivities toward the *t*-butoxy radical are also added in this table.[10] By Scaiano et al., the rate constants for the reactions of the phenyl radical and *t*-butoxy radical with toluene were evaluated in the form of absolute values with the laser-flash photolysis method, which is believed to be one of the most reliable methods.[11,12] Therefore, the relative rates in Table 2 can be easily converted into the absolute values in the range of 10^3 to 10^7 M^{-1} sec^{-1}. The hydrogen atoms at α-carbon adjacent to the sulfide group are more reactive toward the attacking radicals than those of the sulfoxide and sulfone groups. The reactivities of sulfides are slightly higher than toluene.

The effect of the ether groups is intermediate between the sulfide group and the sulfoxide or sulfone group; the difference between the sulfoxide and sulfone groups is not clear in Table 2. Although the polar character of the transition state may also affect the reactivities, it is confirmed that the phenyl radical is fairly free of polar effects. A highly electrophilic *t*-BuO$^{\centerdot}$ also shows a similar tendency of the reactivities (Table 2). Thus, the observed rates reflect the stabilities of the produced-radicals. The order of the delocalized-stabilization is as follows.

Table 2
RELATIVE REACTIVITIES OF PHENYL RADICAL
(REACTION 2 AT 60°C) AND *t*-BUTOXY RADICAL
(AT 130°C) PER HYDROGEN ATOM (UNDERLINED)

	Ph[a]		*t*-BuO[b]
X =	$k_H(CH_3X)$	$k_H(CH_3CH_2X)$	$k_H(CH_3X)$
−SR	17	54	
−SPh	13	26	2.12
−S(O)R	0.7		
−S(O)$_2$R		11	0.05
−OR	4.7	33	
−OPh	3.1	14	1.44
−R	1.0	9	0.1
−Ph	9.0	41	(1.0)
	$(1.7 \times 10^6)^c$		$(2.3 \times 10^5)^d$

[a]   Reference 9.
[b]   Reference 10.
[c,d]   In unit of $M^{-1}\ sec^{-1}$ in References 11 and 12, respectively.

$$\overset{\cdot}{C}HR-SR \simeq \overset{\cdot}{C}HR-Ph > \overset{\cdot}{C}HR-OR > \overset{\cdot}{C}HR-S(O)R \simeq \overset{\cdot}{C}HR-S(O_2)R \simeq \overset{\cdot}{C}HR-R$$

The reactivities of the ether groups in Table 2 seem to be higher than those expected from Table 1.

Two or three substituents increase the effects; i.e., toward *t*-BuO· (toluene = 1.0; see Table 2), $k_H = 16.3$ for $(EtS)_2CH_2$, $k_H = 10.0$ for $(PhS)_2CH_2$, and $k_H = 13.9$ for $(PhS)_3CH$.[10] Diphenysulfide derivative is less reactive than diethylsulfide one; some steric hindrance effects of the phenylsulfide substitutions may be suggested.

3. Addition to Alkenes

The Q-e scheme which is evaluated from the copolymer compositions gives information about both polar nature and resonance stabilization of vinyl monomers.[13] In Table 3, some of the Q-e values are summarized.[14] The Q-values of vinyl sulfides are larger than those of vinyl sulfoxide, sulfone, and ethers, indicating that the propagating carbon-centered radicals ($\sim PCH_2\overset{\cdot}{C}HX$) are stabilized by the adjacent sulfur atom.

$$\sim P^{\cdot} + CH_2{=}CHX \overset{k_p}{\longrightarrow} \sim PCH_2\overset{\cdot}{C}HX \tag{3}$$

Although vinyl sulfides have considerably smaller Q values than styrene (Q = 1.00), they belong to conjugated monomer (Q > 0.3). The Q values of vinyl sulfoxide and sulfone are as small as vinyl ethers which belong to the nonconjugated monomers (Q ≪ 0.3).

The e-values are a measure of the electron density of the C=C double bond, which is assumed to be equal to the polar nature of the propagating radical. The large negative e-values of vinyl sulfide and vinyl ethers indicate the high electron density of the double bonds caused by the strong electron-releasing powers of the substitutents. The positive e-values of vinyl sulfoxides and sulfones indicate the electron-poor double bond caused by their strong electron-withdrawing powers. Thus, the actual reactivities (1/r) are mainly determined by above both effects; i.e., quite high reactivity of vinyl sulfides toward the electrophilic polyacrylonitrile radical can be reasonably interpreted by Q-e scheme.

Table 3
ADDITION RATE PARAMETERS; Q-e VALUES[14] AND RATE PARAMETERS OF REACTION 4[17]

X in CH$_2$=CHX	Q	e	1/r(St)[a]	1/r(AN)[b]	$k_1{}^c \times 10^{-5}$	K[d]	$\rho^+(k_1)$
–SR	0.37	–1.2			260	3.4	2.23
–SPh	0.34	–1.4	2	25	48	1.7	1.99
–S(O)R	0.13	0.69	(1)	(1)	0.14	0.01	1.20
–S(O$_2$)R	0.07	1.2	3	3	0.16	0.004	0.81
–OR	0.03	–1.2			0.76	0.026	1.67
–OPh	0.08	–1.2	0.05	0.05	1.0	0.028	1.84

^a Polystyryl radical.
^b Polyacrylonitrile radical.
^c Units of M^{-1} sec^{-1}.
^d K = k_1/k_{-1} calculated from Kk_2 assuming $k_2 = 10^9$ M^{-1} sec^{-1}.

The substitution by two sulfide groups increases the delocalization; the Q value of CH$_2$ = C(SEt)$_2$ is reported to be 2.70 which is about 10 times greater than that of the monosubstituted one.[15] For phenylthio derivative [CH$_2$=C(SPh)$_2$] Q = 0.42, which suggests steric effects between two phenylthiyl moieties.[16]

The addition rates of these vinyl monomers toward the phenylthiyl radicals have been measured by applying the xenon-flash photolysis method.[17] There are some difficulties in the application of this method to the kinetic study for the radical reactions. One of them is the assignment of the transient absorption bands; for phenylthiyl radicals, the absorptions appeared in the visible region (ca. 500 nm) have been assigned, since both diphenyl disulfide and phenylthiol gave the identical absorption band.[18] Another difficulty is caused by the reversibility of the addition process of the thiyl radicals to alkenes, which appears in the decay profiles of the thiyl radicals in microsecond time scale; i.e., the decay rates of the thiyl radicals are not accelerated even by the addition of the reactive alkenes to the systems. In order to shift the equilibrium (Reaction 4), it is necessary to add the third species, which is selectively reactive to the carbon-centered radicals (PhSCH$_2$ – $\cdot$CHY) but not to the phenylthiyl radicals; in this case oxygen molecule was found to be appropriate. Thus, the rate constants can be determined by taking such reversibility into consideration in the form of the absolute values.

$$\text{PhS}^\cdot + \text{CH}_2\text{=CHX} \underset{k_{-1}}{\overset{k_1}{\rightleftharpoons}} \text{PhSCH}_2\text{–}\overset{\cdot}{\text{C}}\text{HX} \xrightarrow[+\text{O}_2]{k_2} \text{peroxyradical} \qquad (4)$$

Some of the data are summarized in Table 3. The addition rate constants thus obtained for vinyl sulfides are quite higher than those of vinyl ether, vinyl sulfoxide, and vinyl sulfone. The latter three alkenes show similar reactivities to alkyl-substituted alkenes, whereas the reactivities of vinyl sulfides are as high as styrene.

The polar nature can be estimated from the Hammett relation with changing the substituents of the phenylthiyl radicals. The rate parameters are easily obtained with changing the substituents in the attacking radical; this is one of the advantages to measure the rate constants with the fast kinetic spectroscopy. The Hammett reaction constants (ρ^+-values) are also listed in Table 3. Since vinyl sulfides and vinyl ethers show similarly large positive ρ^+-values, it is confirmed that the electron-releasing abilities of both sulfide and ether groups are similar. Therefore, the large difference in the reactivities between them is attributable to the stabilities of the transition states which may be the reflections of the stabilities of the carbon-centered radicals in the products (PhSCH$_2$ – $\dot{}$CHY). The equilibrium constants (K = k_1/k_{-1} in Reaction 4), which can be estimated from the flash photolysis data as relative values, support this consideration. The K values for vinyl sulfides are about 10^2 larger than those of vinyl ethers, suggesting that the carbon-centered radicals produced from vinyl sulfides (PhSCH$_2$ – $\dot{}$CHSR) are more stable than those from vinyl ethers (PhSCH$_2$ – $\dot{}$CHOR). It is also confirmed from the K values that the stabilities of the carbon-centered radicals attached to the sulfoxide and sulfone groups are as low as the ether group.

The smaller ρ^+ values of vinyl sulfoxides and sulfones than those of vinyl ethers and sulfides indicate the electron-poor double bond of the former alkenes; thus, the reactivities also change with the polar nature of the attacking radicals. The ρ^+ values of vinyl sulfoxide and sulfone show the same sign to those of sulfides and ethers, although the e-values change the sign; this reason is interpreted by the linear free-energy relationship originated in the substituent effect on the stabilities of the substituted phenylthiyl radicals in the reactants. Therefore, actual ρ^+ value for the reactivity consists of both the polar nature of the transition state and the substituent effect on the thermodynamic stabilities of the phenylthiyl radicals.

The order of the stabilities of the carbon-centered radicals, PhSCH$_2$ – $\dot{}$CHY, derived from the phenylthiyl radical-addition reactions is qualitatively consistent with those from decomposition data and from the hydrogen-abstraction data, although there seems some exceptions such as the oxygen atom-effect for hydrogen-abstraction. In all data described above, difference between the alkyl and phenyl groups attached to the sulfur and oxygen atoms appears to be small; this problem is not yet reasonably interpreted.

4. ESR Parameters

ESR data also give useful information about the α-substituent effect. Carton et al. reported that the magnitude of the α-proton hyperfine coupling constant (hfc) in MeS$\dot{\text{C}}$H$_2$ is smaller than those of the sulfoxide and sulfone groups, in which the latters are similar to that of ethyl radical.[19]

MeS$\dot{\text{C}}$H$_2$	MeS(O)$\dot{\text{C}}$H$_2$	MeS(O$_2$)$\dot{\text{C}}$H$_2$	MeO$\dot{\text{C}}$H$_2$	Me$\dot{\text{C}}$H$_2$
hfc — 16.5 G	20.0 G	22.3 G	17.1 G	22.4 G

Decrease in the α-proton hfcs implies the delocalization of the unpaired electron onto adjacent atoms.[19] Although the decrease in hfc by the oxygen atom seems to be similar to that of sulfur atom, the stronger electronegativity of the oxygen atom (X_p = 3.5) than that of the sulfur atom (X_p = 2.5) may be responsible to this observation.

Barriers to rotation about X–CH$_2$ bond in RX–$\dot{\text{C}}$H$_2$ radicals were evaluated with analysis of the ESR spectra showing exchange broadenings; the barrier about S–CH$_2$ bond is greater than that about O–CH$_2$ bond by about 1 to 2 kcal/mol.[20,21] The higher barriers in the sulfur atom substituted radicals suggest that there is greater delocalization of the unpaired electron onto the sulfur atom than that of the oxygen atom.

Kinetic study for the carbon-centered radical substituted by three CF$_3$S groups was reported.[22] The rate constants for dissociation of (CF$_3$S)$_3$C–C(SCF$_3$)$_3$ was determined by scavenging the carbon radical with a stable free radical; the equilibrium constant K = k_d/k_{-d} was measured by ESR to be 7.5×10^{-10} M^{-1} at 30°C.

$$(CF_3S)_3C\text{–}C(SCF_3)_3 \underset{k_{-d} = 6.0 \times 10^4\ M^{-1}sec^{-1}}{\overset{k_d = 4.5 \times 10^{-5}sec^{-1}}{\rightleftharpoons}} 2\ \dot{\text{C}}(SCF_3)_3 \qquad (5)$$

From the activation energy for the C–C bond dissociation (21.0 kcal/mol) and the activation energy for dimerization of the radical (7.3 kcal/mol), the low rate constant for dimerization is attributed to the strong steric interaction (F- and B-strain) of the CF_3S groups in the transition state rather than the delocalized stabilization of the radical $[\dot{C}(SCF_3)_3]$. The ESR parameters indicate that $\dot{C}(SCF_3)_3$ is planar in contrast to the pyramidal radical of $\dot{C}(OCF_3)_3$. A low recombination rate constant ($5 \times 10^5 \ M^{-1} \ sec^{-1}$ at 25°C) was also reported for the nitrogen analog, $N(SCF_3)_2$, in which the facile N–N bond homolysis is interpreted by the sterically and electronically induced destabilization of the N–N bond.[23]

The radicals $R_nMS-\dot{C}(SR)_2$, which are produced by the addition of $R_nM\dot{}$ radicals to trithiocarbonates $[S=C(SR)_2]$, decay with second-order kinetics at the rates of the diffusion-controlled limit.[24]

$$R_nM\dot{} + S{=}C(SR)_2 \rightarrow R_nMS-\dot{C}(SR)_2 \qquad (6)$$

In contrast, thiylaminyl radicals $\dot{N}(SR)_2$ are stable and persistent; some thermodynamic data such as the enthalpies of dissociation of the dimers were also reported.[25] The sulfonamidyl radicals $(RSO_2\dot{N}R)$ are rather short-lived ones.

5. Reaction with Oxygen

The absolute rate constants for the reactions of the carbon-centered radicals substituted by the sulfur atoms with oxygen were reported. Asmus et al. observed the transient absorption bands attributable to $RS-\dot{C}R_2$ at approximately 280 nm by pulse radiolysis of sulfides in aqueous solution. The radicals are produced by the $\dot{O}H$-addition to sulfides followed by H_2O elimination ($CH_3SCH_3 + \dot{O}H \rightarrow CH_3\dot{S}(OH)CH_3 \rightarrow \dot{C}H_2SCH_3 + H_2O$).[26] The decay rates of the radicals increased with the addition of oxygen, from which the rate constants were evaluated.

$$RS\dot{C}R_2 + O_2 \xrightarrow{k_o} \text{peroxyradical} \qquad (7)$$

The rate constants for $\dot{C}H_2SCH_3$ is $4.4 \times 10^8 \ M^{-1} \ sec^{-1}$ at room temperature is significantly smaller than even that of the resonance stabilized-benzyl radical ($2.5 \times 10^9 \ M^{-1} \ sec^{-1}$).[27]

B. β-Position

The most prominent characteristic of the carbon-centered radicals containing the sulfur atoms at the β-position is easy cleavage of the C–S bond. This phenomenon is common to the sulfide, sulfoxide, and sulfone groups. Ueno et al. found that the carbon-centered radicals (3) produced by the addition of the alkylthin radical to suitable dithioacetals (2) dissociate at the C–S bond (Reaction 8).[28]

$$
\begin{array}{c}
\text{O} \\
\parallel \quad /\text{SR}^1 \\
\text{Ph--C--C} \quad \xrightarrow{+R_3Sn\dot{}} \\
\text{H} \quad \backslash \text{SR}^2 \\
\textbf{2}
\end{array}
\qquad
\begin{array}{c}
\text{OSnR}_3 \\
\mid \quad /\text{SR}_1 \\
\text{Ph--C--C} \\
\text{H} \quad \backslash \text{SR}^2 \\
\textbf{3}
\end{array}
\qquad
\begin{array}{c}
\text{OSnR}_3 \\
\mid \\
\text{Ph--C=CH--SR}^2 + \dot{}\text{SR}^1 \\
\nearrow \\
\\
\searrow \\
\text{OSnR}_3 \\
\mid \\
\text{Ph--C=CH--SR}^1 + \dot{}\text{SR}^2
\end{array}
\qquad (8)
$$

From the eliminating radicals ($\dot{}SR^1$ or $\dot{}SR^2$), the leaving abilities which are proportional to the stabilities of the radicals of each pair are estimated as follows:

$$PhS^{\bullet} > PhSO_2^{\bullet}, \square\ PhSO^{\bullet} > PhSO_2^{\bullet}, \square\ MeSO^{\bullet} > MeS^{\bullet}$$

This result is qualitatively consistent with Kice's suggestion.[4] For the various thiyl radicals, a following order for the stabilities was obtained.

The addition reactivities of some of these thiyl radicals toward an alkene give information about the stabilities of the thiyl radicals; i.e., for styrene, the addition rate constants of the thiyl radicals **6** and **7** were evaluated by the flash photolysis method; the reactivity of **5** was too low to measure by this method.[29,30]

$$CH_2=CHPh \begin{cases} + \mathbf{5} & \xrightarrow{10^3\ M^{-1}sec^{-1}} & (9) \\ + \mathbf{6} & \xrightarrow{2.0 \times 10^7\ M^{-1}sec^{-1}} & (10) \\ + \mathbf{7} & \xrightarrow{2.5 \times 10^7\ M^{-1}sec^{-1}} & (11) \end{cases}$$

Although these rate constants are compatible with the order of the stabilities of the thiyl radicals estimated from Reaction 8, difference in stabilities between **5** and **6** presumed from the rate parameters looks greater than that evaluated by Reaction 8.

A more systematic comparison was reported by Wagner et al. measuring the product-ratios (**12/13**) of homolytic cleavages in diradicals (**11**) generated by the photochemical excitation of δ-substituted valerophenones (**10**);[31]

The relative rates are listed in Table 4. The sulfinyl radicals are the most stable both in the alkyl and phenyl series. The alkylsulfonyl radical is more stable than alkylthiyl radical, whereas the phenyl substitution inverts the order, indicating that the unpaired electron on the sulfur atom of the thiyl radical delocalizes onto the phenyl ring. An increase in the relative rates by the phenyl substitution for the thiyl radicals is about 700 times, whereas those for sulfinyl and sulfonyl radicals are 8 and 15 times, respectively. Apparently, the

X	k_{-x}	X	k_{-x}
−SBu	0.16	−SPh	110
−S(O)Bu	74	−S(O)Ph	630
−S(O$_2$)Bu	0.46	−S(O$_2$)Ph	6.8

[a] k_{-x} for X = Cl is 1.0.

phenylthiyl radical belongs to the π-radicals. In the alkylsulfinyl radicals the unpaired electron delocalizes sufficiently onto the oxygen-atom. In the phenylsulfinyl radical, some further delocalizations onto the phenyl ring can be recognized. The sulfinyl radicals have also been classified as π-radical by their great stabilization, but the interaction between the unpaired-electron orbital of S=O moiety and phenyl π-orbitals may be quite different from that of the phenylthiyl radical. According to the reactivity data and spectroscopic data such as ESR and optical studies, the sulfonyl radicals are attributed to the σ-radical,[32,33] however, the phenyl substitution considerably increases the stability.

The absolute rate of k_{-x} for − SBu (Reaction 12) can be calculated to be 2.7×10^5 sec^{-1} on the basis of the reported k_s: for the fastest case (PhSO$^\cdot$), the rate constant is approximately 10^9 sec^{-1}. By the flash photolysis, the reverse rate constants (k_{-1}) of the addition process (Reaction 4) can be estimated as a ratio of k_2; the values of k_{-1} for the nonconjugated vinyl monomers are 10^6 to 10^7 sec^{-1} for PhS$^\cdot$,[29] which are closed to the $k, _{-1}$ value of 10^8 sec^{-1} for PhS$^\cdot$ in Table 4.

In the stereochemical studies of the free-radical eliminations of the β-phenylsulfinyl radicals (Reaction 13), alkenes form in a stereoselective manner; from erythro derivatives of **14**, *cis*-alkenes form. This stereoselectivity is the result of rapid loss of phenylsulfinyl radical before appreciable rotation about the central C–C bond in intermediate radical (**15**).[34]

CHBrR−CHRS(O$_n$)Ph $\xrightarrow{\ +^\cdot\text{SnR}_3\ }$ (15, structure) $\longrightarrow$ PhṠ(O$_n$) + (alkenes) (13)

14   **15**

On the other hand, the β-phenylsulfide and β-phenylsulfone derivatives form alkenes in a nonstereospecific manner, suggesting that the elimination is slower than the rotation and that the stabilization by sulfur bridging in the intermediate radicals (**15**) is unimportant. These findings are consistent with the results in Table 4.

In the gas phase, 2,3-dimethylthiirane reacts with attacking radicals such as hydrogen atom, methyl radical, and phenyl radical yielding butene-2 via the intermediates such as **15**. From *cis*-2,3-dimethylthiirane an excess amount of *cis*-butene-2 (70 to 90%) was found.[35] This shows that the release of the thiyl radicals in the intermediates is faster than the rotation of the central C–C bond. This observation indicates that in gas-phase reactions other factors such as excess energy may influence the reactivity.

In the reaction from **14** to **15**, the relative rates are a measure of the activation of the C–Br bond by the β-substituents of PhS(O)$_n$, however, appreciable anchimeric assistance was not found.[34]

Table 5

σ$_\alpha^{\cdot}$ VALUES FOR X–C$_6$H$_4$–$\overset{\cdot}{C}$H$_2^{36}$

X	σ$_\alpha^{\cdot}$	X	σ$_\alpha^{\cdot}$
4-SMe	0.063	4-OMe	0.018
4-SPh	0.058	4-OPh	0.018
4-S(O)Ph	0.026	4-S(O)OMe	0.016
4-S(O)Me	0.018	4-S(O$_2$)OMe	0.013
4-S(O$_2$)Ph	0.018	H	0.000

C. Others

The sulfur substituents on the 4-position of the benzyl radical affect on hfc of the benzylic hydrogen. The substituent constants, σ$_\alpha^{\cdot}$, were defined from the hfcs which reflect the spin delocalization in the substituted benzyl radical relative to the unsubstituted radical.[36] In Table 5, the positive values of σ$_\alpha^{\cdot}$ imply the stabilization ability. The order of the effects of the substituents is surprisingly compatible with the directly bonded α-effect.

In the case of the Q values, 4-RS-substituted styrene shows a considerably positive Q value, whereas for 4-RO-substituted styrene the Q value is similar to unsubstituted styrene (Q = 1.0).[15]

RS—⟨○⟩—CH=CH$_2$ RO—⟨○⟩—CH=CH$_2$
Q=3.29, e=-1.65   Q=1.0, e=-1.0

This finding is consistent with the σ$_\alpha^{\cdot}$ parameters for the sulfide and ether groups listed in Table 5.

The hydrogen-atom donor ability of 2-methylthiophene to the phenyl radical is greater than those of toluene and 2-methylfuran;[9] i.e., 2-methylthiophene:2-methylfuran:toluene = 14.7:8.6:9.0 and these values can be converted to the rate constants by the use of the data in Table 2.

$$Ph^{\cdot} \; + \; Me\text{—}\langle\text{thiophene}\rangle \xrightarrow{\; k_H \;} \; PhH \; + \; {}^{\cdot}CH_2\text{—}\langle\text{thiophene}\rangle \tag{14}$$

This was interpreted by the greater stabilization which comes from greater participation of the stable resonance structures. The carbon radical produced from 2-methylthiophene is more stable than that of 3-methyl; i.e., the rate constant of 3-methylthiophene is similar to that of toluene.

Methyl affinity of 2-vinylthiophene is about twice greater than that of styrene.[37] The Q value of 2-vinylthiophene (Q = 3.0) is also greater than that of styrene by a factor of 3.

Kinetic studies for direct attack of free radicals toward the thiophene and furan ring were reported by laser flash photolysis.[38] The rate constants for the phenyl radical are 6.4 × 10^6 and 2.7 × 10^6 M^{-1} sec^{-1} at 25°C, respectively. Those for the triethylsilyl radical are 5.0 × 10^6 and 1.4 × 10^6 M^{-1} sec^{-1} at 25°C, respectively; for both radicals, thiophene is more reactive than furan. In the case of (EtO)$_2\overset{\cdot}{P}$O this order of the reactivities is temperature dependent.[38] The adduct radicals formed by the radical attack at 2-position were detected by ESR at low temperature.

$$R^\cdot \ + \ \ \ \ \ \longrightarrow \ \ \ \ \ R\text{—} \quad\quad (15)$$

The kinetic studies for OH radical reactions in gas phase using time-resolved resonance fluorescence spectroscopy revealed that thiophene is less reactive than furan over a wide temperature range.[39] Some changes in the reaction mechanism may be suggested.

III. SULFUR-CENTERED RADICALS

A. Thiyl Radicals
1. Reactions with Small Molecules
a. Oxygen

The rate constant for the reaction of the thiyl radical with oxygen was determined with the pulse radiolysis method by Asmus et al.;[26] from the decay rates of the transient absorption at 330 nm which was attributable to the penicilliaminethiyl radical (PenS$^\cdot$) in the presence of oxygen, the rate constant was measured at room temperature.

$$\text{PenS}^\cdot \ + \ O_2 \xrightarrow[\ k_o = 4.0 \times 10^7 \ M^{-1}\text{sec}^{-1}\ ]{k_o} \text{PenSOO}^\cdot \quad\quad (16)$$

In general, even resonance stabilized carbon-centered radicals react with oxygen with quite high rate constants close to the diffusion-controlled limits.[27] Thus, the alkylthiyl radical is less reactive to oxygen than the carbon-centered radicals by a factor of 10^2, and similarly the oxyradicals also show low reactivity toward oxygen. Such low reactivities of these thiyl and oxyradicals are interpreted by the smaller s-character of the sulfur and oxygen bonds.

This low reactivity of the thiyl radical to oxygen was used for the oxidative addition of the thiyl radicals to alkenes in solution containing oxygen.[40] The products of oxidation at sulfur atom were not formed on the addition of an amine which acts as a hydrolysis reagent for the peroxy intermediate radical.

$$\text{RS}^\cdot \ + \ CH_2\!=\!CHR \xrightarrow{+O_2} \overset{\displaystyle OO\cdot}{\underset{\displaystyle |}{\text{RSCH}_2\text{–CHR}}} \xrightarrow{+\,\text{amine}} \overset{\displaystyle OH}{\underset{\displaystyle |}{\text{RSCH}_2\text{–CHR}}} \quad\quad (17)$$

Similar reaction-scheme was established in the gas-phase reactions involving the alkylthiyl radicals, alkenes, and oxygen. Low reactivity of the thiyl radical with oxygen in the absence of alkenes was also confirmed by the product analysis.[41]

By applying the ESR method, a similar reaction-scheme is presumed.[42] The thiyl radical (**8**) can be expected to be reactive from the stability-order estimated from Reaction 8. The ESR signals of the thiyl radical were not significantly depleted in the presence of either oxygen or alkenes, however, when both oxygen and alkenes were present, the ESR signals from the thiyl radical were rapidly replaced by the signals attributable to a peroxy radical structure.

$$\underset{\mathbf{8}}{\overset{\displaystyle S}{\underset{\displaystyle \|}{(RO)_2PS^\cdot}}} \ + \ CH_2\!=\!CHR \rightleftharpoons \overset{\displaystyle S}{\underset{\displaystyle \|}{(RO)_2PSCH_2^\cdot CHR}} \xrightarrow{+O_2} \begin{array}{l}\text{peroxy-}\\\text{radical}\end{array} \quad\quad (18)$$

In the flash photolysis study of the phenylthiyl radicals and some conjugated thiyl radicals such as **7**, the decay rates of the transient absorptions in the visible region, which can be attributed to the thiyl radicals, were not affected in the presence of either oxygen or alkene. Similar to the ESR observation above, only when both reagents were present, the decay rates of the thiyl radicals were accelerated. Therefore, on the basis of the reaction scheme (Reactions 4 and 18), the addition rate constants of many phenylthiyl radicals toward various alkenes were obtained with varying both concentrations of alkenes and oxygen.[43] Contrary to the visible region, in the UV region the transient absorption bands attributable to the adduct radicals were found. This will be described in the next section.

Dialkyl disulfides also gave the transient absorption bands in the UV region attributable to the species produced by the C–S bond fission in addition to the thiyl radicals. Kuntz et al. reported that the flash photolysis of some dialkyl disulfides with the light of 250 to 370 nm in degassed solution gave the perthiyl radicals (RSS·, Reaction 20) in addition to the ordinary S–S bond cleavage.[44]

$$\text{RSSR} \xrightarrow{\text{h}\nu} 2 \text{ RS}^{\bullet} \tag{19}$$

$$\text{RSSR} \xrightarrow{\text{h}\nu} \text{RSS}^{\bullet} + \text{R}^{\bullet} \tag{20}$$

The penicillamineperthiyl radical (380 nm) reacts with oxygen with a rate constant of $4 \times 10^7 \ M^{-1} \ \text{sec}^{-1}$.[44] This value is similar to that with the pencillaminethiyl radical (Reaction 16), although the perthiyl radicals are anticipated to be extraordinarily resonance stabilized; such characteristics are not prominent in the rapid reaction with oxygen. The C–S bond cleavage by the sensitized photolysis was also confirmed by the CIDNP method.[45]

b. Sulfur Compounds

Bonifačić and Asmus attributed the transient band at 380 nm that appeared by the pulse radiolysis of dialkyl disulfides to the adduct radical (**16**) produced by homolytic substitution reaction (S_H2) as shown in Reaction 21.[46]

$$\text{RS}^{\bullet} + \text{RSSR} \underset{2.3 \times 10^4 \ \text{sec}^{-1}}{\overset{3.8 \times 10^6 \ M^{-1}\text{sec}^{-1}}{\rightleftharpoons}} \overset{\displaystyle \text{SR}}{\underset{\displaystyle \underset{\bullet}{\text{RS–SR}}}{\big|}} \tag{21}$$

16

The addition rate constant of the methylthiyl radical to disulfide is intermediate between the phenyl radical ($1.4 \times 10^8 \ M^{-1} \ \text{sec}^{-1}$)[47] and the methyl radical ($6.0 \times 10^4 \ M^{-1} \ \text{sec}^{-1}$),[48] and similar to the trialkyltin radical.[49] The rate constant for the phenyl radical can be calculated by the combination of the relative and absolute rates.[11]

From the laser photolysis and modulation spectroscopy for diphenyl disulfides, Burkey and Griller[50] found the absorption bands at 340 nm attributable to the adduct radicals under the condition of high concentration of disulfides. In addition to the adduct bands, the phenylthiyl radical-bands appear at 300 and 500 nm. The 500 nm bands have been used by Ito and Matsuda for the evaluations of the absolute rate parameters.[43] The optical studies for the intermediate radicals by the photolysis in rigid matrix were also reported,[51,52] however, there are some discrepancies between the absorption maxima observed with both methods.

Addition abilities of the thiyl radicals to thiolate anions are also high. By the pulse radiolysis of aqueous solution containing dialkyl disulfides or thiolate anions, the transient absorption bands attributable to the disulfide radical anions (RSSR⁻·) are identified and kinetic data were obtained.[53,54] The radical anions were produced by the direct attachment of the hydride electron or by some other processes such as the reverse process in Reaction 22.

$$RSSR^{-\bullet} \; \underset{8 \times 10^5 \; sec^{-1}}{\overset{4.9 \times 10^9 \; M^{-1}sec^{-1}}{\rightleftharpoons}} \; RS^- + RS^\bullet \tag{22}$$

The radical anions exist in the equilibrium with the thiyl radicals and the thiolate anions. Some kinetic parameters evaluated for cysteine are shown in Reaction 22.

By the flash photolysis of the alkyl thiolate anions, the transient absorption bands attributable to the radical anions of disulfides were also observed.[55,56] In this case, the radical anions are formed via the thiyl radicals produced by the photoejection of the electron (Reaction 23) followed by the reverse process of Reaction 22.

$$RS^- \xrightarrow{h\nu} RS^\bullet + e^- \tag{23}$$

For phenylthiolate anions, Caspari and Granzow[55] observed the transient absorption bands attributable to the diphenyldisulfide radical anions, however, Thyrion presented the spectroscopic data which could not show the radical anion formations.[18] With photolysis of the mixture of the phenylthiolate anion and aromatic hydrocarbons, the colored radical anions of aromatic hydrocarbons are formed. The lifetimes of the colored radical anions depend upon the reduction potentials of the aromatic hydrocarbons.[57]

2. Hydrogen-Atom Abstraction
a. Abstraction Ability

Many reaction products formed by the hydrogen-atom abstraction of the thiyl radicals from the reactive hydrogen have been reported. Recently, quantitative kinetic data have also been accumulated. From many data, it can be pointed out that there is a difference in the hydrogen abstraction abilities between alkylthiyl and phenylthiyl radicals. In the pyrolysis of aryldiazothiolates (Ph–N=N–SAr; kinetic data for Reaction 24 are shown in Table 6), Zwet and Kooyman[58] reported that the phenylthiyl radicals are considerably unreactive for hydrogen abstraction from tetraline at 80°C, whereas the alkylthiyl radicals can easily abstract the hydrogen atoms from ordinary hydrogen donors.

$$Ph-N=N-SAr \rightarrow Ph^\bullet + N_2 + ArS^\bullet \tag{24}$$

Unexpected observation that the hydrogen abstraction of the phenylthiyl radical occurred in inert solvents such as chlorobenzene was reasonably interpreted by the idea that thiols are produced with the assistance of the phenyl-radical addition to the solvent. This reaction may yield the cyclohexadienyl-type radical which possess an extremely active hydrogen.

In the studies of photoreduction of aromatic ketone in the presence of amines and alcohols, Cohen et al. found that the aliphatic thiols catalytically convert alkylaminyl radicals (HC–N$^\bullet$) to α-aminoalkyl radicals ($^\bullet$C–NH) according to Reaction 25.[59]

Table 6
ACTIVATION ENERGIES (E_a) AND RATES (k_H)
FOR THE REACTIONS FOR $Ph_3C^{\cdot} + X-C_6H_4SH^a$
AND DECOMPOSITION RATES (k_d) OF
$PhN_2SC_6H_4-X^b$

X	E_a^* (kcal/mol)	k_H (at 10°C) ($M^{-1} sec^{-1}$)	$k_d \times 10^5$ (55.5°C) (sec^{-1})
4-OMe	7.28	28.8	66
4-t-Bu	10.88	5.19	9.2
H	9.25	6.11	5.0
3-Cl	10.25	7.09	
4-Cl	10.35	9.20	
4-CF$_3$	11.49	3.91	
4-NO$_2$			0.3

[a] Reference 75.
[b] Reference 58.

$$\left[Ph_2C{=}O^*(T_1) + \overset{|\,|}{\underset{|}{HCNH}} \rightarrow \overset{|\,|}{Ph_2C{-}OH} + \overset{|\,|}{\underset{|}{HCN^{\cdot}}} \right]$$

$$\overset{|\,|}{\underset{|}{HC{-}N^{\cdot}}} + RSH \xrightarrow{10^5\ M^{-1}sec^{-1}} \overset{|\,|}{\underset{|}{HC{-}NH}} + RS^{\cdot} \underset{10^3\ M^{-1}sec^{-1}}{\overset{}{\rightleftharpoons}} \overset{|\,|}{\underset{|}{^{\cdot}C{-}NH}} + RSH \quad (25)$$

The rate parameters for aliphatic thiols were estimated as shown above, however, the catalysis by aromatic thiols is less effective than that by aliphatic thiols. Since the donor abilities of the hydrogen atom of aromatic thiols are higher than that of alkylthiols, it is presumed that the hydrogen abstraction abilities of the phenylthiyl radicals from amines are lower than those of the alkylthiyl radicals.

The absolute rate constant for the alkylthiyl radical with ethylbenzene has been estimated to be $1 \times 10^6\ M^{-1}\ sec^{-1}$ from the competitive reaction with $(RO)_3P$ (Reaction 27).[60]

$$RS^{\cdot} + R_2CH{-}C_6H_4{-}X \rightarrow RSH + R_2\overset{\cdot}{C}{-}C_6H_4{-}X \quad (26)$$

The latter rate constant (Reaction 27) was estimated by Walling et al. from the competitive reaction with styrene,[61] which was evaluated by Sivertz[62] with the rotating sector method (Reaction 28).

$$RS^{\cdot} + (RO)_3P \xrightarrow{1.2 \times 10^7\ M^{-1}sec^{-1}} RS{-}\overset{\cdot}{P}(OR)_3 \quad (27)$$

$$RS^{\cdot} + CH_2{=}CHPh \xrightarrow{8 \times 10^8\ M^{-1}sec^{-1}} RSCH_2\overset{\cdot}{C}HPh \quad (28)$$

This rate constant seems to lead to a conclusion that the ability of hydrogen abstraction of the alkylthiyl radical is as high as the alkoxy radical ($1.1 \times 10^6\ M^{-1}\ sec^{-1}$).[12] The abilities of hydrogen-atom abstraction of various free radicals from methane are summarized in Table 7;[63] the rate constant for methylthiyl radical is 1/5000 of that for t-butoxy radical. The rate constants for hydrogen-atom abstraction from trialkyltin hydride are also collected in Table

Table 7
THE ABSOLUTE RATE CONSTANTS FOR THE H-ATOM ABSTRACTION ABILITIES OF VARIOUS RADICALS

Radical	XH	$k, M^{-1} sec^{-1}$	Temp. (°C)	Ref.
$CH_3S^{\cdot}$	CH_4	1	164	63
$CH_3O^{\cdot}$	CH_4	5.0×10^3	164	63
$CH_3^{\cdot}$	CH_4	1.3×10	164	63
$CH_3NH^{\cdot}$	CH_4	8.0	164	63
$PhC(O)S^{\cdot}$	Et_3SnH	9×10^3	60	64
$PhC(O)O^{\cdot}$	Et_3SnH	6×10^4	50	64
$t\text{-}BuO^{\cdot}$	Et_3SnH	2×10^8	RT	65
$CH_3^{\cdot}$	Et_3SnH	1×10^7	RT	66
$t\text{-}Bu^{\cdot}$	Et_3SnH	1.8×10^6	RT	66

Note: RT, room temperature.

6.[64-66] It is found that the benzoylthiyl radical abstracts the hydrogen atom with slower rate than the benzoyloxy radical.[64] Although the direct comparison is not possible because of a lack of the directly measured rate data, the rate constant of $1.0 \times 10^6\ M^{-1}\ sec^{-1}$ for Reaction 26 looks to be over estimated.

The rate constant for the hydrogen-atom abstraction of ethylthiyl radical from diethylsulfide in gas phase was obtained by the combination of the product analysis and computer simulation method to be $4.3 \times 10^6\ M^{-1}\ sec^{-1}$ at 25°C.[67] However, this value is greater than the rate constant of the t-butoxy radical with $RSCH_3$ which can be calculated from the data in Table 2 to be $5 \times 10^5\ M^{-1}\ sec^{-1}$. The former rate constant in gas phase cannot be directly compared with the latter value in solution.

The flash photolysis experiments of the phenylthiyl radicals also support their low reactivities for hydrogen abstraction. Even in cumene as a solvent, the decay rate of the phenylthiyl radical was not accelerated and shows that recombination of the phenylthiyl radical occurs predominantly.[43]

b. Hydrogen-Donor Ability of Thiols

There are many kinetic data showing the high hydrogen-donor abilities of thiols. Many reliable rate constants were obtained by the pulse radiolysis method (Table 8).[68-73] Greig and Thynne[63] observed the thermodynamic parameters for the quite efficient hydrogen abstraction by the methyl radical from methane thiol (Reaction 29).[63]

$$^{\cdot}CH_3\ +\ CH_3SH \rightarrow CH_4\ +\ CH_3S^{\cdot} \tag{29}$$

Reaction 29 is exothermic by about 14 kcal/mol.

The rate constant for the reverse process of Reaction 26 was reported by Burkhart with the rotating sector method.[69] The rate constant for the hydrogen-atom abstraction from alkane thiol (Reaction 30) is shown below (further low values are calculated from the chain-transfer constants as listed in Table 8).

$$Ph\dot{C}H_2\ +\ RSH \xrightarrow{\ 2.2\ \times\ 10^4\ M^{-1}sec^{-1}\ } PhCH_3\ +\ RS^{\cdot} \tag{30}$$

Such slow down of Reaction 30 is reasonably interpreted by taking the resonance stabilization of the benzyl radical (ca. 13 kcal/mol) into consideration;[74] the reaction becomes only

Table 8
HYDROGEN-DONOR ABILITIES OF THIOLS

R$\cdot$	RSH	k, M^{-1} sec^{-1}	Temp. (°C)	Ref.
$\cdot$OH	CySH	4.9×10^9	RT	70
$\cdot$CH$_2$OH	CySH	6.8×10^7	RT	70
CH$_3$CHOH	CySH	1.7×10^8	RT	71
CH$_3$C(OH)CH$_3$	CySH	3.3×10^8	RT	71
$\cdot$CH$_3$	CH$_3$SH	7.4×10^7	RT	68
PhCH$_2\cdot$	PhCH$_2$SH	2.2×10^4	25	69
c-C$_6$H$_{11}\cdot$	c-C$_6$H$_{11}$SH	3.9×10^5	25	69
C$_3$H$_7\cdot$	C$_3$H$_7$SH	2.9×10^6	25	69
PolySt$\cdot$	n-BuSH	4.4×10^3	60	14
PolyMMA$\cdot$	n-BuSH	9.0×10	60	14
PolyVAc$\cdot$	n-BuSH	9.8×10^4	60	14
Ph$\cdot$	PhSH	1.9×10^5	45	72
ROO$\cdot$	PhSH	5.1×10^3	30	73
Ph$_3$C$\cdot$	PhSH	2.0×10	10	75

Note: CySH, cysteine.

approximately 1 kcal/mol exothermic. It would be anticipated that the reverse process of Reaction 30, which is Reaction 26, is slower than the forward process of Reaction 30. The low hydrogen-abstraction ability of the phenylthiyl radicals can also be predicted from the thermodynamic data. Since the phenylthiyl radical is more stabile than the alkylthiyl radical by 9.6 kcal/mol,[74] the hydrogen-atom abstraction of the phenylthiyl radical from toluene is yet about 10 kcal/mol endothermic.

Colle and Lewis reported the absolute rates and Arrhenius parameters for the reactions of the triarylmethyl radicals with substituted phenylthiols (Reaction 31).[75]

$$Ph_3C\cdot + HS-C_6H_4-X \rightarrow Ph_3CH + \cdot S-C_6H_4-X \tag{31}$$

Some of the data are shown in Table 6. A modest correlation with Taft's σ_R suggests that the resonance in the produced-thiyl radicals is more important than the special polar transition state. The activation energies of the reactions increase with the electron-withdrawing substituents of the phenylthiols. This indicates that the phenylthiyl radicals with electron-releasing substituents are more stable than those with electron-withdrawing ones (Table 6). A similar tendency was found for the substituent effect of the decomposition rates of aryldiazothiolates (Reaction 24);[58] great stabilization of the phenylthiyl radical by the substitution of the electron-donating substituents such as 4-methoxy group is confirmed (Table 6).

3. Addition to Alkenes and Alkynes
a. Relative Rates

Addition of the thiyl radicals to alkenes is a typical reversible process; it is hard to determine the rate constants even in relative ones. By the same reason, many interesting phenomena have been found.

A ratio of the addition rates of one thiyl radical to a pair of alkenes (P) can be expressed by taking the reversible process into consideration.[1,76]

$$RS^{\cdot} + M_1 \underset{k_{-a1}}{\overset{k_{a1}}{\rightleftharpoons}} RS\text{-}M_1^{\cdot} \xrightarrow[+[RSH]]{k_{d1}} RSM_1H + RS^{\cdot} \qquad (32)$$

$$RS^{\cdot} + M_2 \underset{k_{-a2}}{\overset{k_{a2}}{\rightleftharpoons}} RS\text{-}M_2^{\cdot} \xrightarrow[+[RSH]]{k_{d2}} RSM_2H + RS^{\cdot} \qquad (33)$$

$$P = -d[M_1]/d[M_2] = \frac{k_{a1}k_{d1}(k_{-a2} + k_{d2}[RSH])}{k_{a2}k_{d2}(k_{-a1} + k_{d1}[RSH])} \qquad (34)$$

For addition to conjugated vinyl monomers such as substituted styrenes, which produce the stable benzyl-type radicals ($RS\text{-}CH_2\text{-}\overset{\cdot}{C}HPh$), the reverse rates are expected to be slow. Thus, Equation 34 is simplified into ordinary equation form.

$$P = k_{a1}/k_{a2} \qquad (35)$$

By this equation the substituent effects for styrene derivatives were found by Walling et al.[76] Then, Cadogan and Sadler,[77] and Church and Gleicher[78] confirmed the substituent effects. The Hammett plots against the σ^+ constants show linear correlations yielding negative ρ^+ values in the range of -0.2 to -0.4; the values are summarized in Table 9. The negative ρ^+-values suggest the polar transition states in which the electron (charge)-transfer from the olefinic double bond to the sulfur atom (**17a** and **17b**) accelerates the addition rates.

$$\left[\begin{array}{c} RS^{\cdot} \leftarrow \quad \begin{array}{c} CH_2 \\ \| \\ CHY \end{array} \end{array} \right] \longleftrightarrow \left[\begin{array}{c} RS^{-}, \quad \begin{array}{c} CH_2^{+\cdot} \\ \| \\ CHY \end{array} \end{array} \right]$$

The ρ^+ values obtained from the flash photolysis method are in good agreement with those obtained from the relative rates measurements, supporting above procedure to obtain the relative rates by neglecting the reversible process.[79] The ρ^+ value for α-methylstyrenes is more negative than that of styrenes, which is reasonable because of the higher electron density of the double bond of the former. When the electron-withdrawing substituents such as chloro-atom are substituted to the phenylthiyl radicals, the ρ^+ values become more negative than those of the unsubstituted phenylthiyl radicals, indicating an increase in the electrophilicity.

Sato and Otsu found an elegant method to determine the relative reactivities of various vinyl monomers toward the radicals by using ESR with the combination of the spin-trapping method.[80] They found that in the presence of the vinyl monomers the carbon-centered radicals ($PhSCH_2\text{-}\overset{\cdot}{C}HY$) were trapped with nitroso compound producing the adducts (**18**). Thus, the ratios of the concentrations of the adduct radicals yielded the relative addition rates of the thiyl radical to various vinyl monomers.

$$PhS^{\cdot} + CH_2\text{=}CXY \underset{k_{-1}}{\overset{k_1}{\rightleftharpoons}} PhSCH_2^{\cdot}CXY \xrightarrow[+[t\text{-}BuN\text{=}O]]{k_2} PhSCH_2\overset{\overset{\displaystyle X}{|}}{\underset{\underset{\displaystyle Y}{|}}{C}}\text{-}\overset{\overset{\displaystyle t\text{-}Bu}{|}}{N}\text{-}O^{\cdot}$$

$$\textbf{18} \qquad (36)$$

Table 9
HAMMETT ρ^+ VALUES FOR ADDITION REACTIONS OF THIYL RADICALS TO SUBSTITUTED STYRENES

RS	Styrenes	(Ref.)	α-Methyl-styrenes	(Ref.)	Stilbenes	(Ref.)
HO$_2$CCH$_2$S˙					−0.4	(77)
PhS˙			−0.38	(78)		
	−0.26	(79)	−0.30	(79)		
4-ClC$_6$H$_4$S˙	−0.56	(79)	−0.63	(79)		

Table 10
ADDITION RATES OF THIYL RADICALS TO ALKENES AND ALKYNES

	k_r(PhS˙)^a	k_{abs}(PhS˙)^b	k_{abs}(RS˙)^{c,d}
CH$_2$=CHPh	12.0	2.7 × 10^7	8.8 × 10^8
(CH$_2$=CH)$_2$			6.9 × 10^7
CH$_2$=C(Me)CO$_2$Me	(1.0)	5.4 × 10^6	
CH$_2$=CH(CN)	0.077	4.6 × 10^5	
CH$_2$=CH(OBu)	0.0058	1.8 × 10^5	
CH$_2$=CH[OC(O)Me]	0.0042	4.6 × 10^4	
CH$_2$=CHMe		1.5 × 10^4	7.0 × 10^7
CH$_2$=CH$_2$			4.8 × 10^5
CH≡CH			7.9 × 10^4

a   Reference 80.
b   4-ClC$_6$H$_4$S˙ (Reference 43).
c   BuS˙ (Reference 62).
d   MeS˙ (Reference 41).

The reverse process of Reaction 36 is neglected under the condition of $k_{-1} \ll k_2[t\text{-BuN}=O]$. Their relative rates are listed in Table 10 with the absolute values obtained from the flash photolysis and rotating sector methods.[43,81] In this table, one can find a tendency that conjugative monomers are more reactive than nonconjugative ones and that in each monomer group, the electron-rich double bonds are more reactive than the electron-poor ones. The rate constants for the phenylthiyl radicals obtained by the flash photolysis method are compatible to the relative ones from the ESR method. The slopes in the plots of the rate constants with e-values of vinyl monomers will be a measure of the polar character of the attacking radicals like the Hammett plots. Negative slopes are observed for the thiyl radicals and oxyradicals, suggesting the high electrophilicity of these radicals. The slopes have been evaluated for various thiyl radicals such as highly electrophilic benzoylthiyl radical,[82] modest 4-chlorobenzenethiyl radical,[43] and the less electrophilic 4-aminobenzenethiyl radical.[83]

b. Absolute Rates

It is interesting to compare the rate constants for the phenylthiyl radicals obtained from the flash photolysis method with those for the alkylthiyl radicals obtained with rotating-sector method (Table 10). The butanethiyl radical adds to styrene and 1-pentene more rapidly than the phenylthiyl radical by factors of 30 and 500, respectively; the former difference is smaller than the latter. This may be reasonable because the rate constant for the reaction system of the butanethiyl radical-styrene (8.8 × 10^8 M^{-1} sec^{-1}) is close to the diffusion-controlled limit. Both methods may yield reliable rate constants.

For nonconjugated alkenes the relative rate measurements are quite difficult since we have to use Equation 34. For a pair of cyclooctene and cyclohexene, the P value depends upon the concentrations of thiols, which acts as chain carriers; this suggests the high reversibility of addition step with cyclooctene.[84] The P value of this pair observed at extremely low temperature is most compatible with the rate constants observed by the flash photolysis method; at low temperature the reverse process may be frozen.[85]

By the ESR measurements, Lunazzi and Placucci found that the ESR signals attributable to the addition of the methylthiyl radical to cyclopentene (**19**) are observable at low temperature, whereas at higher temperature the ESR signals are replaced by the cyclopentenyl radical (**20**) produced by the hydrogen-atom abstraction.[86]

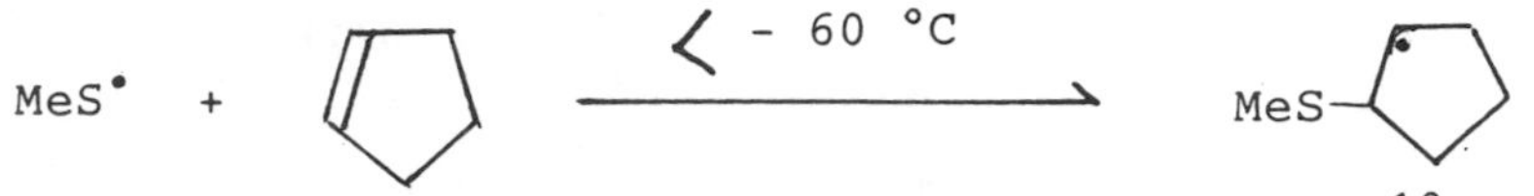

$$\text{MeS}^{\bullet} + \quad \xrightarrow{\quad < -60\ ^\circ\text{C} \quad} \quad \text{MeS} \underset{19}{\bigcirc}^{\bullet} \qquad (37)$$

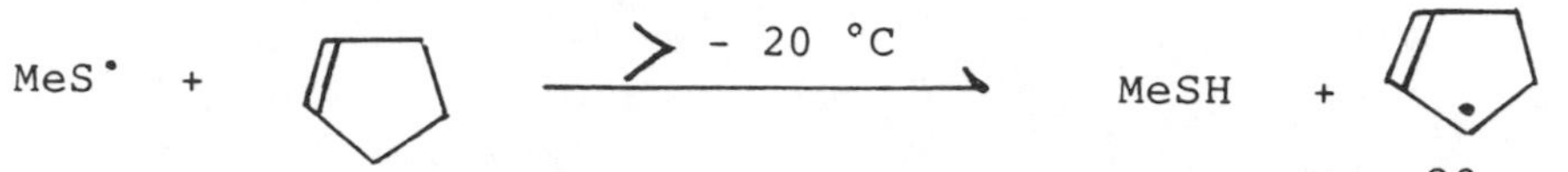

$$\text{MeS}^{\bullet} + \quad \xrightarrow{\quad > -20\ ^\circ\text{C} \quad} \quad \text{MeSH} + \underset{20}{\bigcirc}^{\bullet} \qquad (38)$$

This indicates that at low temperature the reverse process of the addition process is frozen and that at higher temperature although the addition occurs reversibly, the cyclopentenyl radical (**20**) is detectable because of the higher activation energies of the abstraction reaction and because of irreversible process.

c. Substituent Effects on Arylthiyl Radicals

One of the big problems in radical chemistry is to compare the reactivities among the attacking radicals, which is quite difficult by the relative rate measurements. For example, the substituent effect of the phenylthiyl radicals in the addition reaction to α-methylstyrene observed by the relative rate measurements is significantly different from that by absolute rate measurements (Reaction 39).[87,88] From the MO calculation (CNDO/2), Lee and Cheun indicated that the observed relative rates reflect the rates other than the addition process.[89]

$$\text{X–C}_6\text{H}_4\text{–S}^{\bullet} + \text{CH}_2=\text{CHY} \underset{k_{-1}}{\overset{k_1}{\rightleftharpoons}} \text{X–C}_6\text{H}_4\text{–S–CH}_2^{\bullet}\text{CHY} \qquad (39)$$

The substituent effects on the rate constants for above reaction observed by the flash photolysis are listed in Table 11.[88] The substituent effect looks to be greater than that expected from the polar nature of the transition state evaluated from the reactions for one thiyl radical with varying the substituents of styrene (Table 9). From the activation energy of the hydrogen-atom abstraction by the triarylmethyl radical from substituted thiols (Reaction 31), it is found that the stabilities of the arylthiyl radicals are strongly influenced by the substituents on the phenyl ring (Table 6). This finding is supported by the MO calculation.[89] Both methods reveal that the phenylthiyl radicals are stabilized by the electron-releasing substituents and vice versa for electron-withdrawing substituents. This tendency for the arylthiyl radicals is the same for the substituted phenoxy radicals reported by Mohany and Darooge,[90] who measured kinetics for hydrogen abstraction by the use of kinetic ESR technique.

The flash photolysis method gives the equilibrium constants as the relative values in addition to the absolute addition rates. The equilibrium constants are inversely proportional to the stabilities of the arylthiyl radicals in the reactants. They are summarized in Table 11.

Table 11
SUBSTITUENT EFFECTS OF X–C$_6$H$_4$–S˙ ON
ADDITION RATE CONSTANTS (k_1) AND
EQUILIBRIUM CONSTANTS (K) TO α-
METHYLSTYRENE (23°C) AND ON RESONANCE
ENERGIES (ΔE$_R$) AND K FOR PHENOXYRADICALS
(60°C)

X	$k_1{}^a$ (M^{-1} sec^{-1})	K^a (M^{-1})	ΔE$_R{}^b$ (kcal/mol)	K × 10^{-3c}
4-OMe	3.3×10^6	0.1	-9.9	0.079
4-Me	1.6×10^7	0.5	-0.6	
4-t-Bu	1.6×10^7	0.5		3.1
H	7.1×10^7	2	0.0	44.0
4-Cl	1.4×10^8	4	5.4	
3-CO$_2$Et				500
3-CF$_3$			11.5	

ᵃ   Reference 88.
ᵇ   Reference 89.
ᶜ   Reference 90; equilibrium constants for the reaction PhO˙ + XC$_6$H$_4$OH
    ⇌ PhOH + XC$_6$H$_4$O˙.

The arylthiyl radicals with electron-releasing substituents are more stable than the electron-withdrawing substituents. This tendency is in good agreement with the results obtained by the other methods described above.

The substituent effects with varying X under constant Y in Reaction 39 can be represented by Equation 40 which is derived by taking both linear free-energy relationship and charge-transfer in the transition state (CTTS) into consideration.[91]

$$\Delta \log k_1 = \alpha \Delta \log K + \gamma \Delta(\text{CTTS}) \tag{40}$$

In the form of reaction constants of the Hammett equation, Equation 41 can be derived.

$$\rho^+(k_1) = \alpha \rho^+(K) + \gamma \rho^+(\text{CTTS}) \tag{41}$$

The sign of the proportional constant α is positive, since the activation energies for addition reactions decrease with the electron-withdrawing substituents on the phenylthiyl radicals in the reactant. With the electron-withdrawing substituents on the phenylthiyl radicals, the participation of the CTTS from olefinic double bond to the thiyl radical increases. This causes a decrease in the barriers of the transition state. This is the same trend of the linear-free energy (LFE) relationship.

In Table 12, the reported ρ^+ values are summarized. The observed ρ^+ values can be reasonably interpreted by the above theory; i.e., large positive ρ^+ value in the reactivities for the reaction system of arylthiyl radicals — α-methylstyrene ($\rho^+ = 1.72$) comes from two factors; one is large Δ log K [ρ^+ (K)] and another is a large positive γ value. The α values may vary with the reactivity-selectivity principle and the γ values may be dependent upon the electron-density of the double bond of alkenes. With changing alkenes (in Table 12), the ρ^+ values vary mainly with the e-values of vinyl monomers, but not with the Q values. This finding indicates that the α values are not variable within alkenes studied and that γ values depend upon the amount of the contribution of CTTS. If the α value can be evaluated, more detailed knowledge may be given; i.e., for the electron-poor double bond

Table 12
HAMMETT ρ^+ VALUES FOR $X-C_6H_4-S^{\cdot}$ +
$CH_2=CRY$[91]

R	Y	ρ^+	e-Value	Q value
–Me	–Ph	1.72	– 1.27	0.98
–H	–Ph	1.37	– 0.8	1.0
–Me	–CO$_2$Me	1.05	0.40	0.74
–H	–CN	0.46	1.20	0.60
–H	–OBu	1.43	– 1.77	0.023
–H	–OC(O)Me	1.31	– 0.22	0.026

such as acrylonitrile, the direction of the charge-transfer cannot be determined. There remains an important problem.

d. Solvent Effects

It is expected that the rates of the addition reactions of the thiyl radicals are influenced by solvents, since the transition states are polar and the thiyl radicals are highly electrophilic. Kuntz et al. found that the absorption and emission maxima of the 4-aminobenzenethiyl radical change with the solvent dipolarity. From the observed large bathochromic shifts, the dipole moment for the ground and excited states are evaluated to be 4.3 and 7.3 D, respectively.[92]

$$H_2N-\!\!\left\langle\!\!\bigcirc\!\!\right\rangle\!\!-S^{\cdot} \quad \longleftrightarrow \quad H_2\overset{+}{N}\!\!=\!\!\left\langle\!\!\bigcirc\!\!\right\rangle\!\!=S^{\bar{\cdot}}$$

In the rate parameters such as addition rate constants and equilibrium constants, large variations were observed with the solvent polarity.[93,94] In the case of the 4-dimethylaminobenzenethiyl radical, linear correlations were found in the plots of these rate parameters with Kirkwood's solvent-polarity parameter. From the slopes of the plots, the dipole moments of the transition state can be evaluated. In going from the reactant to the transition state, a prominent reduction of the dipole moment is caused by polar solvents, which decreases the solvation-stabilization of the transition state resulting in a retardation of the addition rates in polar solvents. Such reduction in the dipole moment of the transition state suggests the presence of CTTS, which directs from alkene to the thiyl radical (Reaction 42). This direction of CTTS is the same as that presumed by the substituent effects described above.[93-95]

$$R_2N-\!\!\left(\!\!\bigcirc\!\!\right)\!\!-S^{\cdot} + \begin{matrix} CH_2 \\ \| \\ CHY \end{matrix} \rightleftharpoons \left[R_2N-\!\!\left(\!\!\bigcirc\!\!\right)\!\!-\dot{S}\!\leftarrow\!\begin{matrix} CH_2 \\ \| \\ CHY \end{matrix}\right] \rightleftharpoons R_2N-\!\!\left(\!\!\bigcirc\!\!\right)\!\!-S-CH_2\dot{C}HY \quad (42)$$

$\mu=4.3$ D　　　　　　　　$\mu=2.6$ D

For other arylthiyl radicals, the effects of solvent-polarity were not observed in the absorption maxima, indicating that the dipole moments of the arylthiyl radicals except the amino-substituted radicals are small. For such nonpolar radicals, it is expected that the polar transition states caused by CTTS may accelerate the addition rates in polar solvents. However, significant solvent effect was not observed; some problems to be solved may remain.

For such ordinary thiyl radicals, however, the effects of solvent-polarity were observed for the addition reaction to the polar double bonds such as nitrones (**21**) which have dipole moments of approximately 3 D.[96] These rates decrease in the polar solvents, suggesting that

CTTS exists in the direction so as to reduce the dipole moment of nitrones (i.e., **22**). In this case, solvents having hydrogen-bonding ability also retard the addition rates to great extent.

$$\text{(43)}$$

Observed CTTS implies high electrophilicities of the thiyl radicals to the olefinic double bond. Thus it is anticipated that the addition rates are retarded in solvents having the π-electron-donor ability. When the electrophilicity of the thiyl radical is enhanced by the electron-withdrawing substituents such as 4-chlorobenzenethiyl radical, the sharp absorption peak in the visible region in the noninteracting solvent is broadened in the π-electron donor solvent such as mesitylene and benzene, suggesting weak charge-transfer interaction between the radical center and the aromatic π-systems. This interaction interferes to approach alkenes. The addition rates of the thiyl radical to styrenes in mesitylene is about a half of those in cyclohexane.[97]

e. Addition to Alkynes

Many interesting stereospecific syntheses were known in the thiyl radical addition to alkynes, however, kinetic studies were quite few. Recently, the rate constants were measured by the flash photolysis method.[98] In Table 13, some of the addition rate constants are summarized. By the comparison with the corresponding alkenes, the structures of the adduct vinyl-type radicals can be elucidated. Alkyl-substituted alkyne, which shows a similar reactivity to alkyl-substituted alkene, forms the σ type (sp²-orbital) vinyl radical, although the orbital in which the unpaired electron exists different from 2p-orbital, the stabilities of the adduct-radicals are similar as far as they are localized. The rate constant for alkyne substituted by the ester group is smaller than that of the corresponding alkene by a factor of 1/300. This indicates that the unpaired electron of the vinyl radical does not conjugate with the π-bond of the ester group, whereas the electron in $2p_z$ of the radical derived from alkene is stabilized by the conjugation with the carbonyl π-bond. This consideration leads to the conclusion that this vinyl radical is σ-radical, however, this is not in agreement with the presumption from the ESR and MO studies.

$$\text{(44)}$$

Reactivity of phenylacetylene suggests that the produced-vinyl radical is in the middle between the π- and σ- character. Further investigation needs to conclude this problem.

B. Sulfinyl Radicals

Characteristic of the sulfinyl radicals is their extraordinary stabilities; α-disulfoxides (**24**) are not detectable under the ordinary conditions, whereas disulfides and α-disulfones are known to be stable.[4] Only when thiosulfinates (**23**) were oxidized at low temperature, α-disulfoxides (**24**) could be detected with ¹H- and ¹³C-NMR (Reaction 45).[99]

Table 13
THE RATE CONSTANTS (M^{-1} sec^{-1}) FOR ADDITION
OF PhS˙ TO ALKYNES AND ALKENES AT 24°C[98]

Alkyne	k_1	Alkene	k_1
HC≡C–R	2.0×10^4	H$_2$C=CH–R	1.0×10^4
HC≡C–CO$_2$Me	8.3×10^3	H$_2$C=CH–CO$_2$Me	2.7×10^5
HC≡C–Ph	7.9×10^5	H$_2$C=CH–Ph	2.0×10^7

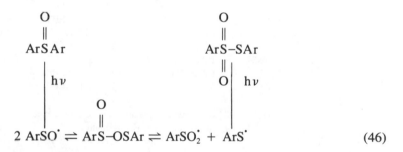

$$\underset{\textbf{23}}{\overset{\overset{\textstyle O}{\|}}{R\text{–}S\text{–}S\text{–}R}} \xrightarrow{\text{[O]}} \underset{\textbf{24}}{[\overset{\overset{\textstyle O}{\|}\;\overset{\textstyle O}{\|}}{R\text{–}S\text{–}S\text{–}R}]} \to \underset{\textbf{25}}{\overset{\overset{\textstyle O}{\|}}{R\text{–}S\text{–}O\text{–}SR}} \to \underset{\textbf{26}}{\overset{\overset{\textstyle O}{\|}}{\underset{\underset{\textstyle O}{\|}}{R\text{–}S\text{–}S\text{–}R}}} \tag{45}$$

Interesting findings about the intermediate radicals produced by the photolysis of sulfoxides and thiosulfonates (**26**) were reported by Chatgilialoglu et al.[100] They measured the ESR spectra in the absence and in the presence of spin-trapping reagents (Reaction 46).

$$2\ ArSO^{\cdot} \rightleftharpoons \overset{\overset{\textstyle O}{\|}}{ArS\text{–}OSAr} \rightleftharpoons ArSO_2^{\cdot} + ArS^{\cdot} \tag{46}$$

Without the spin-trapping reagent, the sulfinyl radical is the only one species detectable with ESR from both starting radical sources. This finding indicates that the stability of the phenylsulfinyl radical is quite higher than those of the phenylthiyl and sulfonyl radicals. On the other hand, in the presence of spin traps, the latter two radicals are detectable as spin adducts. The sulfinyl radical is not trapped by spin-trapping reagents such as nitroso compounds. This is caused by the difference in the addition reactivities among three radicals. The arylthiyl radicals react with 2-nitrosopropane (*t*-BuNO) and nitrones (**21**) with the rate constants of about $10^7\ M^{-1}$ sec^{-1}.[96] The sulfonyl radicals are also known to be reactive to the double bond. On the other hand, it is revealed from this observation that the sulfinyl radicals are unreactive to the spin-trapping reagents having quite reactive double bonds.

Iino and Matsuda[101] observed the isomerization of deuterio-styrene during the decomposition of sulfinyl-radical sources. The rapid addition-elimination mechanism of the sulfinyl radical to styrene was proposed. Although it is possible to consider the participation of the thiyl and sulfonyl radicals, which may be present as an equilibrium with the sulfinyl radicals as shown in Reaction 46, this possibility was excluded because the polymerization of styrene which will be initiated by the sulfonyl radical was not recognized. A more detailed kinetic study for the reactivities of the sulfinyl radicals is desired.

ESR study for the phenylsulfinyl radical indicates that the unpaired electron exists mainly in the S–O π-orbital which conjugates with the phenyl π-system.[102] The hfcs for various phenyl substituted radicals are listed in Table 14.[103] The hfcs at *para*- and *ortho*-hydrogens

Table 14
HYPERFINE COUPLING CONSTANTS (IN GAUSS) FOR PHENYL SUBSTITUTED SULFUR RADICALS AND RELATED RADICALS

	PhSO˙ [a]	PhSO$_2$˙ [b]	PhCH$_2$˙ [c]	PhCO˙ [d]
a_H^{ortho}	2.5	1.06	5.1	>0.1
a_H^{meta}	0.7	0.33	1.6	1.16
a_H^{para}	2.5	0.50	6.3	>0.1

[a] Reference 102.
[b] Reference 33.
[c] Reference 36.
[d] Reference 103.

Table 15
RECOMBINATION RATE CONSTANTS (IN UNITS OF $M^{-1}\ sec^{-1}$)

Radical	k_r	Temp. (°C)	Method	Ref.
t-BuSO˙	6.0×10^7	−100	KEPR	104
R$_2$NSO˙	1.1×10^9	−100	KEPR	107
PhCO˙	1.0×10^9	0	KEPR	106
PhS˙	1.3×10^9	23	KAB	50
t-BuSS˙	2.0×10^8	−83	KEPR	105
R$_2$NS˙	1.3×10^9	−100	KEPR	107
RSO$_2$˙	4.6×10^9	23	KEPR	108

Note: KEPR, kinetic electron paramagnetic resonance method; KAB, kinetic absorption spectroscopy method.

of the phenylsulfinyl radical are greater than that at *meta*-hydrogens. This trend is rather similar to the benzyl and phenoxy radicals, whereas the benzoyl radical, a typical σ-radical, exhibits a larger hfc at *meta*-hydrogens than those at *para*- and *ortho*-hydrogens. On this basis, the phenylsulfinyl radical can be assigned as a π-type radical.

Although the sulfinyl radicals can be produced by the direct fission of the bond by photolysis and pyrolysis,[102] indirect methods are also useful to produce the sulfinyl radicals. The reactions with the hydroxyradical produced by the Ti^{III}-H$_2$O$_2$ system and radiolysis in aqueous solution are especially efficient.[102] For example, ˙OH reacts with RSSR yielding RSO˙ and RSH. At first, the adduct radical [RS˙(OH)SR] is formed and then it dissociates into sulfenic acid (RSOH) and RS˙ followed by the hydrogen-atom transfer from the former to the latter.

Other kinetic data which show low reactivities of the sulfinyl radicals are found in the recombination rate constants. In Table 15, the recombination rate constants for the radicals including the sulfur atom and the related radicals are listed. These data are mainly determined with the kinetic ESR method, which is the most reliable method for this purpose.[103-108] The recombination rate of t-BuṠ=O is considerably slower than the diffusion-controlled limit. The rate of R$_2$N−Ṡ=O is fast; the lone pair electrons on the nitrogen atom may neutralize the positively charged sulfur atom of the $^+$S−O$^-$ bond, which results in an acceleration of the recombination rate. The perthiyl radical is just below the diffusion-controlled limit; similarity in the electronic structures between the sulfinyl and perthiyl radicals is suggested.

Table 16

**HALOGEN ABSTRACTION RATE CONSTANTS (M^{-1} sec^{-1})
OF Et$_3$Si˙ FROM SULFONYL HALIDES AT ROOM
TEMPERATURE**

RX[a]	$k \times 10^{-9}$	RX[a]	$k \times 10^{-6}$	RX[b]	$k \times 10^{-5}$
MeSO$_2$Cl	3.18	MeSO$_2$F	13.0	MeCl	1.2
PhSO$_2$Cl	4.56	PhSO$_2$F	8.9	PhCH$_2$Cl	200
PhCH$_2$SO$_2$Cl	5.73				

[a]   Reference 108.
[b]   Reference 110.

The sulfenic acids (RSOH) have been found to be active radical scavengers; the rate constants for the reaction of alkyl sulfinic acid with peroxyradical (Reaction 47) were reported at least $10^7 M^{-1}$ sec^{-1}. This rate constant is greater than that of phenylthiol with peroxyradical (Table 8). The high efficiencies of sulfenic acids as hydrogen-atom transfer agents are a consequence of the appreciable stabilities of the sulfinyl radicals.[109]

$$\text{ROO}^˙ + \text{RSOH} \xrightarrow{\;>10^7\,M^{-1}\text{sec}^{-1}\;} \text{ROOH} + \text{RSO}^˙ \qquad (47)$$

C. Sulfonyl Radicals

Kinetic study for the sulfonyl radicals using laser-flash photolysis was recently reported by Chatgilialoglu et al.[108] The halogen abstraction rate constants from sulfonyl halides by the triethylsilyl radical are evaluated (Table 16).

$$\text{Et}_3\text{Si}^˙ + \text{RSO}_2\text{X} \rightarrow \text{Et}_3\text{SiX} + \text{RSO}_2^˙ \qquad (48)$$

Both for sulfonyl chlorides having rate constants close to diffusion-controlled limit and for less reactive sulfonyl fluorides, the difference in the reactivities between alkylsulfonyl halides and phenylsulfonyl halides is small compared with that between alkyl and benzyl halide.[110] This indicates that the sulfur-halogen bond strength is not affected by the nature of the alkyl and phenyl substituents. In the phenylsulfonyl radical in the products of Reaction 48, the unpaired electron on the sulfonyl moiety does not delocalize onto even the neighboring aromatic ring. This implies the σ-radical character of the phenylsulfonyl radical.

The ESR parameters afford an evidence that the sulfonyl radicals are pyramidal with respect to the sulfur atom. Rotation about the C–S bond is hindered as indicated by nonequivalent hfcs of two *ortho*-hydrogens in the *para*-substituted phenylsulfonyl radicals.[33,106] The absorption maxima of the optical absorption spectra of the arylsulfonyl radicals observed by xenon-flash[32] and laser-flash photolysis (ca. 340 nm)[108] are the same to those of the alkylsulfonyl radicals observed by pulse radiolysis.[112] This also supports the idea that the unpaired electron localizes mainly within the sulfonyl moiety and that the adjacent phenyl π-system does not appreciably interact with the radical center.

In this respect, it is expected that the substituent effect of the arylsulfonyl radicals is also small. This was confirmed by Kobayashi et al. in the dissociation rates of the azosulfones (Reaction 49).[113]

$$\text{Ph–N=N–S(O)}_2\text{Ar} \rightarrow \text{Ph}^˙ + \text{N}_2 + \text{ArSO}_2^˙ \qquad (49)$$

In the halogen-abstraction reactions by the carbon-centered radicals from substituted phenylsulfonyl halides, the substituent effect was quite small.[114]

$$R^{\cdot} + ArSO_2X \rightarrow RX + ArSO_2^{\cdot} \tag{50}$$

Although a considerable difference in the stabilities between the alkylsulfonyl and phenylsulfonyl radicals is found for the dissociation of δ-valerophenones as shown in Table 4, for other reactions such as halogen-donor abilities and addition abilities, the similarity of both sulfonyl radicals has been emphasized.

A considerable substituent effect was observed in the electrophilicities of the arylsulfonyl radicals. On the addition reactions of each arylsulfonyl radical to substituted styrenes, the negative ρ^+ value was observed. With changing substituents on the phenylsulfonyl-radical side, the ρ^+ values vary as shown below.[115]

$$4\text{-MeO--}C_6H_4SO_2^{\cdot} + \text{Sub. styrenes} \ (\rho^+ = -0.35) \tag{51}$$

$$4\text{-Me--}C_6H_4\text{--}SO_2^{\cdot} + \text{Sub. styrenes} \ (\rho^+ = -0.50) \tag{52}$$

$$3\text{-NO}_2\text{--}C_6H_4\text{--}SO_2^{\cdot} + \text{Sub. styrenes} \ (\rho^+ = -0.88) \tag{53}$$

Electrophilic nature of the arylsulfonyl radicals varies with substituents, thus, it is expected that in the reaction where the polar arylsulfonyl radicals play important roles in the transition state, some appreciable substituent effects can be found. The ρ^+ values observed above are only slightly more negative than the arylthiyl radicals (-0.26 to -0.56 as shown in Table 9). The polarization of the S–O bond does not so much alter the unpaired electron density of the sulfur atom as would be anticipated from neutral sulfone molecules.

The measurements of the absolute rate constants for the loss of SO_2 from the sulfonyl radicals have been tried by the laser-flash photolysis.[108,116] In the case of the benzylsulfonyl radical, the rate for the cleavage of the S–C bond is so fast that the sulfonyl radical cannot be detected even immediately after the laser pulse.

$$PhCH_2SO_2^{\cdot} \rightarrow Ph\dot{C}H_2 + SO_2$$

$$k > 10^8 \ sec^{-1} \quad \text{at} \quad 23°C$$

$$k > 2 \times 10^6 \ sec^{-1} \quad \text{at} \quad -83°C \tag{54}$$

The lower limits of the rate constants were evaluated from the laser pulse duration (20 nsec) or time required for generation of the sulfonyl radical (300 nsec). The activation energy for this reaction is estimated to be lower than 5.0 kcal/mol. The rate for the loss of SO_2 is faster than that for the loss of CO from $PhCH_2CO^{\cdot}$ ($6 \times 10^6 \ sec^{-1}$ at room temperature). When the alkyl radicals are unstable, the rates of the dissociation are quite slow at room temperature. The equilibrium constant for Reaction 54 was estimated to be $9 \times 10^3 \ M$ at 150°C for R = Me.[117]

The reverse process of Reaction 54 is the radical-trapping reaction by SO_2.

$$R^{\cdot} + SO_2 \rightarrow RSO_2^{\cdot} \tag{55}$$

Gas-phase data show that the methyl and ethyl radicals react with the rate constants of 5×10^6 and $5 \times 10^5 \ M^{-1}sec^{-1}$ at 25°C, respectively, which are considerably slower than those with oxygen ($>10^9 \ M^{-1}sec^{-1}$).[118] By flash photolysis in gas phase, the rate constant of $1.8 \times 10^8 \ M^{-1}sec^{-1}$ at 22°C was reported by James et al.[119] They pointed out that the kinetic behavior is similar to the reaction system of the methyl radical with oxygen; some ambiguous points may remain.

The rate constant for the addition reaction of the *t*-butoxy radical to sulfur dioxide was estimated by the ESR method.[33]

$$t\text{-BuO}^\bullet + SO_2 \xrightarrow[10^7 \text{ to } 10^8 \ M^{-1}\text{sec}^{-1}]{} t\text{-BuO–SO}_2^\bullet \tag{56}$$

The rate constant obtained from the competition method is greater than those of the alkyl radicals, thus, the rate constants for Reaction 55 lower than 10^7 to $10^8 \ M^{-1} \ \text{sec}^{-1}$ may be reasonable.

Direct measurement of the absolute rate constant for the reaction of peroxyradical ($ROO^\bullet$) with sulfur dioxide was reported by Calvert et al. with using kinetic absorption spectroscopy.[120] The rate constant $6.4 \times 10^6 \ M^{-1} \ \text{sec}^{-1}$ at 23°C was obtained for this reaction. Both the O-atom transfer path (Reaction 57) and addition path (Reaction 58) are considered.

$$MeOO^\bullet + SO_2 \xrightarrow{6.4 \times 10^6 \ M^{-1}\text{sec}^{-1}} MeO^\bullet + SO_3 \tag{57}$$

$$\rightarrow MeOOSO_2^\bullet \tag{58}$$

It is hard to investigate quantitatively the hydrogen-atom abstraction abilities of the sulfonyl radicals, since the produced sulfinic acids (RSO_2H) are quite unstable. The rate constants for Reaction 59 were measured by Horowitz et al. by using liquid-phase γ-radiolysis for free-radical chain reactions.[121] One of values is shown below.

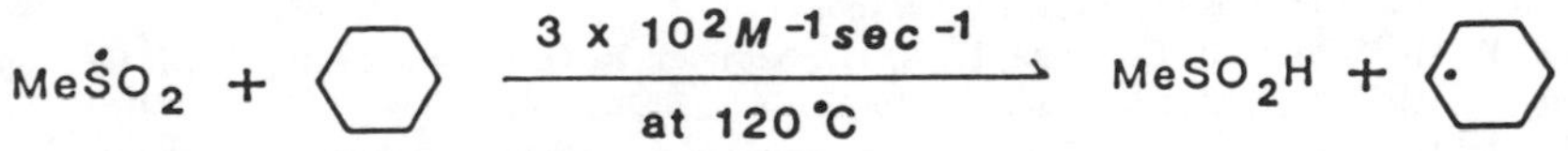

$$Me\dot{S}O_2 + \bigcirc \xrightarrow[\text{at } 120\,^\circ C]{3 \times 10^2 M^{-1} sec^{-1}} MeSO_2H + \langle\cdot\rangle \tag{59}$$

Hydrogen donor abilities of sulfinic acids are quite high. The rate constants for Reaction 60 reported by Gilbert et al. are greater than those of the opposite process (Reaction 59) by a factor of approximately 10^3.[122] The rate constants of approximately $10^6 \ M^{-1} \ \text{sec}^{-1}$ are about 10^2 smaller than those of thiols (Table 8).[122]

$$Me^\bullet + MeSO_2H \xrightarrow{10^6 \ M^{-1}\text{sec}^{-1}} MeH + MeSO_2^\bullet \tag{60}$$

The Cl-atom transfer reaction by the methyl radical (Reaction 61) is slower than by triethylsilyl radical (Reaction 48) by a factor of 10^5.[119]

$$Me^\bullet + MeSO_2Cl \xrightarrow[2 \times 10^4 \ M^{-1}\text{sec}^{-1} \text{ at } 150°C]{} MeCl + MeSO_2^\bullet \tag{61}$$

Recombination rate constants of the alkyl and phenylsulfonyl radicals were measured with kinetic ESR spectroscopy (Table 15),[108] which are close to the rate constants of the diffusion controlled-process. The polarization of the S–O bonds in the sulfone group does not inhibit the proximity of two sulfonyl radicals. This may be related to the pyramidal structure.

REFERENCES

1. **Walling, C.**, *Free Radicals in Solution,* Wiley, New York, 1957.
2. **Pryor, W. A.**, *Mechanisms of Sulfur Reactions,* McGraw Hill, New York, 1962.
3. **Kellog, R. M.**, Thiyl radicals, in *Method in Free-Radical Chemistry,* Huyser, R. M., Ed., Marcel Dekker, New York, 1969, chap. 1.
4. **Kice, J.**, Sulfur-centered radicals, in *Free Radicals,* Kochi, J. K. Ed., John Wiley, New York, 1972, chap. 24.
5. **Block, E.**, *Reactions of Organic Compounds,* Academic Press, New York, 1978, chap. 5.
6. **Ohno, A. and Ohnishi, Y.**, Resonance participation of sulfur and oxygen in radicals. Decomposition of azobis(2-propane) derivatives, *Tetrahedron Lett.,* 50, 4405, 1969.
7. **Bernard, F., Epioties, N. D., Cherry, W., Schlegel, H. B., Whangbo, M.-H., and Wolfe, S.**, A molecular orbital interpretation of the static, dynamic, and chemical properties of CH_2X radicals, *J. Am. Chem. Soc.,* 98, 469, 1976.
8. **Price, C. C., and Oae, S.**, *Sulfur Bonding,* Ronald Press, New York, 1962.
9. **Bridger, R. F. and Russell, G. A.**, Directive effects in the attack of phenyl radicals on carbon-hydrogen bonds, *J. Am. Chem. Soc.,* 85, 3754, 1963.
10. **Ueyama, K., Namba, H., and Oae, S.**, Hydrogen abstraction reactions of α-heteroatom substituted compounds by *t*-butoxy radical, *Bull. Chem. Soc. Jpn.,* 41, 1928, 1968.
11. **Scaiano, J. C. and Stewart, L. C.**, Phenyl radical kinetics, *J. Am. Chem. Soc.,* 105, 3609, 1983.
12. **Paul, H., Small, R. D., and Scaiano, J. C.**, Hydrogen abstraction by *t*-butoxy radicals. A laser photolysis and ESR study, *J. Am. Chem. Soc.,* 100, 4520, 1978.
13. **Alfrey, T. and Price, C. C.**, Relative reactivities in vinyl copolymerization, *J. Polymer Sci.,* 2, 101, 1947.
14. **Brandrup, J. and Immergut, E. H.**, *Polymer Handbook,* Wiley, New York, 1975.
15. **Tagaki, W., Tada, T., Nomura, R., and Oae, S.**, 3d-Orbital resonance divalent sulfides. XIV, *Bull. Chem. Soc. Jpn.,* 41, 1696, 1968.
16. **Otsu, T., Tsuda, K., Fukunaga, T., and Inoue, H.**, Preparation and polymerization of *bis*(phenylthio)ethanone, *J. Chem. Soc. Jpn.,* 89, 72, 1968.
17. **Ito, O., Furuya, S., and Matsuda, M.**, Rates of addition of phenylthio radicals to vinyl sulfide, sulphoxide, sulphone,and ether groups, *J. Chem. Soc. Perkin Trans.,* 2, 139, 1984.
18. **Thyrion, F. C.**, Photolysis of aromatic sulfur molecules, *J. Phys. Chem.,* 77, 1478, 1973.
19. **Carton, P. M., Gilbert, B. C., Laue, H. A. H., Norman, R. O. C., and Sealy, R. C.**, Electron spin resonance studies. XIVII, *J. Chem. Soc. Perkin Trans.,* 2, 1245, 1975.
20. **Biddles, I., Hudson, A., and Wiffen, J. T.**, A comparative study of some oxygen and sulfur substituted alkyl radicals, *Tetrahedron,* 28, 867, 1972.
21. **Griller, D., Nonhebel, D., and Walton, J. C.**, An electron spin resonance study of 1-alkylthioallyl, 3-allylthiopropynyl, and alkylthioalkyl radicals, *J. Chem. Soc. Perkin Trans. 2,* 1817, 1984.
22. **Hass, A., Schlosser, K., and Steenken, S.**, C–C bond homolysis in $(CF_3S)_3C-C(SCF_3)_3$ at room temperature, *J. Am. Chem. Soc.,* 101, 6282, 1979.
23. **Schlosser, K. and Steenken, S.**, Reversible N–N bond homolysis in $(CF_3S)_3N-N(SCF_3)_3$ at room temperature, *J. Am. Chem. Soc.,* 105, 1504, 1983.
24. **Forrest, D. and Ingold, K. U.**, An electron paramagnetic resonance study of free-radical addition to trithiocarbonates and of the formation and destruction of tetrafulvalen by free-radical process, *J. Am. Chem. Soc.,* 100, 3868, 1978.
25. **Miura, Y. and Kinoshita, M.**, Electron spin resonance study of *N*-(arylthio)alkylaminyl radicals generated by photolysis of *N,N-bis* (arylthio)alkylamines, *J. Org. Chem.,* 49, 2724, 1984.
26. **Schafer, K., Bonifacic, M., Bahnemann, D., and Asmus, K.-D.**, Addition of oxygen to organic sulfur radicals, *J. Phys. Chem.,* 82, 2777, 1978.
27. **Maillard, B., Ingold, K. U., and Scaiano, J. C.**, Rate constants for the reactions of free radicals with oxygen in solution, *J. Am. Chem. Soc.,* 105, 5095, 1983.
28. **Ueno, Y., Miyano, T., and Okamura, M.**, A new method for the evaluation of the relative stability of organosulfur radicals by competitive elimination technique, *Tetrahedron Lett.,* 23, 443, 1982.
29. **Ito, O. and Matsuda, M.**, Evaluation of addition rates of the thiyl radicals to vinyl monomers by flash photolysis. II, *J. Am. Chem. Soc.,* 101, 5732, 1979.
30. **Ito, O., Nogami, K., and Matsuda, M.**, Flash photolysis study on initiation of radical polymerization, *J. Phys. Chem.,* 86, 1365, 1981.
31. **Wagner, P. J., Sedon, J. M., and Lindstrom, M. J.**, Rates of radical cleavage in photogenerated diradicals, *J. Am. Chem. Soc.,* 100, 2579, 1978.
32. **Thoi, H. H., Ito, O., Iino, M., and Matsuda, M.**, Studies for sulfonyl radicals. IV, *J. Phys. Chem.,* 82, 314, 1978.

33. **Davies, A. G., Robert, B. P., and Sanderson, B. R.,** An electron spin resonance study of some sulphonyl radicals in solution, *J. Chem. Soc. Perkin Trans. 2,* 626, 1973.
34. **Booth, T. E., Greene, J. L., Shevlin, P. B.,** Stereochemistry of free-radical elimination β-phenylsulfonyl radicals, *J. Org. Chem.,* 45, 794, 1980.
35. **Nishino, J. and Yamada, M.,** The reaction of *cis*-2,3-dimethylthiirane with hydrogen atoms in the gas phase, *J. Chem. Soc. Jpn. p.* 1735, 1982.
36. **Wayner, D. D. M. and Arnold, D. R.,** Substituted effects on benzyl radical hyperfine coupling constants. II, *Can. J. Chem.,* 62, 1164, 1984.
37. **Carrock, F. and Szwarc, M.,** Methyl affinity of substituted styrene, their homologs and analogs, *J. Am. Chem. Soc.,* 81, 4138, 1959.
38. **Burkey, T. J., Griller, D., Lunazzi, L., and Nazran, A. S.,** Homolytic substitution at furan and thiophene, *J. Org. Chem.,* 48, 3704, 1983.
39. **Wine, P. H. and Thompson, R. J.,** Kinetics of OH reactions with furan, thiophene and THF, *Int. J. Chem. Kinet.,* 16, 867, 1984.
40. **Oswald, A. A. and Wallance, T. J.,** Anionic oxidation of thiols and cooxidation of thiols with olefins, in *Organic Sulfur Compounds,* Vol. 2, Kaharash, N. and Meyer, C. Y., Eds., Pergamon Press, Oxford, 1966, chap 8.
41. **Graham, D. M. and Soltys, J. F.,** Photo-induced reaction of thiols and olefins. V, *Can. J. Chem.,* 48, 2173, 1970.
42. **Gilbert, B. C., Kelsall, P. A., Sexton, M. D., McConnacchie, G. D. G., and Symons, M. C. R.,** Electron resonance studies of the photolysis and reaction of *O,O'*-dialkylhydrogen phosphorodithioates, their salts, corresponding disulfides, and related compounds, *J. Chem. Soc. Perkin Trans. 2,* 629, 1984.
43. **Ito, O. and Matsuda, M.,** Evaluation of addition rates of *p*-chlorobenzenethiyl radical to vinyl monomers by means of flash photolysis, *J. Am. Chem. Soc.,* 101, 1815, 1979.
44. **Morine, G. H. and Kuntz, R. R.,** Observations of C–S bond and S–S bond cleavage in the photolysis of disulfides in solution, *Photochem. Photobiol.,* 33, 1, 1981.
45. **Rosenfeld, S. M., Lawler, R. G., and Ward, H. R.,** Photo-CIDNP from carbon-sulfur cleavage of alkyl disulfides, *J. Am. Chem. Soc.,* 94, 9255, 1972.
46. **Bonifacic, M. and Asmus, K.-D.,** Adduct formation and absolute rate constants in the displacement reaction of thiyl radicals with disulfides, *J. Phys. Chem.,* 88, 6286, 1984.
47. **Pyror, W. A. and Smith, K.,** The mechanism of the substitution reaction on sulfur atoms by radicals or nucleophiles, *J. Am. Chem. Soc.,* 92, 2731, 1970.
48. **Suama, M. and Takazaki, Y.,** Reaction of methyl radicals with dimethyl disulfide, *Bull. Inst. Chem. Rev. Kyoto Univ. Jpn.,* 40, 229, 1962.
49. **Spanswick, J. and Ingold, K. U.,** Tri-*n*-butyltin hydride with disulfides, *Int. J. Chem. Kinet.,* 11, 157, 1970.
50. **Burkey, T. J. and Griller, D.,** Miceller systems as derived for enhancing the lifetimes and concentration of free radicals, *J. Am. Chem. Soc.,* 107, 246, 1985.
51. **Feher, F., Gladden, T., and Kunz, D.,** Spectroscopic investigation of sulfur radicals at low temperature, *Z. Naturforsch.,* 25b, 1215, 1970.
52. **Adam, F. C. and Elliot, A. J.,** Intermediates in the photolysis of alkyl sulfides and disulfides in dilute glass matrices, *Can. J. Chem.,* 55, 1546, 1977.
53. **Adams, G. E., Armstromg, R. C., Charlesby, A., Michael, D. E., and Wilson, R. L.,** Pulse radiolysis of sulfur compounds. III., *Trans. Faraday Soc.,* 65, 732, 1969.
54. **Packer, J. E.,** The radiation chemistry of thiols, in *The Chemistry of The Thiol Group,* Patai, S. Ed., John Wiley, New York, 1974, chap. 11.
55. **Caspari, C. and Granzow, A.,** The flash photolysis of mercaptans in aqueous solution, *J. Phys. Chem.,* 74, 836, 1970.
56. **Tung, T. L. and Stone, A.,** Flash photolysis of aqueous solutions of cysteine, *J. Phys. Chem.,* 78, 1130, 1974.
57. **Tagaya, H., Aruga, T., Ito, O., and Matsuda, M.,** Determination of rate constants for electron transfer from radical anions of aromatic compounds to diaryl disulfides, *J. Am. Chem. Soc.,* 103, 5484, 1981.
58. **van Zwet, H. and Kooyman, E. C.,** Sulfur compounds in free radical reactions. II, *Rec. Trav. Chim.,* 86, 1143, 1967.
59. **Stone, P. G. and Cohen, S. G.,** Effects of structure on rates and quantum yields in photoreduction of fluorenone by amines, *J. Am. Chem. Soc.,* 104, 3435, 1982.
60. **Pryor, W. A., Gojon, G., and Church, D. F.,** Relative rate constants for hydrogen atom abstractions by cyclohexanethiyl and benzenethiyl radicals, *J. Org. Chem.,* 43, 793, 1978.
61. **Walling, C. and Pearson, M. S.,** Some radical reactions of trivalent phosphorus derivatives with mer-captans, peroxides, and olefins. A new radical cyclization, *J. Am. Chem. Soc.,* 86, 2262, 1964.
62. **Sivertz, C.,** Studies of the photoinitiated addition of mercaptans to olefins. IV, *J. Phys. Chem.,* 63, 34, 1959.

63. **Greig, G. and Thynne, J. C. J.,** Hydrogen and deuterium atom abstraction from trideuteromethyl mercaptan by methyl radicals, *Trans. Faraday Soc.,* 62, 379, 1966.

64. **Lehnig, M., Schwindt, J., and Neumann, W. P.,** Kinetics of SH$_2$ reaction of sulfur, *Chem. Ber.,* 108, 1355, 1975.

65. **Scaiano, J. C.,** Determination of trialkylstanyl radicals using laser flash photolysis, *J. Am. Chem. Soc.,* 102, 5399, 1980.

66. **Chatgilialoglu, C., Ingold, K. U., and Scaiano, J. C.,** Rate constant and Arrhenius parameteres for the reactions of primary, secondary, and tert alkyl radicals with tri-*n*-butyl hydride, *J. Am. Chem. Soc.,* 103, 7739, 1981.

67. **Ekwenchi, M. M., Safarik, I., and Straus, O. P.,** Reaction of hydrogen atom with diethyldisulfide and ethylmethyldisulfide, *Int. J. Chem. Kinet.,* 13, 799, 1981.

68. **Karmann, W., Grazov, A., Meussner, G., and Henglein, A.,** Pulse radiolysis of simple mercaptans in air-free aqueous solution, *Int. J. Radiat. Phys. Chem.,* 1, 395, 1969.

69. **Burkhart, R. D.,** Radical-radical reactions in different solvents. Propyl, cyclohexyl, and benzyl radicals, *J. Phys. Chem.,* 72, 2703, 1969.

70. **Adams, G. E., McNaughton, G. S., and Michael, B. D.,** Pulse radiolysis of sulfur compounds, *Trans. Faraday,* 64, 902, 1968.

71. **Wolfenden, B. S. and Willson, R. L.,** Radical-cations as reference chromogens in kinetic studies of one-electron transfer reactions, *J. Chem. Soc. Perkin Trans. 2,* 805, 1982.

72. **Kryger, R. G., Lorand, J. P., Stevens, N. R., and Herron, N. R.,** Radicals and scavengers. VII, *J. Am. Chem. Soc.,* 99, 7589, 1977.

73. **Chenier, J. H. B., Furimsky, E., and Howard, J. A.,** Arrhenius parameters for reaction of the *tert*-butylperoxy and 2-ethyl-2-propyl radicals with some nonhindered phenols, aromatic amines, and thiophenols, *Can. J. Chem.,* 52, 3682, 1974.

74. **Colussi, A. J. and Benzon, S. W.,** The very low-pressure pyrolysis of phenyl and methyl sulfide and benzyl methyl sulfide. The enthalpy of formation of the methylthio and phenylthio radicals, *Int. J. Chem. Kinet.,* 9, 295, 1977.

75. **Colle, T. and Lewis, E.,** Hydrogen atom transfer from thiophenols to triarylmethyl radicals, *J. Am. Chem. Soc.,* 101, 1810, 1979.

76. **Walling, C. and Helmreich, W.,** Reactivity and reversibility in the reaction of thiyl radicals with olefins, *J. Am. Chem. Soc.,* 81, 1144, 1959.

77. **Cadogan, J. I. G. and Sadler, I. H.,** Quantitative aspects of radical addition. IV, *J. Chem. Soc. B,* p. 1191, 1966.

78. **Church, D. F. and Gleicher, G. J.,** The effect of substituents on the addition of thiophenol to α-methylstyrene, *J. Org. Chem.,* 40, 536, 1975.

79. **Ito, O. and Matsuda, M.,** Polar effects in addition reactions of benzenethiyl radicals to substituted styrenes and α-methylstyrenes determined by flash photolysis, *J. Am. Chem. Soc.,* 104, 1701, 1982.

80. **Sato, T., Abe, M., and Otsu, T.,** Application of spin trapping technique to radical polymerization. XVI, *Makromol. Chem.,* 180, 1165, 1979.

81. **Graham, D. M. and Soltys, J. F.,** Photo-initiated reaction of thiols and olefins. III, *Can. J. Chem.,* 47, 2529, 1969.

82. **Sato, T., Abe, M., and Otsu, T.,** Application of spin trapping technique to radical polymerization. XV, *Makromol. Chem.,* 178, 1951, 1977.

83. **Natarajan, L. V., Lembke, R. R., and Kuntz, R. R.,** Kinetics of addition of *p*-aminophenylthiyl radicals to vinyl monomers, *J. Photochem.,* 15, 13, 1981.

84. **Gaile, L. H.,** Free-radical addition reactions with cycloolefins. Role of medium ring size effect, *J. Org. Chem.,* 34, 81, 1969.

85. **Ito, O. and Matsuda, M.,** Reactivities of cycloalkenes toward phenylthio radicals, *J. Org. Chem.,* 49, 17, 1984.

86. **Lunazzi, L. and Placucci, G.,** Temperature-controlled addition vs. abstraction by methylthiyl radical with cyclopentene, *J. Chem. Soc. Chem. Commun.,* p. 533, 1979.

87. **Geers, B. N., Gleicher, G. J., and Church, D. F.,** Additions of substituted phenylthiyl radicals to substituted α-methylstyrenes, *Tetrahedron,* 36, 997, 1979.

88. **Ito, O. and Matsuda, M.,** Polar effect in addition rates of substituted benzenethiyl radicals to α-methylstyrene determined by flash photolysis, *J. Org. Chem.,* 47, 2261, 1982.

89. **Lee, I. and Cheun, Y. G.,** Determination of reactivity by MO theory. XXII. Mo studies of substitutent effects on rates of phenylthiyl radical addition to α-methylstyrene, *J. Korean Chem. Soc.,* 26, 1, 1982.

90. **Mohaney, L. R. and DaRooge, M. A.,** The kinetic behavior and thermodynamic properties of phenoxy radicals, *J. Am. Chem. Soc.,* 47, 4722, 1975.

91. **Ito, O. and Matsuda, M.,** Evaluation of addition rates of thiyl radicals to vinyl monomers by flash photolysis. III, *J. Am. Chem. Soc.,* 103, 5871, 1981.

92. **Morine, G. H. and Kuntz, R. R.,** Spectral shifts of the *p*-aminophenylthiyl radical absorption and emission in solution, *Chem. Phys. Lett.,* 67, 552, 1978.

93. **Ito, O. and Matsuda, M.,** Solvent effects on rates of free radical reactions, *J. Am. Chem. Soc.,* 104, 568, 1982.

94. **Ito, O. and Matsuda, M.,** Solvent effects of free-radical reaction. II, *J. Phys. Chem.,* 88, 1002, 1984.

95. **Fong, C. W., Kamlet, M. J., and Taft, R. W.,** Linear solvent energy relationship. XXIV, *J. Org. Chem.,* 48, 832, 1983.

96. **Ito, O. and Matsuda, M.,** Flash photolysis study for substitutent and solvent effects on spin-trapping rates of phenythiyl radicals with nitrones, *Bull. Chem. Soc. Jpn.,* 57, 1745, 1984.

97. **Ito, O. and Matsuda, M.,** Flash photolysis study of solvent effects in addition reactions of thiyl radicals to styrene or α-methylstyrene, *J. Phys. Chem.,* 86, 2076, 1982.

98. **Ito, O., Omori, R., and Matsuda, M.,** Determination of addition rates of benzenethiyl radicals to alkynes by flash photolysis, *J. Am. Chem. Soc.,* 104, 3934, 1982.

99. **Freeman, F. and Angeletakis, C. N.,** α-Disulfoxide formation during the *m*-chloroperoxybenzoic acid oxidation of S-(2,2-dimethylpropyl)2,2-dimethylpropanethiosulfinate, *J. Am. Chem. Soc.,* 104, 5766, 1982.

100. **Chatgilialoglu, C., Gilbert, B. C., Gill, B., and Sexton, M. D.,** Electron spin resonance studies of radicals formed during the thermolysis and photolysis of sulphoxides and thiosulphonates, *J. Chem. Soc. Perkin Trans. 2,* p. 1141, 1980.

101. **Iino, M. and Matsuda, M.,** Studies of sulfinyl radicals. II, *J. Org. Chem.,* 26, 3108, 1983.

102. **Gilbert, B. C., Kirk, C. M., Norman, R. O. C., and Laue, H. A. H.,** Electron spin resonance studies. LI. Alphatic and aromatic sulfinyl radicals, *J. Chem. Soc. Perkin Trans.,* p. 497, 1977.

103. **Kursic, P. J. and Rettig, T. A.,** Electron spin resonance study of benzoyl radicals in solution, *J. Am. Chem. Soc.,* 92, 7221, 1970.

104. **Howard, J. A. and Furimsky, E.,** An electron spin resonance study of the *tert*-butylsulfinyl radical, *Can. J. Chem.,* 55, 555, 1974.

105. **Bennett, J. E. and Brunton, G.,** Electron spin resonance study of *t*-butylperthyl radical, *J. Chem. Soc. Chem. Commun.,* p. 62, 1979.

106. **Huggenberger, C., Lipscher, J., and Fischer, H.,** Self-termination of benzoyl radicals to ground- and excited-state benzil. Symmetry control of a radical combination, *J. Phys. Chem.,* 84, 3467, 1980.

107. **Baban, J. A. and Robert, B. P.,** An electron spin resonance study of dialkylaminothiyl, dialkylamino-sulfinyl, and alkyl(sulfinyl)aminyl radicals. Radical addition to *N*-sulfinylamines, *J. Chem. Soc. Perkin. Trans. 2,* p. 678, 1978.

108. **Chatgilialoglu, C., Lunazz, L., and Ingold, K. U.,** Kinetic studies on the formation and decay of some sulfonyl radicals, *J. Org. Chem.,* 48, 3588, 1983.

109. **Koelewijn, P. and Berger, H.,** Mechanism of the antioxidant action of dialkyl sulfoxides, *Rec. Trav. Chim. Pays Bas.,* 91, 1275, 1972.

110. **Chatgilialoglu, C., Ingold, K. U., and Scaiano, J. C.,** Absolute rate constants for the reaction of triethylsilyl radicals with organic halides, *J. Am. Chem. Soc.,* 104, 5123, 1982.

111. **Chatgilialoglu, C., Gilbert, B., and Norman, R. O. C.,** Investigations of structure and conformation. XII. The structure of aromatic sulfonyl radicals: an ESR and INDO molecular orbital study, *J. Chem. Soc. Perkin Trans. 2,* p. 771, 1979.

112. **Eriksen, T. E. and Lind, J.,** Optical absorption spectra of alkylsulfonyl radicals formed by pulse radiolysis of alkylsulfonyl chlorides, *Radiochem. Radioanal. Lett.,* 25, 11, 1976.

113. **Kojima, M., Minato, H., and Kobayashi, M.,** Kinetic study on homolysis of ary arylazo sulfones, *Bull. Chem. Soc. Jpn.,* 45, 2032, 1972.

114. **da Silva Correa, C. M. M. and Oliveira, M. A. B. C. S.,** Reaction of arensulphonyl halides with free radicals. II, *J. Chem. Soc. Perkin Trans.,* p. 711, 1983.

115. **da Silva Correa, C. M. M. and Waters, W. A.,** Evidence for electrophilic character of the toluene-*p*-sulfonyl free radical: relative rates of its addition to some substituted styrenes, *J. Chem. Soc. Perkin Trans. 2,* p. 1575, 1972.

116. **Gould, I. R., Tung, C., Turro, N. J., Givens, R. S., and Matuszewski, B.,** Mechanistic studies of the photodecomposition of arylmethyl sulfones in homogenious and micellar solutions, *J. Am. Chem. Soc.,* 106, 1789, 1984.

117. **Horowitz, A.,** Liquid phase kinetic study of the formation and decomposition of methylsulfonyl and cyclohexylsulfonyl radicals in the cyclohexane-MeSO$_2$Cl-SO$_2$ system at 150°C, *Int. J. Chem. Kinet.,* 7, 927, 1975.

118. **Good, A. and Thynne, J. C. J.,** Reaction of free radicals with sulfur dioxide. II, *Trans. Faraday Soc.,* 63, 2720, 1967.

119. **James, F. C., Kerr, J. A., and Simons, J.,** Direct measurement of the rate of reaction of the methyl radical with sulphur dioxide, *J. Chem. Soc. Faraday Trans.,* p. 2124, 1973.

120. **Kan, C. S., Mcquigg, R. D., Whitbech, M. R., and Calvert, J. G.,** Kinetic flash spectroscopy study of the CH$_3$OO–CH$_3$OO and CH$_3$–SO$_2$ reaction, *Int. J. Chem. Kinet.,* 11, 921, 1979.

121. **Horowitz, A. and Rajbenbach, L. A.,** The free radical mechanism of the decomposition of alkylsulfonyl chlorides in liquid cyclohexane, *J. Am. Chem. Soc.,* 97, 10, 1975.
122. **Gilbert, B. G., Norman, R. O. C., and Sealy, R. C.,** Electron spin resonance studies. XLIII. Reaction of dimethylsulphoxide with the hydroxyl radical, *J. Chem. Soc. Perkin Trans. 2,* p. 303, 1975.

Chapter 16

REACTIONS OF SILYL RADICALS

Takaaki Dohmaru

TABLE OF CONTENTS

I. INTRODUCTION

Silyl radicals occupy an important position in the silicon chemistry which is expanding with its wide variety of practical use: from silicone polymers on one hand to silicon microchips on the other.

Recent reviews have surveyed the silyl radical reactions from the kinetic[1,2] and the mechanistic[3-5] point of view. These reviews give a good background to silyl radical chemistry in general. The purpose of this chapter is to provide the present state of knowledge of quantitative kinetics of silyl radicals in the gas phase as well as in the liquid phase. The radicals to be described here are limited to *small organic* silyl radicals, so that principally the $Me_3Si^{\cdot}$ and the $Et_3Si^{\cdot}$ radicals and, to the lesser extent, the $H_3Si^{\cdot}$ and the $Cl_3Si^{\cdot}$ radicals will be surveyed.

The formation of the radicals is one of the important subjects in radical chemistry, but as it has been extensively described recently,[4,5] it will be omitted here.

II. THERMOCHEMISTRY OF SILICON COMPOUNDS

A. Bond Energies

Thermochemical data of the silicon-containing species are still very limited compared with the carbon analogs which have been thoroughly studied both experimentally and theoretically.[6] The major data sources for thermodynamic properties of silicon compounds have been CATCH tables,[7] JANAF tables,[8] and the NBS compilation.[9] There have also been a number of experimental[10] and theoretical[11-16] papers on the thermochemistry of silicon compounds.

Walsh[10] obtained the bond dissociation energies of Si–H and Si–C bonds by the iodination method[17] and reviewed and discussed the results both for their intrinsic interest and for their bearing on questions of stability and reactivity of intermediate species. O'Neal and Ring have developed methods for calculating thermochemical quantities of silicon compounds using the bond additivity law[11] and later the group additivity scheme.[12] These methods may not be of good accuracy, but are sometimes convenient to use for calculating thermochemical properties of a compound whose experimental values are not available.[18] Davidson[13] calculated heat of formations of methylsilanes employing Benson's electrostatic model.[19] Bell et al.[14] used the MOBI (Molecular Orbital Bond Index) method to compute heats of formation of silicon compounds. Recently Walsh[15] compiled and critically evaluated the heat of formation of silicon-halogen compounds from the standpoints of bond additivity and general chemical reactivity of the species involved. Ho and Coltrin[16] calculated the heats of formation of H–Si–Cl compounds by *ab initio* electronic structure calculations. Some of the representative bond dissociation energies derived from these studies are listed in Table 1.

B. Hydrogen Abstraction from Hydrosilanes

The hydrogen abstraction from a Si–H bond is the most fundamental reaction in the chemistry involving silyl radicals. This reaction not only serves as the most convenient method for producing silyl radicals (Equation 1)[4,5]

$$R^{\cdot} + X_3SiH \rightarrow RH + X_3Si^{\cdot} \tag{1}$$

but also is encountered in almost any sequences of reactions that involve silyl radicals. Thus, it is a matter of course that extensive kinetic studies have been concentrated on this subject.[20,21,27-39]

Arthur and Bell[27] have given both a comprehensive compilation and a critical evaluation of the experimental data up to the middle of 1977. A number of kinetic studies have also

Table 1
BOND DISSOCIATION ENERGIES OF BONDS INVOLVING
REPRESENTATIVE SILYL GROUPS[a,b,c]

R	D(R–H)	Ref.	D(R–CH₃)	Ref.	D(R–Cl)	Ref.	D(R–R)	Ref.
H₃Si·	378.0*	20	377.3*	23	485.5	14	309.6	10
	388.7	14	369.0	10	456.1	16	330.9	14
	383.7	16	385.4	14				
Me₃Si·	376.0*	25	355.0*	24	485.3*	23	337.0*	26
	377.8*	21	384.0	22	472.8	10	336.1	14
	377.1	14	374.0	10	462.8	14		
			355.0	14				
Cl₃Si·	382.0*	22	358.1*	23	435.1*	23	322.2	23
	383.7*	23	377.0	22	464.4	10	326.3	14
	384.7	14	371.7	14	415.0	14		
	387.4	16			447.7	15		
					464.0	16		

a $kJ \, mol^{-1}$.
b The starred values are obtained experimentally and the nonstarred ones from thermochemical or theoretical calculations.
c Tenths of $kJ \, mol^{-1}$ are shown for consistency, not to claim spurious accuracy.

appeared since then.[20,21,28-39] Arrhenius parameters for hydrogen abstraction from some representative silanes by the methyl or the trifluoromethyl radicals are shown in Table 2.

A factors for the H-abstraction from silanes by radicals are almost constant; the majority fit within the range $\log(A/cm^3 \, mol^{-1} \, sec^{-1}) = 11.9 \pm 0.3$,[27] corresponding to a variation in the rate constant of a factor of 4. The activation energy is clearly the major factor in determining the radical-substrate reactivity. The activation energies for attack by the $CH_3^·$ are clearly higher than by the $CF_3^·$ in cases of SiH_4 and Me_3SiH. These differences can be rationalized by the difference of $D(CF_3–H)–D(CH_3–H) = 9.2 \, kJ \, mol^{-1}$.[27] Polar effects seem to be important to elucidate the reactivity of the $CF_3^·$ radical toward Cl_3SiH. But the quantitative discussion on this matter is hampered at present due to the lack of the accurate data of H-abstractions from Cl_3SiH by fluorinated methyl radicals.

III. STRUCTURE OF SILYL RADICALS

A. IR Spectra

The reported IR spectra of $^·SiH_3$,[42] $^·SiCl_3$,[43] and $^·SiF_3$[44] radicals indicate a pyramidal shape for these radicals. In studies of the vacuum-ultraviolet (VUV) photolysis of SiH_4,[42] $HSiCl_3$,[43] or $HSiF_3$[44] in inert matrices at 14 K, evidence has been obtained indicating that H atom detachment occurs and that the $^·SiH_3$ radical, the $^·SiCl_3$ radical, or the $^·SiF_3$ radical is stabilized, respectively. A detailed consideration of the observed vibrational fundamentals indicates that an angle between the threefold axis and each of the Si–Cl bonds or the Si–F bonds is $72° \pm 5°$ or $71° \pm 2°$, respectively. These angles are close to the value characteristic of sp^3 hybridization of the valence electrons at the central atom (70.5°).

B. ESR Spectra

The structure of silyl radicals have been discussed in terms of the extent of the deviation from planarity based on the isotropic hyperfine coupling (hfc) to α-protons and to ^{29}Si of the respective silyl radical. Table 3 illustrates ESR parameters of some simple silyl radicals.

Cochran[45] first studied the $^·SiH_3$ radical by ESR spectroscopy and reported the proton hfc of 7.6 G, remarkably smaller than 23.0 G of the $^·CH_3$ radical. The difference of the two values was discussed in terms of the deviation from the planarity of the $^·SiH_3$ radical.[46]

Table 2

**ARRHENIUS PARAMETERS FOR HYDROGEN ABSTRACTION FROM
REPRESENTATIVE HYDROSILANES BY THE METHYL OR THE
TRIFLUOROMETHYL RADICAL**

Attacking radical	Substrate	Temp. log (k)	log A (cm³ mol⁻¹ sec⁻¹)	E (kJ mol⁻¹)	log k(400 K) (cm³ mol⁻¹ sec⁻¹)	Ref.
CH_3	SiH_4	301—486	11.89 ± 0.11	29.2 ± 0.8	8.08	27
CF_3	SiH_4	303—413	11.97 ± 0.19	21.4 ± 1.3	9.18	27
CH_3	Me_3SiH	345—526	11.10 ± 0.14	30.2 ± 1.0	7.16	27
CF_3	Me_3SiH	323—476	12.28 ± 0.09	23.3 ± 0.7	9.23	40
CH_3	Cl_3SiH	333—443	10.84 ± 0.11	18.1 ± 0.8	8.48	41
CF_3	Cl_3SiH	323—567	11.82	26.11	8.41	29

Table 3

ESR PARAMETERS FOR SMALL SILYL RADICALS

Silyl radical	Matrix	Temp. (K)	g Value[a]	Coupling constant[a,b] $a_{\alpha-H}$	$a_{\beta-H}$	$a_{^{29}Si}$	Ref.
SiH_3	Ar	−269		7.6			45
	Ar	−269	2.0036	7.53			57
	Kr	−269		−8.1		266	47
	Kr	−269	2.0013	7.63			57
	Xe	−269	2.003			190	48
	N_2	−269	2.0013	7.65			57
	Silica gel	−150	2.003	7.9		182	54
	Solution/C_2H_6	−153		+7.96			49
	Solution/SiH_4	−70	2.0032	+7.84			51
SiD_3	Silica gel	−150	2.003	1.2(D)		182	54
$MeSiH_2$	Solution/C_2H_6	−121		−11.82	7.98		49
	Solution/$MeSiH_3$	−70	2.0032	−12.11	8.21		51
	$MeSiH_3$	−196	2.0032	+11.8	8.0	181	52, 53
Me_2SiH	Solution/C_2H_6	−123		−16.99	7.19	183.05	49
	Solution/Me_2SiH_2	−70	2.0031	−17.29	7.30		51
Me_3Si	Solution/C_2H_6	−120			6.28	181.14	49
	Solution/Me_3SiH	−70	2.0031		6.42		50
	Solution/Me_3SiH	−70	2.0031		6.34	183	51
	Me_3SiH	−196	2.0029		6.3	181	52
	Me_4Si	−103			6.32	172.5	55
	Me_3SiCl	−88			6.2	129	55

[a] Isotropic.
[b] Values in Gauss.

Gordy et al.[47] studied the ˙SiH_3 radical formed in Kr matrix at 4.2 K and reported the isotropic ^{29}Si hyperfine splitting (hfs) of 266 G. This large value leads to the pyramidal structure of the ˙SiH_3 radical with HSiH angle of 110.6°. The orbital occupied by an unpaired electron was calculated to have 22% s character. Later, however, they obtained the isotropic ^{29}Si hfs of 190 G of the ˙SiH_3 radical in Xe matrix at 4.2 K.[48]

Krusic and Kochi[49] and Hudson et al.[50,51] generated some methylsilyl radicals in fluid solution via hydrogen abstraction from parent hydrosilanes by the *t*-BuO˙ radical photochemically produced. Both groups obtained the similar values of 181 to 183 G for the isotropic ^{29}Si hfs of the Me$_2$SiH$^{\cdot}$ radical[49] and the Me$_3$Si$^{\cdot}$ radical.[49,51] Combined with the value of the $^{\cdot}$SiH$_3$ radical in Kr matrix (266 G),[47] they concluded that a flattening advances along the series H$_3$Si$^{\cdot}$ to Me$_3$Si$^{\cdot}$ although the Me$_3$Si$^{\cdot}$ radical remains pyramidal in structure. α-Coupling constants obtained by both groups are in good agreement, i.e., 12 G for the MeSiH$_2$$^{\cdot}$ radical and 17 G for the Me$_2$SiH$^{\cdot}$ radical. They gave negative signs to these and a positive sign to the α-coupling constant of the $^{\cdot}$SiH$_3$ radical, so that the splitting constant might decrease as the flattening proceeded. This tendency would be substantiated by the fact that in a planar radical the α-proton hfc constant is expected to arise via spin polarization of the Si–H bond and to be negative in sign, and that a bending in the radical introduces s character into the orbital containing the unpaired electron and a positive contribution to the proton coupling constant.

Symons et al.[52,53] also generated the MeSiH$_2$$^{\cdot}$ and Me$_3$Si$^{\cdot}$ radicals in matrices at 77 K. The isotropic ^{29}Si hfs of both radicals were found equal (181 G). This value was in good agreement with those in fluid solution[49-51] but the conclusion they came to was quite different. They prefered the value of ^{29}Si hfs of the $^{\cdot}$SiH$_3$ radical in Xe matrix (190 G),[48] and proposed that the methylsilyl radicals should have the same configuration at the central silicon as does the $^{\cdot}$SiH$_3$ radical. The ^{29}Si hf tensors showed that these radicals were pyramidal (bond angle, 113° to 114°), with $3p/3s = 5.7$. Symons et al. obtained 11.8 G for the α-proton coupling of the MeSiH$_2$$^{\cdot}$ radical in excellent agreement with those by other groups.[49,51] On the basis of the consideration of a linewidth effect and of a relationship between p/s ratio and the magnitude of proton coupling constants in various radicals, they gave a positive sign to the proton coupling constant for the MeSiH$_2$$^{\cdot}$ radical. Combined with the observed tendency that β-proton coupling constants of methylsilyl radicals decreased on methylation, Symons et al. suggested that, on methylation, there is a slight overall deviation from the planarity but a small delocalization of spin density due to hyperconjugation on the methyl groups just offsets this. This is the opposite conclusion to that of the studies in fluid solutions.[49-51] The apparent discrepancy in the foregoing argument seems to arise entirely from the experimental values of the ^{29}Si hfs of the $^{\cdot}$SiH$_3$ radical which differ too much in the different matrices. Thus Katsu et al.[54] reinvestigated the ^{29}Si hfs of the $^{\cdot}$SiH$_3$ radical and obtained the well-resolved ESR spectra of ^{29}SiH$_3$ and ^{29}SiD$_3$ radicals trapped on the surface of silica gel. The value for ^{29}Si hfs of these radicals was 182 G which is close to that in Xe matrix.[48] This value is also equal to those of the methylsilyl radicals obtained in fluid solutions[49-51] and in solid matrices.[52,53] Katsu et al. suggested that the mean bond angles at the silicon atom of those radicals are, to a first approximation, unaffected by the successive methyl substitutions as has been suggested by Symons and Sharp.[52]

There seems to be a general agreement in that the silyl and the methylsilyl radicals described here are nonplanar, the extent of which, however, is controversial at present.

For the interested readers references are available for ESR studies of Me$_2$SiCl$^{\cdot}$,[52,55] MeSiCl$_2$$^{\cdot}$,[55] $^{\cdot}$SiCl$_3$,[55,58] $^{\cdot}$SiH$_2$F,[56] $^{\cdot}$SiHF$_2$,[56] $^{\cdot}$SiF$_3$,[56,59,60] and Me$_2$SiF$^{\cdot}$ radicals.[56]

C. Theoretical Calculations

The quantum mechanical calculations of various levels have been performed on the molecular geometry and hfs constants of silyl radicals since around 1970. Although a few of them are concerned with $^{\cdot}$SiF$_3$[64,71,78] and Me$_x$H$_{3-x}$Si$^{\cdot}$ [66,78] radicals, they are mostly on the $^{\cdot}$SiH$_3$ radical.[61-78]

The semi-empirical calculations[61-71] of the HSiH angle of the optimized geometry for the $^{\cdot}$SiH$_3$ radical gave a variety of values with complete[69,70] and nearly[71] planar configurations on the one hand and a more bent structure than the tetrahedral one[64] on the other. All the *ab initio* calculations indicate a pyramidal structure for the $^{\cdot}$SiH$_3$ radical.[72-78] The spectrum of HSiH angles by *ab initio* calculations is narrower than that by the semi-empirical methods

Table 4
CALCULATED HYPERFINE COUPLING CONSTANTS (G) FOR THE $H_3Si^·$ RADICAL

a_H		a_{29Si}		Ab initio	
Static	Dynamic[a]	Static	Dynamic[a]	calc. level	Ref.
− 0.98		− 177.61		DZ, CGTO	75
6.08		− 197.82		DZ + P, CGTO	75
3.70	2.20	− 201.01	− 194.33	DZ + P, CGTO	76
− 10.1	− 12.3	− 173.8	− 164.8	DZ + P, CGTO (PO theory)	77

[a]　The vibrational effect is allowed for.

and is in relatively close agreement with the experimental values of 110.6° (47), 113 to 114° (52), and 112.8° (79).

The most recent values of the calculated hfs constants for the $H_3Si^·$ radical are shown in Table 4. The ^{29}Si hfs values seem self-consistent and are in agreement with the lower value by ESR measurement.[48,54] As shown in Table 3, experimental ^1H hfs values are in good agreement with each other in magnitude but are controversial in its sign. Table 4 shows, however, that detailed *ab initio* calculations have not settled this problem yet.

IV. REACTIONS OF SILYL RADICALS

A. Recombination and Disproportionation

Thynne[80] first calculated the rate constant for self-combination of the $Me_3Si^·$ radical using the thermochemical relation $A_f/A_r = \exp(\Delta S°/R)$.[81] Based upon the experimental value[82] of $A_r = 10^{13.5}$ cm^3 $mol^{-1}sec^{-1}$ and the estimated values of $S°(Me_3Si^·) = 336$ J deg^{-1} mol^{-1} and $S°(Me_3SiSiMe_3) = 429$ J deg^{-1} mol^{-1},

$$2Me_3Si^· \overset{f}{\underset{r}{\rightleftarrows}} Me_3SiSiMe_3 \qquad (2)$$

Thynne obtained $k_f = 10^{5.5}$ cm^3 mol^{-1} sec^{-1}. This value itself is too small compared with those of carbon centered radicals (see Table 5), but it inspired the interest of a number of researchers on this subject. Table 5 summarizes the recent rate constants for self-combination of $Me_3Si^·$ and $H_3Si^·$ radicals, and $Me_3C^·$ and $CH_3^·$ radicals, for comparison. The activation energy for recombination of silyl radicals has been found to be zero or very small.[83] The liquid-phase rate constants have been measured by the kinetic ESR spectroscopy,[83-85] and are in satisfactory agreement with each other. On the other hand, gas-phase rate constants have been obtained by a variety of methods. Cadman et al.[86,87] measured the rate constant for combination of the $Me_3Si^·$ radical by a rotating sector technique in the gas phase. But the value of $10^{14.25}$ cm^3 mol^{-1} sec^{-1} has been questioned by a number of authors[26,88-91] as being unreasonably high. The rest of the gas-phase values varies from $10^{10.7}$ to $10^{13.0}$, so that one may conclude that the correct value lies close to the liquid-phase values and also to those for the corresponding carbon radicals.[92,93] The value for the $H_3Si^·$ radical by Ring et al.[94] is clearly too small compared with those for the $Me_3Si^·$ radical.

Until very recently the disproportionation of the $Me_3Si^·$

Table 5
RATE CONSTANTS FOR SELF-COMBINATION OF Me₃Si· AND H₃Si· RADICALS

Radical	Temp. (K)	log(k/cm³ mol⁻¹ sec⁻¹)		Method	Ref.
		Gas phase	Liquid phase		
Me₃Si·	298	5.5		Thermochemical	80
	298	~13[a]	12.34 ± 0.14	Kinetic ESR	84
	319—399	14.25 ± 0.3		Rotating sector	86, 87
	298		12.74 ± 0.02	Kinetic ESR	85
	293		12.48 ± 0.03[b]	Kinetic ESR	83
	770—872	13.0		Kinetic + thermochemical	26
	298	12.2		Estimation	88
	298	12.48		Estimation	91
	473—623	13.38/11.3 ± 0.5		Kinetic/thermochemical	89
	295	<13.5/10.7 ± 1.3		Kinetic/thermochemical	90
H₃Si·	733	~9		Kinetic	94
	123—153		12.8[c]	Kinetic ESR	83
Me₃C·	700	11.8 ± 0.3		VLPP[d]	92
	298	12.1		Modulation spectroscopy	93
H₃C·	438	13.57		Rotating sector	95

[a]   Estimated value.
[b] $A = 10^{12.9 \pm 0.2}$ cm³ mol⁻¹ sec⁻¹, $E_a = 4.2 \pm 0.84$ kJ mol⁻¹ (191 to 293 K).
[c]   Extrapolated to 298 K. $A = 10^{13.2 \pm 0.5}$ cm³ mol⁻¹ sec⁻¹, $E_a = 2.51 \pm 2.51$ kJ mol⁻¹.
[d]   Very low pressure pyrolysis.

$$\text{Me}_3\text{Si}^{\cdot} + \text{Me}_3\text{Si}^{\cdot} \begin{array}{l} \overset{c}{\nearrow} \text{Me}_3\text{SiSiMe}_3 \\ \underset{d}{\searrow} \text{Me}_3\text{SiH} + \text{Me}_2\text{Si}{=}\text{CH}_2 \end{array} \tag{3}$$

radical to a silaolefin, 2-methyl-2-silapropene, has been considered to be a minor process compared to the radical combination. Incidentally, in Table 5, the gas-phase rate constants represent k_c and the liquid-phase ones represent $k_c + k_d$, since the formation of Me₃SiSiMe₃ has been monitored in the gas-phase studies while the decay of the Me₃Si· radical has been monitored in the liquid-phase studies. No allowance for the contribution of k_d has been added to the data in Table 5. Some groups[88,91,96] have measured the formation of Me₃SiH to obtain k_d and others[91,97,98] have analyzed Me₂Si=CH₂ by trapping it with alcohols.[100] Table 6 summarizes the recent values of k_d/k_c

$$\text{Me}_2\text{Si}{=}\text{CH}_2 + \text{ROH} \rightarrow \text{Me}_3\text{SiOR} \tag{4}$$

for the Me₃Si· radical. The values of k_d/k_c scatter widely from ~0[91] to 0.48,[96] but there is no disagreement with respect to the occurrence of the disproportionation of the Me₃Si· radical.

Gammie et al.[88] have studied some *cross*-disproportionation of the Me₃Si· radical with Me₂SiH· (0.3), MeSiH₂· (0.5), and SiH₃· (0.8) radicals where the values in parentheses represent k_d/k_c at 298 K.

B. Addition Reactions
1. Addition to Olefins
The addition of silanes to olefins (Equation 5) is called "hydrosilylation" and has been one of the three major methods

Table 6

VALUES OF k_d/k_c FOR THE $Me_3Si^{\cdot}$ RADICAL

Temp. (K)	Phase	k_d/k_c	Method	Ref.
		0.046^a		99
303	Gas	0.48	Kinetic	96
298	Gas	0.31 ± 0.8	Trapping by MeOH	97
298	Liquid	0.19 ± 0.05	Trapping by Me_3COH	98
298	Gas	0.05 ± 0.01	Kinetic	88
298	Gas	$0.007/<0.13$	Kinetic/trapping by MeOH	91

^{a} Experimental details are not given.

$$R_3SiH + \; \text{\Large$>$}C=C\text{\Large$<$} \; \rightarrow R_3SiC-C-H \tag{5}$$

(together with "direct synthesis" and Grignard synthesis) for the production of organosilicon compounds.[101] The addition has been promoted most frequently by transition-metal-complex-catalysis.[102] Free radical hydrosilylations initiated by peroxides, UV or gamma irradiation or heat have also been extensively studied,[103] but since these studies were performed almost from the synthetic point of view, the kinetic data to be described here are still rather limited.

The free radical chain mechanism was proposed in 1947 by Sommer et al.[104] who first studied the peroxide-initiated addition of Cl_3SiH to 1-octene, viz., Equations 6 to 9. This mechanism is presumably valid in free-radical hydrosilylations in general.[3-5]

$$\text{Initiator} \rightarrow In^{\cdot} \tag{6}$$

$$In^{\cdot} + R_3SiH \rightarrow InH + R_3Si^{\cdot} \tag{7}$$

$$R_3Si^{\cdot} + \; \text{\Large$>$}C=C\text{\Large$<$} \; \rightarrow R_3Si-C-C^{\cdot} \tag{8}$$

$$R_3SiC-C^{\cdot} + R_3SiH \rightarrow R_3SiC-C-H + R_3Si^{\cdot} \tag{9}$$

The nature of the initiation step depends upon the initiator used. The most widely used initiator is di-*tert*-butyl peroxide combined with UV irradiation.[50]

The addition step (Equation 8) seems to be irreversible at ordinary temperatures. Bennett et al.[105] studied the addition of the $Me_3Si^{\cdot}$ radical to *cis*-1-deuterio-1-hexene or 4-methyl-2-pentene and observed no isomerization in either olefin up to 413 K. Dohmaru et al.[106,107] studied the gas-phase addition of the $Cl_3Si^{\cdot}$ radical to ethylene competitively to acetone in the higher temperature range and found kinetically that the addition step becomes appreciably reversible around 473 K. Later, Dohmaru and

$$Cl_3Si^{\cdot} + CH_2{=}CH_2 \rightleftarrows Cl_3SiCH_2CH_2^{\cdot} \tag{10}$$

Nagata observed that *cis-trans* isomerization of 2-butene[108] and of 2-pentene[109] was induced by the $Cl_3Si^{\cdot}$ radical. These observations lend further support to the reversibility of Reaction 10.

Pollock et al.[110] have estimated the rate constant for addition of the $D_3SiSiD_2^{\cdot}$ radical to ethylene, based upon the quantum yield measurements of the products in the mercury photo-

$$SiD_3SiD_2^{\cdot} + CH_2{=}CH_2 \rightarrow SiD_3SiD_2CH_2CH_2^{\cdot} \qquad (11)$$

sensitization of a mixture of H_2, Si_2D_6, and C_2H_4, viz., $k_{11} = 4 \times 10^9$ cm^3 mol^{-1} sec^{-1} at 298 K. This value is two to three orders of magnitude larger than the corresponding rate constants for alkyl radical additions to olefins.[111] Dohmaru and Nagata obtained the relative gas-phase Arrhenius parameters for both of the forward and reverse steps of addition of the $Cl_3Si^{\cdot}$ radical to 1-olefins[112,113] and 2-olefins[18] (k_{12} and k_{-12}) by a competitive method employing Reaction 14 as a reference reaction. Those relative values can be put on an absolute basis utilizing the rate

$$Cl_3Si^{\cdot} + \;\;\rangle C{=}C\langle\;\; \rightleftarrows Cl_3SiC{-}\overset{|\;\;|}{\underset{|\;\;|}{C}}{}^{\cdot} \qquad (12)$$

$$\overset{|\;\;|}{\underset{|\;\;|}{Cl_3SiC{-}C}}{}^{\cdot} + Cl_3SiH \rightarrow \overset{|\;\;|}{\underset{|\;\;|}{Cl_3SiC{-}CH}} + Cl_3Si^{\cdot} \qquad (13)$$

$$Cl_3Si^{\cdot} + CH_3COCH_3 \rightarrow (CH_3)_2\overset{\cdot}{C}OSiCl_3 \qquad (14)$$

constants fairly reasonably estimated for Reactions 13 and 14. The rate constants thus obtained are shown in Table 7.

In the liquid phase between 199 and 293 K, Choo and Gaspar[114] obtained the rate constant for addition of the $Me_3Si^{\cdot}$ radical to ethylene using the kinetic ESR method. $Log(A_{15}/cm^3$ mol^{-1} $sec^{-1}) = 10.0 \pm 0.2$, E_{15}/kJ $mol^{-1} = 10.5 \pm 0.8$ and k_{15}/cm^3 mol^{-1} $sec^{-1} = (1.7 \pm 1.0) \times 10^8$

$$Me_3Si^{\cdot} + CH_2{=}CH_2 \rightarrow Me_3SiCH_2CH_2^{\cdot} \qquad (15)$$

at 293 K. Comparison of this Arrhenius parameter with the one for the $CH_3^{\cdot}$ radical addition to ethylene, viz.,[115] $log(A/cm^3$ mol^{-1} $sec^{-1}) = 11.1$ and E/kJ $mol^{-1} = 28.5$ reveals that it is its low activation energy that enhances the rate of addition of the $Me_3Si^{\cdot}$ radical to ethylene. Product stability is suggested[114] to be the dominant factor favoring the silyl radical addition; the stabilization of a carbon-centered free radical by a β-silicon substituent has been shown to be significant.[116,117]

Choo and Gaspar's rate constant was, however, questioned by Chatgilialoglu et al.[118] who recently studied liquid-phase addition of the $Et_3Si^{\cdot}$ radical to a large number of unsaturated compounds by laser-flash photolysis technique. This method is based on the monitoring of the growth of the Et_3Si-benzil adduct radical formed in the laser flash photolysis of DTBPO-Et_3SiH solvent. Rate constants at 300 K and some Arrhenius parameters are shown in Table 8. The reactivities of C=C double bonds have a wide range, e.g., 1.1×10^{12} cm^3 mol^{-1} sec^{-1} for acrylonitrile to 9.4×10^8 cm^3 mol^{-1} sec^{-1} for cyclohexene. Monoalkyl and 1,1-dialkyl olefins are somewhat more reactive than 1,2-dialkyl olefins, and olefins in which

Table 7

ABSOLUTE ARRHENIUS PARAMETERS FOR FORWARD AND REVERSE STEPS OF ADDITION OF THE $Cl_3Si^{\cdot}$ RADICAL TO

$$\text{OLEFINS: } Cl_3Si^{\cdot} + {>}C{=}C{<} \underset{k_r}{\overset{k_f}{\rightleftharpoons}} Cl_3Si{-}\overset{|}{\underset{|}{C}}{-}\overset{|}{\underset{|}{C}}^{\cdot a}$$

Olefin[b]	log(k_f^c/cm³ mol⁻¹ sec⁻¹)	k_f^d	log(k_r/sec⁻¹)	Ref.
$H_2C{=}CH_2$	12.8—9.2/θ[e]	1	15.0—117.2/θ	107
$H_2C{=}CHCH_3$	13.1—7.5/θ	3.1	14.8—108.4/θ	112
$H_2C{=}CHCH_2CH_3$	13.0—6.7/θ	3.4	15.3—113.0/θ	112
$H_2C{=}CH(CH_2)_2CH_3$	13.1—6.7/θ	4.1	14.8—107.1/θ	112
$H_2C{=}C(CH_3)_2$	13.1—2.5/θ	13.4	16.3—116.3/θ	112
cis-$CH_3CH{=}CHCH_3$	13.5—12.6/θ	4.7	15.6—102.1/θ	18
trans-$CH_3CH{=}CHCH_3$	13.4—12.6/θ	3.7	15.5—101.3/θ	18
cis-$\overset{*}{C}H_3CH{=}CHCH_2CH_3$	13.2—12.6/θ	} 6.5	16.9—110.9/θ	18
cis-$CH_3CH{=}\overset{*}{C}HCH_2CH_3$	13.5—12.6/θ		16.9—110.9/θ	18
$\overset{*}{C}H_3CH{=}C(CH_3)_2$	14.0—12.6/θ	} 16.7	16.7—105.9/θ	18
$CH_3CH{=}\overset{*}{C}(CH_3)_2$	12.8—12.6/θ		16.7—103.8/θ	18
Cyclopentene	11.8—2.1/θ	1.4	12.9—86.2/θ	18
$O{=}C(CH_3)_2$	12.6—12.6/θ			18

ᵃ Temperature ranges are 406 to 524 K for 1-olefins and 406 to 455 K for 2-olefins.
ᵇ The starred site indicates the reaction center.
ᶜ Per one site for symmetrical olefins; k_f was calculated based upon the assumed value of k_{14}/cm³ mol⁻¹ sec⁻¹ = 12.6—12.6/θ.
ᵈ The relative rates per bond at 460 K are compared.
ᵉ θ = 2.303 RT in kJ mol⁻¹.

the double bond is activated by conjugation with a neighboring π-electron system are substantially more reactive. The rates of $Et_3Si^{\cdot}$ addition to olefins are greatly enhanced when electron-withdrawing atoms such as chlorine or groups such as $C(O)OCH_3$ attach a double bond. This tendency is opposite to that of the electrophilic[112,119,120] $Cl_3Si^{\cdot}$ radical, which adds to olefins more readily when double bonds are substituted by electron-donating alkyl groups (Table 7). The Arrhenius A factors listed in Table 8 for 1-hexene, styrene, and $CH_2{=}CCl_2$ except ethylene are slightly larger than the "normal" value,[121] viz., $10^{11.5\pm0.5}$ cm³ mol⁻¹ sec⁻¹. This implies that the transition state for the addition of the $Et_3Si^{\cdot}$ radical to ethylene is rather "tight" but those for the addition to the other three olefins are fairly "loose". A factors listed in Table 7 show that the transition states for the addition of the $Cl_3Si^{\cdot}$ radical to olefins are even more "loose".

Very recently, Horowitz[122] studied the photolysis and radiolysis of C_2Cl_4 solutions in Et_3SiH at 298 K. Accurate product analysis lead to the conclusion that the $Et_3Si^{\cdot}$ radical almost exclusively abstracted Cl atom from C_2Cl_4 (Equation 16).

$$Et_3Si^{\cdot} + Cl_2C{=}CCl_2 \rightarrow Et_3SiCl + Cl_2C{=}CCl^{\cdot} \tag{16}$$

This result sheds a new light on the course and mechanism of the reaction of the $Et_3Si^{\cdot}$ radical with chloroethylenes. Chloroethylenes listed in Table 8 have been postulated to react with the $Et_3Si^{\cdot}$ radical solely via addition reaction. Thus the interpretation of these data may have to be reexamined.

Table 8
RATE CONSTANTS AT ~300 K AND SOME ARRHENIUS PARAMETERS FOR ADDITION OF THE $Et_3Si^{\cdot}$ RADICAL TO OLEFINS[a]

Olefin	Temp. (K)	log A ($cm^3\ mol^{-1}\ sec^{-1}$)	E ($kJ\ mol^{-1}$)	k ($cm^3\ mol^{-1}\ sec^{-1}$)
$H_2C=CH_2$	154—270	11.40 ± 0.60	5.9 ± 3.3	$(2.2 ± 0.4) × 10^{10}$ [b]
$H_2C=CH(CH_2)_3CH_3$	261—332	12.00 ± 0.57	13.3 ± 3.2	$(4.8 ± 0.5) × 10^9$
$H_2C=CHC(CH_3)_3$				$(3.7 ± 0.3) × 10^9$
$H_2C=C(CH_2)_5$				$(7.4 ± 1.3) × 10^9$
$H_2C=CHCN$				$(1.1 ± 0.2) × 10^{12}$
$H_2C=C(CH_3)C(O)OCH_3$				$(4.6 ± 0.8) × 10^{11}$
$H_2C=CHC_6H_5$	236—324	12.35 ± 0.23	5.7 ± 1.2	$(2.2 ± 0.2) × 10^{11}$
$H_2C=CCl_2$	233—327	12.35 ± 0.36	5.2 ± 1.9	$(2.7 ± 0.3) × 10^{11}$
cis-$ClCH=CHCl$				$(2.1 ± 0.4) × 10^{10}$
trans-$ClCH=CHCl$				$(8.9 ± 0.2) × 10^9$
$Cl_2C=CCl_2$				$(1.0 ± 0.2) × 10^{10}$
trans-$CH_3CH=CHC(O)OCH_3$				$(4.3 ± 0.4) × 10^{10}$
trans-$CH_3CH_2CH=CHCH_2CH_3$				$(9.6 ± 1.5) × 10^8$
$HC=CH(CH_2)_3$				$(2.2 ± 0.3) × 10^9$
$HC=CH(CH_2)_4$				$(9.4 ± 1.1) × 10^8$
$HC=CHO(CH_2)_3$				$(1.2 ± 0.1) × 10^9$
Nobornadiene				$(1.5 ± 0.1) × 10^{10}$

[a] From Reference 118. Unless otherwise noted the data were obtained by laser flash photolysis with benzil as the probe in di-tert-butyl peroxide-triethylsilane as solvent.

[b] Obtained by an EPR competition method.

2. Addition to Acetylenes and Nitriles

Silyl radical additions to acetylenes are achieved in the similar manner as the addition to olefins.[4,5] The sole kinetic study on this subject seems to be the one by Chatgilialoglu et al.[118] who measured the absolute rate constants for the addition of the $Et_3Si^{\cdot}$ radical to some unsaturated compounds by flash-photolysis technique with benzil as a probe for the ESR study. The results are shown in Table 9. Acetylenes are only slightly less reactive than the corresponding olefins (c.f. Table 8).

Nitriles are considerably less reactive toward the $Et_3Si^{\cdot}$ radical compared with terminal olefins or terminal acetylenes.

3. Addition to Carbonyl Compounds

There have been a number of studies on the addition of the $Me_3Si^{\cdot}$ (or $Et_3Si^{\cdot}$) radical to ketones.[123-127] The observed products

$$Me_3Si^{\cdot} + R_2CO \rightarrow R_2\dot{C}OSiMe_3 \qquad (17)$$

are rationalized by the formation of an intermediate adduct radical, $R_2\dot{C}OSiMe_3$ (Equation 17), which dimerizes, disproportionates, or abstracts an H atom when hydrogen sources are available.

The Sussex group has carried out ESR studies of the addition of the $Et_3Si^{\cdot}$ (and $Me_3Si^{\cdot}$[130]) radical to a number of ketones,[128-130] esters,[129-131] and an acid.[128] Cooper et al.[130] found the ease of addition of the $Et_3Si^{\cdot}$ radical to carbonyl compounds to decrease in the order diketones > oxalates > ketones > trifluoroacetates > formates > acetates. The results are accounted for in terms of bond energy differences, stabilization of the adduct radical formed, and polar effects.

Table 9
RATE CONSTANTS FOR
THE ADDITION OF THE
Et₃Si· RADICAL TO
ACETYLENES AND
NITRILES AT ~300 Kᵃ

Substrate	$k/cm^3mol^{-1}sec^{-1}$
$HC \equiv CC(CH_3)_3$	$(2.3 \pm 0.2) \times 10^9$
$HC \equiv CC_6H_5$	$(1.0 \pm 0.1) \times 10^{11}$
$N \equiv CC(CH_3)_3$	$\leqslant 3 \times 10^{8}$ [b]
$N \equiv CC_6H_5$	$(3.2 \pm 0.2) \times 10^9$

ᵃ   Reference 118.
ᵇ   Too slow to measure.

Nakao et al. studied the addition of the $Cl_3Si·$ radical to esters,[132,133] lactones,[134] and acetals.[135] The mechanistic study[136] revealed that the γ-initiated reaction of trichlorosilane

$$Cl_3SiH + CH_3CCH_3 \rightarrow CH_3CHOCH_3 \text{ (I)} \qquad (18)$$
$$\qquad\qquad\quad \| \qquad\qquad\quad |$$
$$\qquad\qquad\quad O \qquad\qquad\quad OSiCl_3$$

$$I + Cl_3SiH \rightarrow CH_3CHOCH_3 \text{ (II)} + HSiCl_2OSiCl_3 \qquad (19)$$
$$\qquad\qquad\qquad\qquad |$$
$$\qquad\qquad\qquad\qquad Cl$$

$$II + Cl_3SiH \rightarrow CH_3CH_2OCH_3 + Cl_4Si \qquad (20)$$

with methylacetate is illustrated by three steps (Equations 18 to 20). The first step constitutes a radical chain reaction to form an adduct, and the second step is the ionic chlorination of the acetal-type intermediate to α-chloroether by Cl_3SiH, and the third step involves a radical chain dechlorination of α-chloroether to ether. The intermediates, (I) and (II), were identified by an nmr spectroscopy. The gas-phase kinetic study[137] of the photolytic reaction of Cl_3SiH with acetone showed that the reaction proceeded by a chain reaction, the propagating steps of which constitute Reactions 21 and 22. Based upon the assumed rate constant

$$Cl_3Si· + (CH_3)_2CO \rightarrow (CH_3)_2\overset{\cdot}{C}OSiCl_3 \text{ (III)} \qquad (21)$$

$$III + Cl_3SiH \rightarrow (CH_3)_2CHOSiCl_3 + Cl_3Si· \qquad (22)$$

of $10^{12.5}$ cm^3 mol^{-1} sec^{-1} for the recombination of III, the absolute rate constant for Reaction 22, which is the rate determining step of this chain cycle, can be calculated as in Equation 23.

$$log(k_{22}/cm^3mol^{-1}sec^{-1}) = 11.3 - 28.0 \text{ kJ/2.303RT} \qquad (23)$$

The only experimental rate constants obtained so far for the addition of a silyl radical to carbonyl oxygen are those by Chatogilialoglu et al.[138,139] The rate constants have been measured in solution by using laser-flash photolysis for compounds having rate constants

Table 10
ARRHENIUS PARAMETERS AND ABSOLUTE RATE
CONSTANTS AT ~300 K FOR THE ADDITION OF
TRIETHYLSILYL RADICALS TO CARBONYL COMPOUNDS[a]

Substrate	$\log(A/cm^3\ mol^{-1}\ sec^{-1})$	$E/kJ\ mol^{-1}$	$k/cm^3\ mol^{-1}\ sec^{-1}$
Duroquinone			$(2.2 \pm 0.1) \times 10^{12}$
Fluorenone			$(1.5 \pm 0.1) \times 10^{12}$
$[CF_3CF_2C(O)]_2O$	11.9 ± 0.07	0.84 ± 0.38	$(5.7 \pm 0.7) \times 10^{11}$
Benzaldehyde			$(4.1 \pm 0.9) \times 10^{11}$
Benzil	12.26 ± 0.07	4.27 ± 0.38	$(3.3 \pm 0.3) \times 10^{11}$
Xanthone			$(2.4 \pm 0.2) \times 10^{11}$
p-Anisaldehyde			$(1.7 \pm 0.1) \times 10^{11}$
Benzophenone			$(3.0 \pm 0.6) \times 10^{10}$
Propionaldehyde	10.81 ± 0.31	4.14 ± 1.72	$(1.2 \pm 0.1) \times 10^{10}$
Acetophenone	12.40 ± 0.38	13.22 ± 2.05	$(1.2 \pm 0.1) \times 10^{10}$
$[CH_3CH_2OC(O)]_2$			$(5.7 \pm 0.3) \times 10^{9}$
$CF_3C(O)O(CH_2)_3CH_3$			$(3.5 \pm 1.3) \times 10^{9}$
$C_6H_5C(O)OCH_3$			$(2.8 \pm 0.4) \times 10^{9}$
$[CH_3CH_2C(O)]_2O$	11.03 ± 0.38	10.50 ± 2.13	$(1.6 \pm 0.6) \times 10^{9}$
Cyclopentanone			$(7.2 \pm 1.2) \times 10^{8}$
Cyclohexanone			$(6.6 \pm 1.1) \times 10^{8}$
Cyclobutanone			$(6.5 \pm 1.8) \times 10^{8}$
$(C_2H_5)_2CO$			$(2.8 \pm 0.8) \times 10^{8}$
Ethyl formate	11.27 ± 0.93	21.42 ± 4.27	$(3.5 \pm 2.5) \times 10^{7}$
$CH_3C(O)OCH_3$			$<10^{2}$

[a]   Reference 138 and 139.

$>10^8\ cm^3\ mol^{-1}\ sec^{-1}$ and kinetic ESR spectroscopy for compounds having rate constants $<10^8\ cm^3\ mol^{-1}\ sec^{-1}$. In most measurements, benzil has been used as a probe. The results are shown in Table 10. Now, the reactivity order for carbonyl groups shown by Cooper et al.[130] can be expanded to 1,4-benzoquinone $\geq$ cyclic diaryl ketones, benzaldehyde, benzil, perfluoroacid anhydride $>$ benzophenone $\geq$ alkyl aryl ketone, alkyl aldehyde $>$ oxalate $>$ benzoate, trifluoroacetate, anhydride $>$ cyclic dialkyl ketone $>$ acyclic dialkyl ketone $>$ formate $>$ acetate. Over this series of compounds the rate constants for $Et_3Si^{\cdot}$ radical addition vary by more than six orders of magnitude. Cooper et al.[130] and Nakao et al.[133] reported that no adduct was formed in the reaction of the $Et_3Si^{\cdot}$ radical with methyl acetate at temperatures from 173 to 283 K or 300 K, respectively. These observations are in good agreement with the result in Table 10 (see the last entry).

C. Aromatic Substitution

Homolytic aromatic substitutions by silyl radicals have been extensively studied (Equation 24).[4,5] Principal interest has been

$$R_3Si^{\cdot} \;+\; \bigcirc \;\longrightarrow\; R_3Si{-}\overset{\cdot}{\underset{H}{\bigcirc}} \quad (\text{IV}) \tag{24}$$

$$\text{IV} \;\xrightarrow{-H^{\cdot}}\; R_3Si{-}\bigcirc \tag{25}$$

focused on the distribution of the final products[140-143] and on the ESR spectra of the intermediate hexadienyl radicals (IV).[144] An optimized geometry of (IV) with R=H has also been calculated by *ab initio* UHF-SCF MO calculation.[145]

Recently, Sakurai et al.[146] reported the chemical and ESR spectroscopic evidence for the reversibility of the addition step (Equation 24) for the first time. 3,6-*bis*(Trimethylsilyl)cyclohexa-1,4-diene (V) was mixed with DTBPO and heated at 403 K for 6.5 hr (Equation 26).

$$\text{(V)} \quad \xrightarrow{\text{Bu}^t\text{O}\cdot} \quad \text{Me}_3\text{Si}-\text{(VI)} \quad + \quad \text{Me}_3\text{Si}-\text{(VII, VIII)} \tag{26}$$

The main products were phenyltrimethylsilane (VI), *p*-(VII), and *m-bis*(trimethylsilyl)benzene (VIII) in 62, 17, and 5% yield, respectively. The results may be well substantiated by the detachment of the Me$_3$Si$^\cdot$ radical from the intermediate radical (IX) whose formation from (V) (Equation 27) has already been verified by an ESR experiment.[147] Further, the intermediacy of the Me$_3$Si$^\cdot$

$$\text{V} \quad \xrightarrow{\text{Bu}^t\text{O}\cdot} \quad \text{Me}_3\text{Si}-\overset{\text{SiMe}_3}{\underset{\text{H}}{\text{C}}} \quad \longrightarrow \quad \text{VI} \; + \; \text{Me}_3\text{Si}\cdot \tag{27}$$

$$(\text{IX})$$

radical was proved employing a spin-trap [(*t*-Bu)$_2$C=CH$_2$]. It was concluded that the cyclohexadienyl radical (IX) becomes kinetically unstable at temperatures higher than 313 to 333 K and eliminates the Me$_3$Si$^\cdot$ radical.

Kinetic data are very few on these substitution processes. Griller et al.[148] estimated the rate constant for Reaction 28,

$$\text{Et}_3\text{Si}\cdot \; + \; \text{(benzene)} \quad \longrightarrow \quad \overset{\text{Et}_3\text{Si}}{\underset{\text{H}}{\text{C}}} \text{(ring)} \tag{28}$$

as $k_{28} \geq 10^8$ cm^3 mol^{-1} sec^{-1} utilizing an ESR method supplemented by optical modulation spectroscopy. Chatgilialoglu et al.[118] obtained rate constants for the reactions of the Et$_3$Si$^\cdot$ radical with some aromatic and heteroaromatic compounds at about 300 K by a flash-photolysis technique with benzil as a probe. The results are shown in Table 11. For simple aromatic and heteroaromatic compounds the Et$_3$Si$^\cdot$ radical adds to the ring giving rise to a substituted hexadienyl radical, as has been shown by ESR spectroscopy in many instances.[144] For chlorobenzene the addition is much more important than the abstraction of a Cl atom because chlorobenzene is twice as reactive as 1-chloropentane, viz., 3.1×10^8 cm^3 mol^{-1} sec^{-1} (see Table 12). For benzonitrile, the Et$_3$Si$^\cdot$ radical may be presumed to add to the ring in view of the fact that the triphenylsilyl radical adds mainly to the 4-position of the aromatic ring rather than to CN group.[149] For aromatics with a conjugated C–C double or triple bond the rate constant is enhanced by two orders of magnitude relative to simple aromatics. It is clear that addition occurs at the terminus of the multiple bond to form a resonance-stabilized benzylic radical as in Equation 29.

$$\text{Et}_3\text{Si}^\cdot \; + \; \text{H}_2\text{C=CHC}_6\text{H}_5 \rightarrow \text{Et}_3\text{SiCH}_2\overset{\cdot}{\text{C}}\text{HC}_6\text{H}_5 \tag{29}$$

The trend in the relative rate constants in Table 11 shows that reactivity is enhanced by both electron-withdrawing and electron-donating substituents; reactivity is determined mainly by the degree of stabilization of the intermediate cyclohexadienyl radicals.

Table 11
RATE CONSTANTS FOR THE REACTIONS OF THE Et$_3$Si$^.$ RADICAL WITH SOME AROMATIC AND HETEROAROMATIC COMPOUNDS AT ~300 K[a]

Substrate	k/cm^3 mol^{-1} sec^{-1}	k/relative
Benzene	$(4.6 \pm 1.0) \times 10^8$	1
Chlorobenzene	$(6.9 \pm 0.2) \times 10^8$	1.5
Toluene	$(1.2 \pm 0.2) \times 10^9$	2.6
Anisole	$(1.7 \pm 0.1) \times 10^9$	3.7
Benzonitrile	$(3.2 \pm 0.2) \times 10^9$	7.0
1-Methyl-naphthalene	$(5.0 \pm 0.5) \times 10^9$	10.9
Phenylacetylene[b]	$(1.0 \pm 0.1) \times 10^{11}$	217.4
Styrene[c]	$(2.2 \pm 0.2) \times 10^{11}$	478.3
Pyridine	$(1.3 \pm 0.3) \times 10^9$	2.8
Pyrrole	$(6.0 \pm 0.3) \times 10^8$	1.3
Furan	$(1.4 \pm 0.1) \times 10^9$	3.0
Thiophene	$(5.0 \pm 0.2) \times 10^9$	10.9

[a] Reference 118.
[b] Table 9.
[c] Table 8.

Table 12
ABSOLUTE ARRHENIUS PARAMETERS FOR Cl ABSTRACTION BY THE Me$_3$Si$^.$ RADICAL AND RELATIVE RATE CONSTANTS FOR Me$_3$Si$^.$, Cl$_3$Si$^.$, AND F$_3$Si$^.$ RADICALS AT 548 K[a]

RCl	log(A/cm^3 mol^{-1} sec^{-1})	E/kJ mol^{-1}	k(RCl)/k(MeCl) at 548 K		
			Me$_3$Si$^.$[b]	Cl$_3$Si$^.$[c]	F$_3$Si$^.$[e]
CH$_3$Cl	11.02	16.98	1	1	
CH$_3$CH$_2$Cl	10.99	15.93	1.4	2.6	1
CH$_3$CH$_2$CH$_2$Cl	10.90	14.47	1.6	3.5[d]	1.00
(CH$_3$)$_2$CHCl	11.30	15.79	3.0	9.2	9.86
(CH$_3$)$_3$CCl	11.15	12.37	4.5	30.6	41.2

[a] Obtained in the gas phase.
[b] Reference 152.
[c] Reference 151.
[d] Reference 119.
[e] Reference 153.

D. Abstraction Reactions

1. Halogen Abstraction

Halogen abstraction reactions by silyl radicals are generally facile reactions; the formation of a strong Si–X bond is the driving force of this reaction. The recent value of D(Me$_3$Si–Cl) =

$$Y_3Si^. + RX \rightarrow Y_3SiX + R^. \tag{30}$$

477 kJ mol^{-1} (14, 22) combined with D(CH$_3$–Cl) = 349 kJ mol^{-1} (6) gives $\Delta H_{30} = -128$

kJ mol^{-1} for Reaction 30 with Y=R=CH$_3$ and X=Cl. Photolytic dehalogenation of an organic halide in the presence of a hydrosilane is known to proceed via a chain reaction involving Reactions 30 and 31 as chain propagating steps.[150] Kerr et al.[151]

$$R^{\cdot} + Y_3SiH \rightarrow RH + Y_3Si^{\cdot} \qquad (31)$$

and Cadman et al.[152] postulated that the rate-determining step of this propagating cycle is Reaction 30 because they detected no products ascribable to the reactions between alkyl radicals. On the other hand, Hudson and Jackson[128] studied the photolysis of a mixture of (t-BuO)$_2$, HSiEt$_3$, and RBr in an ESR cavity at 233 K. The ESR spectra showed that the only radical present in significant amount was the radical R$^{\cdot}$ indicating that the rate-determining step in this system was Reaction 31. As we will see later (Table 12), k_{30} differs by almost five orders of magnitude depending on the haloalkane employed. Thus it would be safe to consider that the rate-determining step may shift from Reaction 30 to Reaction 31 as a substrate changes from RCl to RBr.

Cadman et al.[152] obtained relative rate constants for Cl abstraction from various organic halides by the Me$_3$Si$^{\cdot}$ radical in the gas phase. They carried out competitive experiments in which the relative rates of formation of the alkanes R^{1}H and R^{2}H from the halides R^{1}X and R^{2}X were measured. An absolute rate constant for Cl abstraction from CH$_3$CH$_2$Cl was obtained by comparing with the rate of formation of (Me$_3$Si)$_2$ and by employing the previous value of log(k_{32}/cm^3 mol^{-1} sec^{-1}) = 14.25 ± 0.3.[87] Absolute Arrhenius

$$Me_3Si^{\cdot} + Me_3Si^{\cdot} \rightarrow Me_6Si_2 \qquad (32)$$

parameters together with the relative rate for Cl$_3$Si$^{\cdot}$[119-151] and F$_3$Si$^{\cdot}$[153] radicals are shown in Table 13. Table 13 shows that A factors are almost constant and well within the *normal* range of 10$^{11.5 \pm 0.5}$ cm^3 mol^{-1} sec^{-1}.[154] The reactivity appears to be controlled by the difference of the activation energies which follow a Polanyi relationship (Equation 33). The relative rate

$$E = \alpha\Delta H + C \qquad (33)$$

constants at 548 K shows that the ease of Cl abstraction from alkyl chlorides by silyl radicals increases in the sequence primary < secondary < tertiary chloride, i.e., it parallels the decreasing bond energy. Comparison of the selectivity among the silyl radicals indicates that the reactivity decreases in the order of Me$_3$Si$^{\cdot}$ > Cl$_3$Si$^{\cdot}$ > F$_3$Si$^{\cdot}$. Cooper[155] has found a similar decrease in reactivity: Me$_3$Si$^{\cdot}$ > Me$_2$ClSi$^{\cdot}$ > MeCl$_2$Si$^{\cdot}$ > Cl$_3$Si$^{\cdot}$ in the competitive study of Cl abstraction in solution. Cooper[155] explained this trend in reactivity by suggesting that d-orbital contribution may give rise to an increase in delocalization of an unpaired electron with increasing chlorine substitution hence making the formation of a new bond more difficult.

The relative activation energies of Cl abstraction by the Cl$_3$Si$^{\cdot}$ radical[119,151] were also found to be reasonably linear with the strength of the bond broken as predicted by the Polanyi relationship. The F$_3$Si$^{\cdot}$ radical,[153] however, exhibits no such relationship. Polar repulsions between the incoming radical and the halogen to be abstracted have been invoked to rationalize the relative reactivities of these radicals.

Aloni et al.[156,157] obtained Arrhenius parameters for the abstraction of Cl atoms by the Et$_3$Si$^{\cdot}$ radical from chloroalkanes in the liquid phase, relative to that from n-pentyl bromide. The results are shown in Table 14. The relative Arrhenius parameters show that variations in both A factors and activation energies are responsible for the reactivity trends. Comparison

Table 13
RELATIVE ARRHENIUS PARAMETERS AND RELATIVE RATE
CONSTANTS FOR Cl ABSTRACTION BY THE Et$_3$Si$^.$ RADICAL[a]

RCl	Temp. (K)	$\log \dfrac{A(RCl)}{A(RBr)}$[b]	E(RCl) − E(RBr)[b] (kJ mol^{-1})	$\dfrac{k(RCl)}{k(CCl_3H)}$[c]	Ref.
CCl$_3$H	353			1	159
	273—423	0.28 ± 0.16	0.79 ± 1.17	1	156
CCl$_2$H$_2$	303—343	0.40 ± 0.10	4.64 ± 0.75	0.32	156
CCl$_3$CH$_3$	353			1.40 ± 0.01	159
CCl$_3$CH$_2$Cl	353			3.07 ± 0.01	159
CCl$_3$CHCl$_2$	353			6.19 ± 0.2	159
	295—448	0.72 ± 0.11	−2.55 ± 0.71	8.00	157
CCl$_3$CCl$_3$	353			7.18 ± 0.3[d]	159
	273—404	1.44 ± 0.08	0.50 ± 0.50	7.40[d]	157
CCl$_4$	353			11.4 ± 0.8	159
	297—398	1.25 ± 0.28	−0.71 ± 0.92	14.4	156
CCl$_3$CN	323—448	1.86 ± 0.12	1.00 ± 1.00	32.7	156
CCl$_2$HCCl$_2$H	292—413	1.06 ± 0.12	5.65 ± 0.79	1.06	157
n-C$_5$H$_{11}$Br	273—448	0	0	0.64	156

[a] Obtained in the liquid phase.
[b] n-C$_5$H$_{11}$Br was used as a reference.
[c] Rate ratio at 353 K.
[d] Value per CCl$_3$ group.

Table 14
ARRHENIUS PARAMETERS AND ABSOLUTE RATE
CONSTANTS AT ~300 K FOR HALOGEN ATOM
ABSTRACTIONS BY THE Et$_3$Si$^.$ RADICAL[a]

Substrate	log(A/cm^3 mol^{-1} sec^{-1})	E/kJ mol^{-1}	k/cm^3 mol^{-1} sec^{-1}
(CH$_3$)$_2$CHI			(1.4 ± 1.0) × 10^{13}
CH$_3$CH$_2$I	13.4 ± 0.5	4.2 ± 2.5	(4.3 ± 0.6) × 10^{12}
CH$_3$I			(8.1 ± 0.4) × 10^{12}
C$_6$H$_5$I			(1.5 ± 0.4) × 10^{12}
CH$_2$=CHCH$_2$Br			(1.5 ± 0.2) × 10^{12}
C$_6$H$_5$CH$_2$Br	13.3 ± 0.4	5.4 ± 2.1	(2.4 ± 0.1) × 10^{12}
(CH$_3$)$_3$CBr	12.7 ± 0.1	3.6 ± 0.5	(1.1 ± 0.05) × 10^{12}
CH$_3$(CH$_2$)$_4$Br	12.3 ± 0.3	3.0 ± 1.5	(5.4 ± 0.1) × 10^{11}
c-C$_3$H$_5$Br			(3.4 ± 0.6) × 10^{11}
C$_6$H$_5$Br			(1.1 ± 0.8) × 10^{11}
CCl$_4$	13.2 ± 0.2	3.3 ± 1.0	(4.6 ± 0.8) × 10^{12}
CH$_2$=CHCH$_2$Cl			(2.4 ± 0.5) × 10^{10}
C$_6$H$_5$CH$_2$Cl	11.9 ± 0.4	8.8 ± 2.1	(2.0 ± 0.6) × 10^{10}
(CH$_3$)$_3$CCl	11.7 ± 0.5	13.4 ± 2.9	(2.5 ± 0.2) × 10^{9}
CH$_3$(CH$_2$)$_4$Cl			(3.1 ± 1.4) × 10^{8}
CH$_3$SO$_2$Cl			(3.18 ± 0.01) × 10^{12}
C$_6$H$_5$SO$_2$Cl			(4.56 ± 0.18) × 10^{12}
C$_6$H$_5$CH$_2$SO$_2$Cl			(5.73 ± 0.28) × 10^{12}
CH$_3$SO$_2$F			(1.30 ± 0.12) × 10^{10}
p-CH$_3$C$_6$H$_5$SO$_2$F			(8.93 ± 1.05) × 10^{9}

[a] References 138, 160, and 161.

with the previous report on Cl$_3$Si$^{\cdot}$ [158] gives the trend in reactivity of Et$_3$Si$^{\cdot}$ > Cl$_3$Si$^{\cdot}$ in agreement with the trends by other groups. [153,155]

Table 14 includes the relative rate constants for Cl abstraction by the Et$_3$Si$^{\cdot}$ radical by Nagai et al. [159] who studied the reactions of Et$_3$SiH and polychloroalkanes in the presence of Bz$_2$O$_2$ at 353 K in the liquid phase. Plots of log(relative rates) vs. Taft σ^* values of the substituents in XCCl$_3$ gave a straight line with the exception of HCCl$_3$. The result was interpreted as indicating that the polar factor was primarily important in determining the reactivity of the Cl atom in XCCl$_3$. For comparison relative rate constants at 353 K were calculated using the Arrhenius parameters by Aloni et al. Two sets of data are in satisfactory agreement with each other.

Recently, Ingold et al. [138,160] measured the absolute rate constants for the reaction of the Et$_3$Si$^{\cdot}$ radical with a number of organic halides in solution by using a laser flash photolysis technique (Table 12). The trends in reactivity for halogen abstraction from RX by the Et$_3$Si$^{\cdot}$ radical are those which would be expected; for a particular R group the rate constants decrease along the series X=I > Br > Cl > F, while for a particular X the rate constants decrease along the series R=ally > benzyl > *tert*-alkyl (>*sec*-alkyl) > primary alkyl > cyclopropyl > phenyl. Detailed comparison between the rate constant for Et$_3$Si$^{\cdot}$ + CCl$_4$ and that for Et$_3$Si$^{\cdot}$ + C$_6$H$_5$CH$_2$Cl revealed that the high reactivity of CCl$_4$ is a consequence of its large A factor, which further indicates the great importance of polar contributions to the transition state for the Cl abstraction; it is proposed [160] that *polar* contributions to the transition state can enhance the reaction rate by increasing the A factor. Table 12 includes absolute rate constants for halogen abstraction from some sulfonyl halides at 300 K. [161] The rate constants for Cl abstraction are all very similar and the kinetic discrimination is not possible among these diffusion-controlled reactions. Thus *slow* halogen abstraction was attempted using sulfonyl fluorides. Again, there is not much difference between the two rate constants for F abstraction reaction, and it is concluded that arylsulfonyls are not much stabilized relatively to alkyl sulfonyl radicals.

Cooper [162] photolyzed a mixture of chlorobenzene and 4 molar excess of Me$_3$SiH and found 11% Me$_3$SiPh and 13% C$_6$H$_6$ in addition to Me$_3$SiCl (Equation 34). The reaction of the Me$_3$Si$^{\cdot}$ radical with

$$\text{PhCl} + \text{Me}_3\text{SiH} \xrightarrow{h\nu} \text{Me}_3\text{SiPh} + \text{C}_6\text{H}_6 + \text{Me}_3\text{SiCl} \qquad (34)$$
$$\phantom{\text{PhCl} + \text{Me}_3\text{SiH} \xrightarrow{h\nu}}\ \ 11\% \qquad\quad 13\%$$

halobenzene involves not only conventionally admitted addition-displacement process (producing Me$_3$SiPh) [163,164] but also abstraction of a Cl atom (producing C$_6$H$_6$).

2. Chlorine Abstraction from Si–Cl Bonds

Atton et al. [165] reported some gas-phase silyl radical reactions in which rapid Cl atom abstraction from a Si–Cl bond was

$$\text{Me}_2\dot{\text{S}}\text{iCl} + \text{Me}_2\text{SiHCl} \rightarrow \text{Me}_2\text{SiCl}_2 + \text{Me}_2\dot{\text{S}}\text{iH} \qquad (35)$$

observed. Davidson and Matthews [166] further attempted to measure kinetics of the unusual abstraction reaction (Equation 36) in the

$$\text{Me}_3\text{Si}^{\cdot} + \text{Me}_2\text{SiCl}_2 \rightarrow \text{Me}_3\text{SiCl} + \text{Me}_2\dot{\text{S}}\text{iCl} \qquad (36)$$

mercury-photosensitization of Me$_3$SiH in the presence of Me$_2$SiCl$_2$ which has a quenching cross-section for Hg* 400 times smaller than that of Me_3SiH.[166] Kinetic data were obtained by measuring the rate of formation of Me_3SiCl competitively with that of $(Me_3Si)_2$. The rate constant for Reaction 36 was given as in Equation 37 based upon $k_{32} = 10^{13}$ cm^3 mol^{-1} sec^{-1} for recombination of the $Me_3Si^{\cdot}$ radical (Table 5).[26] It is interesting to compare k_{36} with the rate

$$\log(k_{36}/cm^3mol^{-1}sec^{-1}) = 9.2 \pm 1.1 - (15.1 \pm 1.4) \text{ kJ mol}^{-1}/2.303 \text{ RT} \quad (37)$$

constants obtained for Cl abstraction from alkyl chloride by the $Me_3Si^{\cdot}$ radical in the gas phase (Table 13). In fact, A factors in Table 13 were based on a *high* value for k_{32} ($10^{14.25}$ cm^3 mol^{-1} sec^{-1})[87] and if they were adjusted to $k_{32} = 10^{13}$ cm^3 mol^{-1} sec^{-1}, they would decrease by $10^{0.63}$ and would range between $10^{10.3}$ and $10^{10.7}$ cm^3 mol^{-1} sec^{-1}. Still these are one power of ten higher than the present A_{36} value. To account for the low A factor, a cyclic

$$\begin{matrix} & Cl & \\ Me_3Si & & SiMe_2 \\ & Cl & \end{matrix}$$

bridged structure was assumed for the transition state for the Cl abstraction from dimethyl-dichlorosilane by the $Me_3Si^{\cdot}$ radical.

3. Hydrogen Abstraction

Three groups[91,96,167] have almost simultaneously investigated the direct photolysis of Me_4Si in the gas phase and reported that the yield of Me_3SiH was pretty high. These authors quite agree in that Me_3SiH formed during the photolysis of Me_4Si is mostly arising from hydrogen abstraction reaction by the $Me_3Si^{\cdot}$ radical (Equation 38) and that the contribution from Reaction 39 to

$$Me_3Si^{\cdot} + Me_4Si \rightarrow Me_3SiH + Me_3SiCH_2^{\cdot} \quad (38)$$

$$2Me_3Si^{\cdot} \rightarrow Me_3SiH + Me_2Si{=}CH_2 \quad (39)$$

Me_3SiH formation is of minor importance (see Table 6). However, the values of k_{38} derived in these studies are in considerable discrepancy.

Gammie et al.[167] reported a high yield (quantum yield = 0.1) of that part of Me_3SiH which arises from a hydrogen uptake by the $Me_3Si^{\cdot}$ radical (Equation 38) in the 147 nm photolysis of Me_4Si. But the readiness of hydrogen abstraction by the $Me_3Si^{\cdot}$ radical was ascribed to a large excess of vibrational energy that the radical carries on the occasion of the initial photochemical split.

Tokach and Koob[96] photolyzed Me_4Si in the presence of a small amount of ethylene as a $Me_3Si^{\cdot}$ radical scavenger, and measured the ratio of $[Me_3SiH]/[Me_6Si_2]$ at varying $[Me_3Si^{\cdot}]$ concentrations. The ratio is expected to follow the relationship shown in Equation 40,

$$[Me_3SiH]/[Me_6Si_2] = (t/k_{41}[Me_6Si_2])^{1/2}k_{38}[Me_4Si] + k_{39}/k_{41} \quad (40)$$

$$2Me_3Si^{\cdot} \rightarrow Me_6Si_2 \quad (41)$$

where t is the photolysis time. An evaluation of the slope of the plot of Equation 40, combined with $k_{41} = 10^{14.25}$ cm^3 mol^{-1} sec^{-1},[87] gives $k_{38} = 1.6 \times 10^7$ cm^3 mol^{-1} sec^{-1} at room temperature. This k_{38} value is many orders of magnitude higher than anticipated based on earlier work ($k_{38} = 10^{-4.6}$ cm^3 mol^{-1} sec^{-1} adjusted to the updated value of D(H–Me$_3$Si) = 377.8 kJ mol^{-1} in Table 1).[168] Thus a more detailed examination of k_{38} was forwarded. Deuterium donors were added to this system and the yields of CH$_3$D, C$_2$H$_6$, Me$_3$SiD, and (Me$_3$Si)$_2$ were measured. Kinetic treatment of the rates of these products

$$CH_3^\cdot + RD \rightarrow CH_3D + R \tag{42}$$

$$Me_3Si^\cdot + RD \rightarrow Me_3SiD + R \tag{43}$$

allows the determination of k_{43}/k_{42}. The k_{43}/k_{42} values thus obtained are 23.4 ± 3.2 (D$_2$S) and 8.5 ± 1.1 (CD$_3$OD). As a test of the conclusion that the Me$_3$Si$^\cdot$ radical abstracts a hydrogen atom more readily than the CH$_3^\cdot$ radical, they photolyzed Me$_4$Si in excess CD$_4$ (which does not absorb light at 147 nm) and found that Me$_3$SiD was produced. Measuring the ratio of [Me$_3$SiD]/[Me$_6$Si$_2$], they obtained k_{43}(CD$_4$) = (3.21 ± 0.58) × 10^3 cm^3 mol^{-1} sec^{-1} which, combined with k_{42}(CD$_4$) in the literature,[169] gives k_{43}/k_{42} = 19.3. To check the effect of the excess energy carried by the Me$_3$Si$^\cdot$ radical formed by the primary photodissociation, Me$_4$Si was diluted with N$_2$ by a factor of 15 and total pressure up to 460 torr. But no effect on k_{43}/k_{42} could be detected. Tokach and Koob[96] concluded that the Me$_3$Si$^\cdot$ radical abstracts a hydrogen atom from a variety of donors approximately 20 times faster than the methyl radical.

This conclusion has been rejected by Bastian et al.[91] who also studied the direct photolysis of Me$_4$Si and obtained log (k_{38}/cm^3 mol^{-1} sec^{-1}) = 11.36 − (52.0 ± 8) kJ mol^{-1}/2.303 RT based on k_{41} = 3 × 10^{13} cm^3 mol^{-1} sec^{-1}. Bastian's value (k_{38} = 1.8 × 10^2 cm^3 mol^{-1} sec^{-1} at 298 K) is five orders of magnitude smaller than the value by Tokach and Koob.[96] The endothermicity of Reaction 38 is 38 ± 9 kJ mol^{-1},[21] which will also be the minimum activation energy, thereby giving a minimum A factor of 2.3 × 10^{13} cm^3 mol^{-1} sec^{-1} based on the value by Tokach and Koob. It was suggested that this value is so high that it can be immediately disregarded.

To shed more light on the hydrogen atom abstraction by the Me$_3$Si$^\cdot$ radical, Ellul et al.[89] employed the simpler system, viz., mercury photosensitization of D$_2$ in the presence of a small amount of Me$_3$SiH. This method has an advantage over the earlier studies[91,96,167] in that the Me$_3$Si$^\cdot$ radical is regarded to be thermally equilibrated as being generated by a chemical reaction as opposed to a photochemical process. Light intensities were kept low to avoid complicating reactions such as radical-radical reactions. The Arrhenius parameters of the Reaction 44 have been determined

$$Me_3Si^\cdot + D_2 \rightarrow Me_3SiD + D \tag{44}$$

relative to Reaction 41, yielding log[($k_{44}/k_{41}^{1/2}$)/cm$^{3/2}$ mol$^{-1/2}$ sec$^{-1/2}$] = 4.89 ± 0.5 − (67 ± 6) kJ mol^{-1}/2.303 RT. This relative value can be put on an absolute basis using the same k_{41} value as in Reference 91: k_{44} = 0.77 cm^3 mol^{-1} sec^{-1} at 298 K. Thus, Elull et al.[89] also rejected the high values by Tokach and Koob in terms of the similar thermochemical consideration as employed by Bastian et al.[91] It is concluded that those high rate constants for hydrogen abstraction by the Me$_3$Si$^\cdot$ radical are not those for a thermally equilibrated system but quite likely to be those for a excited species brought about by the initial photodecomposition.

Gammie et al.[88] studied the photolysis of *bis*(trimethylsilyl)mercury in the presence of a hydrogen donor, and obtained rate constants for hydrogen abstraction from Si–H bonds by the Me$_3$Si$^\cdot$ radical. The results are shown in Table 15. The values of k_{abst} by the Me$_3$Si$^\cdot$

Table 15
RATE CONSTANTS FOR
HYDROGEN ABSTRACTION
BY THE Me$_3$Si˙ RADICAL
AND THE CH$_3$˙ RADICAL

	k_{abstr}/cm^3 mol^{-1} sec^{-1}	
Substrate	**Me$_3$Si˙[a,b,c]**	**CH$_3$[d]**
Si$_2$H$_6$	3.25×10^8	6.9×10^7
SiH$_4$	6.51×10^6	5.0×10^6
MeSiH$_3$	1.95×10^6	2.1×10^6
Me$_2$SiH$_2$	4.73×10^5	1.0×10^6
Me$_3$CH	2.55×10^4	5.1×10^5

[a]   Reference 88.
[b]   Calculated based on $k_{comb} = 1.5 \times 10^{12}$
      cm^3 mol^{-1} sec^{-1} (see Table 5).
[c]   Corrected for cross disproportionation.
[d]   References 170 and 171.

radical are as high as those by the CH$_3$˙ radical. This tendency is rather in agreement with that by Gammie et al.[167] and Tokach and Koob[96] but is surprising in view of the

$$Me_3Si˙ + HSiR_3 \rightarrow Me_3SiH + R_3Si˙ \qquad (45)$$

essential thermoneutrality of the process (Reaction 45). It is suggested that the ease of Reaction 45 reflects a diminished triplet repulsion in the $\geq$Si··H··Si$\leq$ as compared with the $\geq$C··H··Si$\leq$ transition state.

E. Bimolecular Homolytic Substitution (S$_H$2)

The process involving the radical attack on a molecule to

$$R˙ + AB \rightarrow RA + B˙ \qquad (46)$$

replace another radical is termed bimolecular homolytic substitution (S$_H$2) (Equation 46). When the attack is on an univalent atom such as hydrogen or halogen, the process is usually called hydrogen-(halogen-) atom abstraction. The majority of the bimolecular homolytic substitutions are of this type. S$_H$2 reactions at multivalent atoms (most frequently elements of group II-V) have also been a subject of extensive studies. Excellent reviews are available on this subject.[172-174]

In rare cases silicon serves as a substitution center for this process.[4,5,174] Homolytic substitutions by atoms such as I,[175] Br,[176,177] and H[110,178] on disilanes (Equation 47) seem to have

$$X˙ + Me_3SiSiMe_3 \rightarrow Me_3SiX + Me_3Si˙ \qquad (47)$$

been established. Band and Davidson[175] studied the gas-phase reaction between hexamethyldisilane and iodine between 458 and 523 K. The reaction was found first order in hexamethyldisilane and one half order in iodine. The only product was trimethylsilyl iodide. A simple chain sequence (Equation 48) was proposed, from which

$$I_2 \rightleftarrows 2I^{\cdot} \tag{48a}$$

$$I^{\cdot} + Me_3SiSiMe_3 \rightarrow Me_3SiI + Me_3Si^{\cdot} \tag{48b}$$

$$Me_3Si^{\cdot} + I_2 \rightarrow Me_3SiI + I \tag{48c}$$

the Arrhenius parameter for reaction (48b) was given by Equation 49.

$$\log(k_{48b}/cm^3mol^{-1}sec^{-1}) = 11.23 \pm 0.50 - (33.8 \pm 4.5)kJ\ mol^{-1}/2.303RT \tag{49}$$

Pollock[110] obtained the rate constants of Reactions 50 and 51 as $k_{50} = 6.7 \times 10^{11}\ cm^3\ mol^{-1}\ sec^{-1}$ and $k_{51} = 5.0 \times 10^{11}\ cm^3\ mol^{-1}\ sec^{-1}$ at 298 K

$$H + H_3SiSiH_3 \rightarrow H_4Si + H_3Si^{\cdot} \tag{50}$$

$$H + D_3SiSiD_3 \rightarrow D_3SiH + D_3Si^{\cdot} \tag{51}$$

based upon the product yields from mercury photosensitization of hydrogen in the presence of disilane and ethylene. Very recently, Ellul et al.[90] studied the pulsed mercury-sensitized photolysis of hydrogen in the presence of disilane. The pseudo-first-order rate constant of disappearance of the hydrogen atom extrapolated to zero-light intensity, is linearly dependent on hexamethyldisilane concentration. The slope of the plots gives $k_{52} = (2.14 \pm 0.15) \times 10^{10}\ cm^3\ mol^{-1}\ sec^{-1}$ at 295 K, and further, combined with the temperature

$$H + Me_3SiSiMe_3 \rightarrow Me_3SiH + Me_3Si^{\cdot} \tag{52}$$

dependence, the rate constant in the Arrhenius form (Equation 53).

$$\log(k_{52}/cm^3mol^{-1}sec^{-1}) = 12.9 \pm 0.1 - (14.7 \pm 0.4)\ kJ\ mol^{-1}/2.303RT \tag{53}$$

These rate constants are summarized in Table 16. The big difference between k_{48b} and k_{52} stems almost from the difference in the activation energies. Qualitatively bond energy consideration supports this trend, viz., $D(Me_3Si-H) = 378\ kJ\ mol^{-1}$ and $D(Me_3Si-I) = 322\ kJ\ mol^{-1}$.[10]

Two types of S_H2 reactions of the siloxy radical attack on silanes, viz., Reactions 54[179] and 55,[167] and 56[180] have been postulated as one of the chain propagating steps in the reactions between silyl radicals and NO.

$$MeR_2SiO^{\cdot} + MeR_2SiH \rightarrow MeR_2SiOSiMeR_2 + H \tag{54}$$

$$Me_3SiO^{\cdot} + Me_4Si \rightarrow Me_3SiOSiMe_3 + Me^{\cdot} \tag{55}$$

$$H_3SiO^{\cdot} + H_4Si \rightarrow H_3SiOSiH_2^{\cdot} + H_2 \tag{56}$$

S_H2 reactions involving silyl radicals as attacking radicals are known with Ph_2P-PPh_2,[181] $P(OR)_2-OR$,[181] Ph_2P-OR,[181] $R_2CH-SCHR_2$,[182] and $R-OR$[183] where the left-hand groups are to leave. No kinetic study has appeared yet.

Table 16
RATE CONSTANTS FOR S_H2 REACTIONS
INVOLVING DISILANES AT 298 K

Reaction	k/cm³ mol⁻¹ sec⁻¹	Ref.
48b		
$I + Me_3SiSiMe_3 \rightarrow Me_3SiI + Me_3Si^{\cdot}$	2.1×10^5	175
52		
$H + Me_3SiSiMe_3 \rightarrow Me_3SiH + Me_3Si^{\cdot}$	2.1×10^{10}	90
50		
$H + H_3SiSiH_3 \rightarrow H_4Si + H_3Si^{\cdot}$	6.7×10^{11}	110
51		
$H + D_3SiSiD_3 \rightarrow D_3SiH + D_3Si^{\cdot}$	5.0×10^{11}	110

F. Miscellaneous

1. Spin Trapping

A number of reports on the spin trapping of trialkylsilyl radicals have appeared. Interested readers may refer to the recent studies on phenyl *tert*-butyl nitrone,[118,184-187] *tert*-butyl nitroxide,[186-190] nitro compounds,[118,191-193] alkyl isocyanate,[118,194] alkyl isocyanide,[118] *N*-dicyanomethyleneaniline,[195] diphenyl diazomethane,[196] *di-tert*-butyl thioketone,[197] quinone,[186] and kinetics of spin trapping.[118,190,198]

2. Reaction with NO

Nay et al.[179] studied the mercury photosensitization of methylsilanes and silane. The major products were hydrogen and disilane. Smaller amounts of higher polysilanes with four or more silicon atoms were also detected. On addition of a small amount of NO to this system, the formation of disilane and higher polysilanes was completely suppressed. Instead, the simultaneous appearance of the corresponding disiloxane was pointed out. The chain mechanism for the formation of disiloxane have been proposed.

$$MeR_2Si^{\cdot} + NO \rightarrow MeR_2SiON \tag{57}$$

$$2MeR_2SiON \rightarrow [MeR_2SiON=NOSiR_2Me] \tag{58}$$

$$[MeR_2SiON=NOSiR_2Me] \rightarrow 2MeR_2SiO^{\cdot} + N_2 \tag{59}$$

$$MeR_2SiO^{\cdot} + MeR_2SiH \rightarrow MeR_2SiOSiR_2Me + H \tag{60}$$

$$H + MeR_2SiH \rightarrow H_2 + MeR_2Si^{\cdot} \tag{61}$$

Kamaratos and Lampe[180] performed the detailed study on the mechanism of the reaction of silyl radicals with NO. The mercuryphotosensitized reaction of silane or methylsilane with NO was monitored by a mass spectrometer which was directly connected to the reaction cell. It was revealed that the addition of NO caused an induction period for the formation of disilane and that during the induction period, di-, tri-, and tetra-siloxanes were produced. The results are rationalized by the following chain

$$SiH_3 + NO \rightarrow SiH_3ON \tag{62}$$

$$SiH_3ON + NO \rightarrow SiH_3ONNO \tag{63}$$

$$SiH_3ONNO \rightarrow SiH_3O + N_2O \qquad (64)$$

$$SiH_3O + SiH_4 \rightarrow SiH_3OSiH_2 + H_2 \qquad (65)$$

$$SiH_3OSiH_2 + SiH_4 \rightarrow SiH_3OSiH_3 + SiH_3 \qquad (66)$$

mechanism. In this mechanism, Reaction 65 is somewhat speculative because hydrogen to be formed in this reaction was not measured. The siloxysilyl radical, SiH_3OSiH_2 arising via Reaction 65 undergoes reaction with NO and after the similar reaction cycle as Reactions 62 to 66, trisiloxane, $SiH_3OSiH_2OSiH_3$, is formed. The formation of higher siloxanes is elucidated in the same way.

3. Reaction with O_2

Reactions of silyl radicals with O_2 have recently been investigated in the UV-visible ($\geqslant 300$ nm) photolysis of Cl_2–SiH_4[36] or Cl_2–Cl_3SiH[37] mixtures in 700 torr of $N_2 + O_2$ on the basis of FTIR product analysis. The $SiH_3^{\cdot}$ radical was postulated to react with O_2 to form $HSi(=O)OH$ which may readily undergo polymerization (Equations 67 and 68).[36]

$$SiH_3^{\cdot} + O_2 \rightarrow SiH_3OO^{\cdot} \rightarrow HSi(=O)OH + H \qquad (67)$$

$$nHSi(=OH) \rightarrow -(HSi(OH)O)-_n \qquad (68)$$

A radical-chain mechanism was proposed for the oxidation of the $Cl_3Si^{\cdot}$ radical in the gas phase[37] as well as in the liquid phase.[199] The presence of the transient radical, $Cl_3SiOO^{\cdot}$

$$Cl_3Si^{\cdot} + O_2 \rightarrow Cl_3SiOO^{\cdot} \qquad (69)$$

$$2Cl_3SiOO^{\cdot} \rightarrow 2Cl_3SiO^{\cdot} + O_2 \qquad (70)$$

$$Cl_3SiO^{\cdot} + Cl_3SiH \rightarrow Cl_3SiOH + Cl_3Si^{\cdot} \qquad (71)$$

has been evidenced by the trapping technique employing NO_2 which has been well-known as a good trap for alkylperoxy radicals,[200] viz., $Cl_3SiOO^{\cdot} + NO_2 (+M) \rightarrow Cl_3SiOONO_2$ $(+M)$.

4. Reaction with N_2O

In the course of the study of the reaction of $O(^3P)$ with trimethylsilane, generating $O(^3P)$ atoms by mercury photosensitization of N_2O, Hoffmeyer et al.[39] postulated that the $Me_3Si^{\cdot}$ radical, besides combining with itself or other radicals, may undergo Reaction 72. They further verified the occurrence

$$Me_3Si^{\cdot} + N_2O \rightarrow Me_3SiO^{\cdot} + N_2 \qquad (72)$$

of Reaction 72 by the kinetic treatment of the competitive system, and obtained the ratio of $k_{72}/k_{73}^{1/2}$ at room temperature.

$$2Me_2Si^{\cdot} \rightarrow Me_6Si_2 \qquad (73)$$

Assuming $k_{73} = 3.0 \times 10^{12}$ cm^3 mol^{-1} sec^{-1} (Table 5),[91] it follows that $k_{72} = 3.0 \times 10^6$ cm^3 mol^{-1} sec^{-1}.

REFERENCES

1. **Davidson, I. M. T.,** Some aspects of silicon radical chemistry, *Q. Rev.,* 25, 111, 1971.
2. **Davidson, I. M. T.,** Kinetic studies in silicon chemistry, in *Reaction Kinetics,* Vol. 1, Specialist Periodical Reports, The Chemical Society, 1975, chap. 5.
3. **Jackson, R. A.,** Group IVB radical reactions, *Adv. Free Radical Chem.,* 3, 231, 1969.
4. **Sakurai, H.,** Group IVB radicals, in *Free Radicals,* Vol. 2, Kochi, J. K., Ed., Wiley-Interscience, New York, 1973, chap. 25.
5. **Wilt, J. M.,** Radical reactions of silanes, in *Reactive Intermediates,* Vol. 3, Abramovitch, R. A., Ed., Plenum Press, New York, 1983, chap. 3.
6. **Benson, S. W.,** *Thermochemical Kinetics,* 2nd ed., Wiley, New York, 1976, 309.
7. **Pedley, J. B. and Iseard, B. S.,** CATCH Table for silicon compounds, University of Sussex, 1972 and 1977.
8. JANAF Thermochemical Tables, 2nd ed., NSRDS-NBS 37, National Bureau of Standards, Washington D.C., 1971; supplements, *J. Phys. Chem. Ref. Data,* 3, 311, 1974; 4, 1, 1975; 7, 793, 1978; 11, 695, 1982.
9. **Wagman, D. D., Evans, W. H., Parker, V. B., Halow, I., Bailey, S. M., and Schumm, R. H.,** National Bureau of Standards, Tech. Note, 270-3, 1968.
10. **Walsh, R.,** Bond dissociation energy values in silicon-containing compounds and some of their implications, *Acc. Chem. Res.,* 14, 246, 1981.
11. **O'Neal, H. E. and Ring, M. A.,** Bond additivity properties of silicon compounds, *Inorg. Chem.,* 5, 435, 1966.
12. **O'Neal, H. E. and Ring, M. A.,** An additivity scheme for the estimation of heats of formation, entropies, and heat capacities of silanes, polysilanes, and their alkyl derivatives, *J. Organomet. Chem.,* 213, 419, 1981.
13. **Davidson, I. M. T.,** Enthalpies of formation of silanes, *J. Organomet. Chem.,* 170, 365, 1979.
14. **Bell, T. N., Perkins, K. A., and Perkins, P. G.,** Heats of formation and dissociation of methylsilanes and chlorosilanes and derived radicals, *J. Chem. Soc. Faraday Trans. 1,* 77, 1779, 1981.
15. **Walsh, R.,** Thermochemistry of silicon-containing compounds. I. Silicon-halogen compounds, an evaluation, *J. Chem. Soc. Faraday Trans. 1,* 79, 2233, 1983.
16. **Ho, P., and Coltrin, M. E.,** A theoretical study of the heats of formation of SiH$_n$, SiCl$_n$, and SiH$_n$Cl$_m$ compounds, *J. Phys. Chem.,* 89, 4647, 1985.
17. **Golden, D. M. and Benson, S. W.,** Free-radical and molecule thermochemistry from studies of gas-phase iodine-atom reactions, *Chem. Rev.,* 69, 125, 1969.
18. **Dohmaru, T. and Nagata, Y.,** Kinetics of the gas-phase addition reactions of trichlorosilyl radicals. III. Additions to 2-olefins, *J. Chem. Soc. Faraday Trans. 1,* 78, 1141, 1982.
19. **Benson, S. W. and Luria, M.,** Electrostatics and the chemical bond. I. Saturated hydrocarbons, *J. Am. Chem. Soc.,* 97, 704, 1975.
20. **Doncaster, A. M. and Walsh, R.,** Kinetics of the gas-phase reaction between iodine and monosilane and the bond dissociation energy D(H$_3$Si–H), *Int. J. Chem. Kinet.,* 13, 503, 1981.
21. **Doncaster, A. M. and Walsh, R.,** Kinetics of the gas-phase reaction between iodine and trimethylsilane and the bond dissociation energy D(Me$_3$Si–H). II, *J. Chem. Soc. Faraday Trans. 1,* 75, 1126, 1979.
22. **Walsh, R. and Wells, J. M.,** Kinetics of the gas-phase reaction between iodine and trichlorosilane and the bond dissociation energy D(Cl$_3$Si H), *J. Chem. Soc. Faraday Trans. 1,* 72, 1212, 1976.
23. **Potzinger, P., Ritter, A., and Krause, J.,** Massenspektrometrische Bestimmung von Bindungsenergien in silicium-organischen Verbindungen, *Z. Naturforsch.* 30a, 347, 1975.
24. **Baldwin, A. C., Davidson, I. M. T., and Reed, M. D.,** Mechanism of thermolysis of tetramethylsilane and trimethylsilane, *J. Chem. Soc. Faraday Trans. 1,* 74, 2171, 1978.
25. **Walsh, R. and Wells, J. M.,** Kinetics of the gas-phase reaction between iodine and trimethylsilane and the bond dissociation energy D(Me$_3$Si–H), *J. Chem. Soc. Faraday Trans. 1,* 72, 100, 1976.
26. **Davidson, I. M. T. and Howard, A. V.,** Mechanism of thermolysis of hexamethyldisilane and the silicon-silicon bond dissociation energy, *J. Chem. Soc. Faraday Trans. 1,* 71, 69, 1975.
27. **Arthur, N. L. and Bell, T. N.,** An evaluation of the kinetic data for hydrogen abstraction from silanes in the gas phase, *Rev. Chem. Intermed.,* 2, 37, 1978.

28. **Volpe, P. and Castiglioni, M.,** Recoil tritium reactions with CH_3SiD_3: pressure dependent yields, *J. Chem. Soc. Faraday Trans. 1,* 74, 818, 1978.

29. **Arthur, N. L., Christie, J. R., and Mitchell, G. D.,** Reactions of trifluoromethyl radicals. IV. Hydrogen abstraction from dichlorosilane and trichlorosilane, *Aust. J. Chem.,* 32, 1017, 1979.

30. **Doncaster, A. M. and Walsh, R.,** Kinetic determination of the bond dissociation energy $D(SiH_3-H)$ and its implications for bond strengths in silanes, *J. Chem. Soc., Chem. Commun.,* 904, 1979.

31. **Braruch, G. and Horowitz, A.,** Liquid-phase reactions of CCl_3 radicals with trimethylsilane and tri- ethylsilane, *J. Phys. Chem.,* 84, 2535, 1980.

32. **Chatgillialoglu, C., Scaiano, J. C., and Ingold, K. U.,** Absolute rate constants for the reactions of *tert*-butoxyl radicals and some ketone triplets with silanes, *Organometallics,* 1, 466, 1982.

33. **Choe, M. and Choo, K. Y.,** Kinetic and equilibrium study of Br atom reactions with trimethylsilane using the very low pressure reactor technique, *Chem. Phys. Lett.,* 89, 115, 1982.

34. **Park, C. R., Song, S. A., Lee, Y. E., and Choo, K. Y.,** Arrhenius parameters for the *tert*-butoxy radical reactions with trimethylsilane in the gas phase, *J. Am. Chem. Soc.,* 104, 6445, 1982.

35. **Arican, H. and Arthur, N. L.,** Reactions of methyl radicals. IV. Hydrogen abstraction from tetrame- thylsilane by the photolysis of both acetone and azomethane, *Aust. J. Chem.,* 36, 2185, 1983.

36. **Niki, H., Maker, P. D., Savage, C. M., and Breitenbach, L. P.,** An FTIR study of the kinetics and mechanism for the Cl- and Br-atom-initiated oxidation of SiH_4, *J. Phys. Chem.,* 89, 1752, 1985.

37. **Niki, H., Maker, P. D., Savage, C. M., Breitenbach, L. P., and Hurley, M. D.,** FTIR study of the kinetics and mechanism for the Cl-atom-initiated reactions of $SiHCl_3$, *J. Phys. Chem.,* 89, 3725, 1985.

38. **Choo, K. Y. and Choe, M. H.,** A kinetic study of Br atom reactions with trimethylsilane by the VLPR (Very Low Pressure Reactor) technique, *Bull. Korean Chem. Soc.,* 6, 196, 1985.

39. **Hoffmeyer, H., Horie, O., Potzinger, P., and Reimann, B.,** Reaction of $O(^3P)$ with trimethylsilane, *J. Phys. Chem.,* 89, 2901, 1985.

40. **Morris, E. R. and Thynne, J. C. J.,** Hydrogen atom abstraction from silane, trimethylsilane and tetra- methylsilane by trifluoromethyl radicals, *Trans. Faraday Soc.,* 66, 183, 1970.

41. **Kerr, J. A., Stephens, A., and Young, J. C.,** Hydrogen abstraction reactions from organosilicon com- pounds. The reactions of methyl, trifluoromethyl, and ethyl radicals with trichlorosilane, *Int. J. Chem. Kinet.,* 1, 371, 1969.

42. **Milligan, D. E. and Jacox, M. E.,** Infrared and ultraviolet spectra of the products of the vacuum-ultraviolet photolysis of silane isolated in an argon matrix, *J. Chem. Phys.,* 52, 2594, 1970.

43. **Jacox, M. E. and Milligan, D. E.,** Matrix-isolation study of the vacuum-ultraviolet photolysis of trich- lorosilane. The infrared spectrum of the free radical $SiCl_3$, *J. Chem. Phys.,* 49, 3130, 1968.

44. **Milligan, D. E., Jacox, M. E., and Guillory, W. A.,** Matrix-isolation study of the vacuum-ultraviolet photolysis of trifluorosilane. Infrared spectrum of the trifluorosilyl free radicals, *J. Chem. Phys.,* 49, 5330, 1968.

45. **Cochran, E. L.,** 4th International Symposium of Free Radical Stabilization, Washington D.C., D-I-1, 1959.

46. **Symons, M. C. R.,** Electron-spin resonance of some simple oxy radicals, *Adv. Chem. Ser.,* 36, 76, 1962.

47. **Morehouse, R. L., Christiansen, J. J., and Gordy, W.,** ESR of free radicals trapped in inert matrices at low temperature: CH_3, SiH_3, GeH_3, and SnH_3, *J. Chem. Phys.,* 45, 1751, 1966.

48. **Jackel, G. S. and Gordy, W.,** Electron spin resonance of free radicals formed from group-IV and group- V hydrides in inert matrices at low temperature, *Phys. Rev.,* 176, 443, 1968.

49. **Krusic, P. J. and Kochi, J. K.,** Electron spin resonance of organosilyl radicals in solution, *J. Am. Chem. Soc.,* 91, 3938, 1969.

50. **Bennett, S. W., Eaborn, C., Hudson, A., Hussain, H. A., and Jackson, R. A.,** Electron spin resonance spectra of trimethyl-silyl, trimethylgermyl and related free radicals in solution, *J. Organomet. Chem.,* 16, P36, 1969.

51. **Bennett, S. W., Eaborn, C., Hudson, A., Jackson, R. A., and Root, K. D. J.,** Electron spin resonance study of some silyl radicals, *J. Chem. Soc. A,* 348, 1970.

52. **Symons, M. C. R. and Sharp, J. H.,** Unstable intermediates. LXXXI. Electron spin resonance spectra of γ-irradiated methyl silanes: methyl silyl radicals, *J. Chem. Soc. A,* 3084, 1970.

53. **Begum, A., Sharp, J. H., and Symons, M. C. R.,** Electronegativity and structure, *J. Chem. Phys.,* 53, 3756, 1970.

54. **Katsu, T., Yatsurugi, Y., Sato, M., and Fujita, Y.,** Silicon-29 hyperfine splitting of the silyl radical, *Chem. Lett.,* 343, 1975.

55. **Hesse, C., Leray, N., and Roncin, J.,** Structure of methyl-chlorosilyl radicals, *J. Chem. Phys.,* 57, 749, 1972.

56. **Merritt, M. V. and Fessenden, R. W.,** ESR spectra of the fluorinated silyl radicals, *J. Chem. Phys.,* 56, 2353, 1972.

57. **Raghunathan, P. and Shimokoshi, K.,** On the anisotropic electron spin resonance spectroscopic parameters and motional states of matrix-isolated silyl radical at low temperatures, *Spectrochim. Acta Part A,* 36A, 285, 1980.

58. **Roncin, J.,** Electron paramagnetic resonance study of free radicals produced by γ-irradiation at 77 K of globular derivatives of methane and silane, *Mol. Cryst.,* 3, 117, 1967.

59. **Hasegawa, A., Sogabe, K., and Miura, M.,** ESR spectra of trifluorosilyl radicals produced in a single crystal of tetrafluorosilane, *Mol. Phys.,* 30, 1889, 1975.

60. **Symons, M. C. R.,** Unstable intermediates. CLXX. Electron spin resonance studies of trifluorosilyl and related radicals, *J. Chem. Soc. Dalton Trans.,* 1568, 1976.

61. **Pauling, L.,** Structure of the methyl radical and other radicals, *J. Chem. Phys.,* 51, 2767, 1969.

62. **Benson, H. G. and Hudson, A.,** Applications of the INDO (intermediate neglect of differential overlap) method to some radicals containing second row elements, *Theoret. Chim. Acta,* 23, 259, 1971.

63. **Wong, S. K., Hutchinson, D. A., and Wan, J. K. S.,** High-order calculation of the silyl radical ESR spectrum, *Spectrosc. Lett.,* 6, 665, 1973.

64. **Gorlov, Y. I., Ukrainskii, I. I., and Penkovskii, V. V.,** UHF (unrestricted Hartree-Fock)-CONDO/2 study of tetraatomic radicals containing silicon and phosphorus, *Theoret. Chim. Acta,* 34, 31, 1974.

65. **Higuchi, J., Kubota, S., Kumamoto, T., and Tokue, I.,** Contribution of d orbitals in silylidyne radical, silylene radical, silyl radical, and silane, *Bull. Chem. Soc. Jpn.,* 47, 2775, 1974.

66. **Reffy, J.,** Organosilicon radicals. II. Methylsilyl radicals, *J. Organomet. Chem.,* 97, 151, 1975.

67. **Chaillet, M., Arriau, J., Leclerc, D., Marey, T., and Tirouflet, J.,** Theoretical study of titanium complexes. I. Comparison of molecules MX_4, MX_3^+, MX_3, and MX_3^- (M = carbon, silicon and titanium) and study of some cyclopentadienyl complexes, *J. Organomet. Chem.,* 117, 27, 1976.

68. **Marynick, D. S.,** An SCF-CI study of the structures, inversion frequencies of SiH_3, PH_3^+, SiH_3^{++}, *J. Chem. Phys.,* 74, 5186, 1981.

69. **Jordan, P. C.,** Lower electronic levels of the radicals SiH, SiH_2, and SiH_3, *J. Chem. Phys.,* 44, 3400, 1966.

70. **Hartmann, H., Papula, L., and Strehl, W.,** One center model calculations of some silicon hydrides of the type SiH_n, SiH_n^+, and SiH_n^-, *Theoret. Chim. Acta,* 17, 131, 1970.

71. **Biddles, I. and Hudson, A.,** INDO (intermediate neglect of differential overlap) study of the hyperfine coupling constants and equilibrium geometries of some tetraatomic radicals, *Mol. Phys.,* 25, 707, 1973.

72. **Wirsam, B.,** CI (configuration interaction) calculations of the low lying electronic states of the silyl, silyl (+) ion, silyl (−) ion radicals, *Chem. Phys. Lett.,* 18, 578, 1973.

73. **Aarons, L. J., Hillier, I., and Guest, M.,** Theoretical study of the structure of some trigonal radicals, *J. Chem. Soc., Faraday Trans. 2,* 70, 167, 1974.

74. **Gordon, M. S.,** Structure and stability of saline(1+), *Chem. Phys. Lett.,* 59, 410, 1978.

75. **Barone, V., Douady, J., Ellinger, Y., Subra, R., and Pauzat, F.,** Nonempirical calculations on the conformation and hyperfine structure of the silyl radical. Influence of vibrational effects, *Chem. Phys. Lett.,* 65, 542, 1979.

76. **Ellinger, Y., Pauzat, F., Barone, V., Douady, J., and Subra, R.,** Ab initio study of the vibrational dependence of hyperfine coupling constants in the methyl, silyl, and formaldehyde anion radicals, *J. Chem. Phys.,* 72, 6390, 1980.

77. **Ohta, K., Nakatsuji, H., Maeda, I., and Yonezawa, T.,** Ab initio calculation of geometries and hfs constants on methyl, silyl, and germyl radicals, *Chem. Phys.,* 67, 49, 1982.

78. **Cartledge, F. K. and Piccione, R. V.,** A theoretical study of trimethylsilyl radical and related species, *Organometallics,* 3, 299, 1984.

79. **Begum, A., Lyons, A. R., and Symons, M. C. R.,** Unstable intermediates. XC. The radicals AlR_3^-, SiR_3, and PR_3^+: their electron spin resonance spectra and pyramidal character, *J. Chem. Soc. A,* 2290, 1971.

80. **Thynne, J. C. J.,** Rate constants for trimethylsilyl radical reactions, *J. Organomet. Chem.,* 17, 155, 1969.

81. **Benson, S. W.,** *Thermochemical Kinetics,* 2nd ed., Wiley, New York, 1976, 10.

82. **Davidson, I. M. T. and Stephenson, I. I.,** The silicon-silicon bond dissociation energy in hexamethyl-disilane, *J. Chem. Soc. A,* 282, 1968.

83. **Gaspar, P. P., Haizlip, A. D., and Choo, K. Y.,** Disappearance of silyl radicals in silane. Flash photolysis-electron spin resonance kinetic study, *J. Am. Chem. Soc.,* 94, 9032, 1972.

84. **Frangopol, P. T. and Ingold, K. U.,** Rate constant for the self-reaction of trimethylsilyl radicals, *J. Organomet. Chem.,* 25, C9, 1970.

85. **Watts, G. B. and Ingold, K. U.,** Kinetic application of electron paramagnetic resonance spectroscopy. V. Self-reactions of some group IV radicals, *J. Am. Chem. Soc.,* 94, 491, 1972.

86. **Cadman, P., Tilsley, G. M., and Trotman-Dickenson, A. F.,** Recombination rate of trimethylsilyl radicals in the gas phase, *J. Chem. Soc. D,* 1721, 1970.

87. **Cadman, P., Tilsley, G. M., and Trotman-Dickenson, A. F.,** Rate of combination of trimethylsilyl radicals in the gas phase, *J. Chem. Soc. Faraday Trans. 1,* 68, 1849, 1972.

88. **Gammie, L., Safarik, I., Strausz, O. P., Roberge, R., and Sandorfy, C.,** Disproportionation and hydrogen abstraction reaction of trimethylsilyl radicals, *J. Am. Chem. Soc.,* 102, 378, 1980.

89. **Ellul, R., Potzinger, P., Reimann, B., and Camilleri, P.,** Arrhenius parameters for the system $(CH_3)_3Si + D_2 \rightleftharpoons (CH_3)_3SiD + D$. The $(CH_3)_3Si–D$ bond dissociation energy, *Ber. Bunsenges. Phys. Chem.,* 85, 407, 1981.

90. **Ellul, R., Potzinger, P., and Reimann, B.,** Reaction of hydrogen atoms with hexamethyldisilane, *J. Phys. Chem.,* 88, 2793, 1984.

91. **Bastian, E., Potzinger, P., Ritter, A., Schuchmann, H. P., von Sonntag, C., and Weddle, G.,** The direct photolysis of tetramethylsilane in the gas and liquid phases, *Ber. Bunsenges. Phys. Chem.,* 84, 56, 1980.

92. **Choo, K. Y., Beadle, P. C., Piszkiewicz, L. W., and Golden, D. M.,** An absolute measurement of the rate constant for *t*-butyl radical combination, *Int. J. Chem. Kinet.,* 8, 45, 1976.

93. **Parkes, D. A. and Quinn, C. P.,** Ultraviolet absorption spectrum of tert-butyl radicals and the rate constant for their recombination, *Chem. Phys. Lett.,* 33, 483, 1975.

94. **Ring, M. A., Puentes, M. J., and O'Neal, H. E.,** Pyrolysis of monosilane, *J. Am. Chem. Soc.,* 92, 4845, 1970.

95. **Kistiakowsky, G. B. and Roberts, E. K.,** Rate of association of methyl radicals, *J. Chem. Phys.,* 21, 1637, 1953.

96. **Tokach, S. K. and Koob, R. D.,** Photolysis of tetramethylsilane at 147 nm. Reactivity of trimethylsilyl and (dimethylsilyl)methylene, *J. Phys. Chem.,* 83, 774, 1979.

97. **Tokach, S. K. and Koob, R. D.,** Disproportionation of trimethylsilyl at 25°C. Mercury photosensitization of trimethylsilane, *J. Am. Chem. Soc.,* 102, 376, 1980.

98. **Cornett, B. J., Choo, K. Y., and Gaspar, P. P.,** Disproportionation of trimethylsilyl radicals to a sila olefin in the liquid phase, *J. Am. Chem. Soc.,* 102, 377, 1980.

99. **Strausz, O. P., Gammie, L., Theodorakoupoulous, G., Mezey, P. G., and Csizmadia, I. G.,** The ground electronic state of silaethene. An ab initio molecular orbital study of the lower electronic manifold, *J. Am. Chem. Soc.,* 98, 1622, 1976.

100. **Gusel'nikov, L. E., Nametkin, N. S., and Vdovin, V. M.,** Unstable silicon analogs of unsaturated compounds, *Acc. Chem. Res.,* 8, 18, 1975.

101. **Kumada, M.,** Current trends in organosilicon chemistry, *J. Syn. Org. Chem. Jpn.,* 40, 462, 1982.

102. **Lukevits, E. Ya. and Voronkov, M. G.,** *Organic Insertion Reactions of Group IV Elements,* Consultants Bureau, New York, 1966.

103. **Eaborn, C. and Bott, R. W.,** *Organometallic Compounds of the Group IV Elements,* Vol. 1, MacDiarmid, A. G., Ed., Dekker, New York, 1968, part 1.

104. **Sommer, L. H., Pietrusza, E. W., and Whitmore, F. C.,** Peroxide-catalyzed addition of trichlorosilane to 1-octene, *J. Am. Chem. Soc.,* 69, 188, 1947.

105. **Bennett, S. W., Eaborn, C., Jackson, R. A., and Pearce, R.,** Reaction of trimethylsilyl and trimethylgermyl radical with olefins, *J. Organomet. Chem.,* 15, P17, 1968.

106. **Dohmaru, T., Nagata, Y., and Tsurugi, J.,** Kinetic study of trichlorosilyl radical addition to ethylene in the gas phase, *Chem. Lett.,* 1031, 1973.

107. **Dohmaru, T. and Nagata, Y.,** Kinetics of gas phase addition reactions of trichlorosilyl radicals. VIII. Reinvestigation of addition to ethylene, *Bull. Chem. Soc. Jpn.,* 56, 2387, 1983.

108. **Dohmaru, T. and Nagata, Y.,** Kinetics of gas-phase addition reactions of trichlorosilyl radicals. V. *Cis-trans* isomerization of 2-butenes induced by ˙$SiCl_3$ radicals, *J. Phys. Chem.,* 86, 4522, 1982.

109. **Dohmaru, T. and Nagata, Y.,** Kinetics of gas phase addition reactions of trichlorosilyl radicals. VI. Orientation of additions of 2-pentenes, *Bull. Chem. Soc. Jpn.,* 56, 1847, 1983.

110. **Pollock, T. L., Sandhu, H. S., Jodhan, A., and Strausz, O. P.,** Photochemistry of silicon compounds. IV. Mercury photosensitization of disilane, *J. Am. Chem. Soc.,* 95, 1017, 1973.

111. **Cvetanovic, R. J. and Irwin, R. S.,** Rates of addition of methyl radicals to olefins in the gas phase, *J. Chem. Phys.,* 46, 1694, 1967.

112. **Dohmaru, T. and Nagata, Y.,** Kinetics of gas phase addition reactions of trichlorosilyl radicals. II. Additions to 1-olefins, *J. Chem. Soc. Faraday Trans. 1,* 75, 2617, 1979.

113. **Dohmaru, T. and Nagata, Y.,** Kinetics of gas phase addition reactions of trichlorosilyl radicals. IV. Relative rates of additions to 1-alkenes, *Bull. Chem. Soc. Jpn.,* 55, 323, 1982.

114. **Choo, K. Y. and Gaspar, P. P.,** Addition of trimethylsilyl radical to ethylene. Flash photolysis-electron spin resonance kinetic study, *J. Am. Chem. Soc.,* 96, 1284, 1974.

115. **Endrenyi, L. and LeRoy, D. J.,** Kinetics of the addition of methyl, propyl, and acetonyl radicals to ethylene and of abstraction of hydrogen from acetone by methyl and propyl radicals, *J. Phys. Chem.,* 71, 1334, 1967.

116. **Krusic, P. J. and Kochi, J. K.,** Electron spin resonance of group IV organometallic alkyl radicals in solution, *J. Am. Chem. Soc.,* 91, 6161, 1969.

117. **Kawamura, T. and Kochi, J. K.,** Hyperconjugative and p-d homoconjugative effects of silicon, germanium, and tin on alkyl radicals from electron spin resonance studies, *J. Am. Chem. Soc.,* 94, 648, 1972.

118. **Chatgilialoglu, C., Ingold, K. U., and Scaiano, J. C.,** Absolute rate constants for the addition of triethylsilyl radicals to various unsaturated compounds, *J. Am. Chem. Soc.,* 105, 3292, 1983.

119. **Cadman, P., Tilsley, G. M., and Trotman-Dickenson, A. F.,** Reactions of trichlorosilyl radicals with n-propyl chloride, cyclopentyl chloride, and carbon tetrachloride, *J. Chem. Soc. A,* 1370, 1969.

120. **Haszeldine, R. N., Pool, C. R., and Tipping, A. E.,** Polyfluoroalkyl derivatives of silicon. XIV. Reaction of trichlorosilane with 1,3,3,3-tetrafluoropropene and 2-chloro-1,3,3,3-tetrafluoropropene, *J. Chem. Soc., Dalton Trans.,* 2292, 1975.

121. **Benson, S. W.,** *Thermochemical Kinetics,* 2nd ed., Wiley, New York, 1976, 169.

122. **Horowitz, A.,** The mechanism of the reaction between silyl radicals and chloroethylenes: a case study of the triethylsilyl-tetrachloroethylene reaction, *J. Am. Chem. Soc.,* 107, 318, 1985.

123. **Beaumont, A. G., Bott, R. W., Eaborn, C., and Jackson, R. A.,** The reaction of *bis*(trimethylsilyl)mercury with ketones, *J. Organomet. Chem.,* 6, 671, 1966.

124. **Beaumont, A. G., Eaborn, C., and Jackson, R. A.,** Organo-silicon compounds. XLVI. The reaction of *bis*(trimethylsilyl)mercury with ketones, *J. Chem. Soc. B,* 1624, 1970.

125. **Janzen, A. F. and Willis, C. J.,** The reaction of hexafluoroacetone with some silanes, *Can. J. Chem.,* 43, 3063, 1965.

126. **Janzen, A. F., Rodesiler, P. F., and Willis, C. J.,** Electrophilic reactions of hexafluoroacetone, *J. Chem. Soc. D,* 672, 1966.

127. **Sitzki, A. and Ruehlmann, K.,** Reaction of acid derivatives and halosilanes with alkali metals. VII. Reaction of trimethylsilyl radicals and cyclohexanone, *Z. Chem.,* 8, 427, 1968.

128. **Hudson, A. and Jackson, R. A.,** Production of specific radicals for electron spin resonance studies, *J. Chem. Soc. D,* 1323, 1969.

129. **Bowles, A. J., Hudson, A., and Jackson, R. A.,** An electron resonance study of reactions involving silyl radicals, *J. Chem. Soc. B,* 1947, 1971.

130. **Cooper, J., Hudson, A., and Jackson, R. A.,** An electron spin resonance study of the reactions of organosilicon, organogermanium, and organotin radicals with carbonyl compounds, *J. Chem. Soc. Perkin Trans. 2,* 1933, 1973.

131. **Coppin, G. N., Hudson, A., and Jackson, R. A.,** An ESR study of some group IV B free radical adducts of diethylketo-malonate, *J. Organomet. Chem.,* 131, 371, 1977.

132. **Tsurugi, J., Nakao, R., and Fukumoto, T.,** A novel γ-induced reduction with trichlorosilane. Dialkyl ether from alkyl aliphatic carboxylate, *J. Am. Chem. Soc.,* 91, 4587, 1969.

133. **Nakao, R., Fukumoto, T., and Tsurugi, J.,** The reduction of methyl acetate with several hydrosilanes under γ irradiation, *Bull. Chem. Soc. Jpn.,* 47, 932, 1974.

134. **Nakao, R., Fukumoto, T., and Tsurugi, J.,** Reduction with trichlorosilane. III. Cyclic ether from lactone, *J. Org. Chem.,* 37, 76, 1972.

135. **Nakao, R., Fukumoto, T., and Tsurugi, J.,** Reduction with trichlorosilane. IV. Ether from acetal, *J. Org. Chem.,* 37, 4349, 1972.

136. **Nagata, Y., Dohmaru, T., and Tsurugi, J.,** Reduction with trichlorosilane. II. Mechanistic study of reduction of methyl acetate to ethyl methyl ether, *J. Org. Chem.,* 38, 795, 1973.

137. **Dohmaru, T., Nagata, Y., and Tsurugi, J.,** Photolytic addition of trichlorosilane to acetone in the gas phase, *Bull. Chem. Soc. Jpn.,* 45, 2660, 1972.

138. **Chatgilialoglu, C., Ingold, K. U., Scaiano, J. C., and Woynar, H.,** Absolute rate constants for some reactions involving triethylsilyl radicals in solution, *J. Am. Chem. Soc.,* 103, 3231, 1981.

139. **Chatgilialoglu, C., Ingold, K. U., and Scaiano, J. C.,** Absolute rate constants for the addition of triethylsilyl radicals to the carbonyl group, *J. Am. Chem. Soc.,* 104, 5119, 1982.

140. **Eaborn, C., Jackson, R. A., and Pearce, R.,** Aromatic substitution by trimethylsilyl radicals, *J. Chem. Soc. D,* 920, 1967.

141. **Sakurai, H. and Hosomi, A.,** Silyl radicals. VIII. Directive effects and relative reactivities of the pentamethyldisilanyl radical in homolytic aromatic silylation, *J. Am. Chem. Soc.,* 93, 1709, 1971.

142. **Sakurai, H., Hosomi, A., and Kumada, M.,** Silyl radicals. IV. Homolytic aromatic silylation with hydrosilanes, *Tetrahedron Lett.,* 1755, 1969.

143. **Bennett, S. W., Stuart, W., Eaborn, C., Jackson, R. A., and Pearce, R.,** Photolytic reactions of *bis*(trimethylsilyl)-mercury and *bis*(trimethylgermyl)mercury with benzene, toluene, and anisole. Free-radical aromatic silylation and germilation, *J. Organomet. Chem.,* 28, 59, 1971.

144. **Kira, M. and Sakurai, H.,** Chemistry of organosilicon compounds. CIL. Electron spin resonance spectra of 6-silyl-substituted cyclohexadienyl radicals produced by the reaction of silyl radicals with benzene, *Chem. Lett.,* p. 927, 1981. See also Reference 118.

145. **Kira, M. and Sakurai, H.,** The *ab initio* optimized geometry of the 6-silylhexadienyl radical, *Chem. Lett.,* 221, 1982.

146. **Sakurai, H., Kira, M., and Sugiyama, H.,** Chemistry of organosilicon compounds. CLXXVI. Reversibility in the homolytic aromatic substitution with silyl and germyl radicals, *Chem. Lett.,* 599, 1983.

147. **Kira, M. and Sakurai, H.,** Temperature dependent electron spin resonance spectra of cyclohexadienyl and silyl-substituted cyclohexadienyl radicals. On the conformation of the radicals, *J. Am. Chem. Soc.,* 99, 3892, 1977.

148. **Griller, D., Marriott, P. R., Nonhebel, D. C., Perkins, M. J., and Wong, P. C.,** Homolytic addition to benzene. Rate constants for the formation and decay of some substituted cyclohexadienyl radicals, *J. Am. Chem. Soc.,* 103, 7761, 1981.

149. **Alberti, A. and Pedulli, G. F.,** Radical intermediates in homolytic aromatic silylation. ESR evidence, *Gazz. Chim. Ital.,* 109, 395, 1979.

150. **Haszeldine, R. N., and Young, J. C.,** Polyfluoroalkyl compounds of silicon. V. The reaction of trichlorosilane with chlorotrifluoroethylene, and halogen-abstraction by silyl radicals, *J. Chem. Soc.,* 4503, 1960.

151. **Kerr, J. A., Smith, B. J. A., Trotman-Dickenson, A. F., and Young, J. C.,** Abstraction of halogen atoms from alkyl halides by the trichlorosilyl radical, *J. Chem. Soc. A,* 510, 1968.

152. **Cadman, P., Tilsley, G. M., and Trotman-Dickenson, A. F.,** Abstraction of chlorine atoms from alkyl chlorides by the trimethylsilyl radical, *J. Chem. Soc. Faraday Trans. 1,* 69, 914, 1973.

153. **Cadman, P. and Owen, H. L.,** Abstraction of chlorine and bromine atoms from alkyl halides by the trifluorosilyl radical, *J. Chem. Soc. Faraday Trans. 1,* 77, 1913, 1981.

154. **Benson, S. W.,** *Thermochemical Kinetics,* 2nd ed., Wiley, New York, 1976, 156.

155. **Cooper, D.,** Reactions of substituted silyl radicals with chloroalkanes, *J. Organomet. Chem.,* 10, 447, 1967.

156. **Aloni, R., Rajbenbach, L. A., and Horowitz, A.,** Abstraction of chlorine atoms from chloroalkanes. V. The liquid-phase reaction between triethylsilyl radicals and dichloromethane, chloroform, carbon tetrachloride, and trichloroacetnitrile, *Int. J. Chem. Kinet.,* 11, 899, 1979.

157. **Aloni, R., Rajbenbach, L. A., and Horowitz, A.,** Free-radical chain dechlorination of chloroethanes in liquid triethyl-silane, *Int. J. Chem. Kinet.,* 13, 23, 1981.

158. **Aloni, R., Horowitz, A., and Rajbenbach, L. A.,** Abstraction of chlorine atoms from chloroalkanes. III. The kinetics of liquid phase reactions of trichlorosilyl radicals with chloromethanes, *Int. J. Chem. Kinet.,* 8, 673, 1976.

159. **Nagai, Y., Yamazaki, K., Shiojima, I., Kobori, N., and Hayashi, M.,** Homolytic reduction of polychloroalkanes by triorganosilicon hydrides, *J. Organomet. Chem.,* 9, P21, 1967.

160. **Chatgilialoglu, C., Ingold, K. U., and Scaiano, J. C.,** Absolute rate constants for the reaction of triethylsilyl radicals with organic halides, *J. Am. Chem. Soc.,* 104, 5123, 1982.

161. **Chatgilialoglu, C., Lunazzi, L., and Ingold, K. U.,** Kinetic studies on the formation and decay of some sulfonium radicals, *J. Org. Chem.,* 48, 3588, 1983.

162. **Cooper, D.,** Reaction of silyl radicals with halobenzenes, *J. Organomet. Chem.,* 7, P26, 1967.

163. **Birchall, J. M., Daniewski, W. M., Haszeldine, R. N., and Holden, L. S.,** Polyfluoroarenes. VII. The photochemical reactions of trichlorosilane and trimethylsilane with hexafluorobenzene, *J. Chem. Soc.,* 65, 6702, 1965.

164. **Eaborn, C., Jackson, R. A., and Walsingham, R. W.,** Production of organosilicon radicals in solution, *J. Chem. Soc. D,* p. 300, 1965.

165. **Atton, D., Bone, S. A., and Davidson, I. M. T.,** Chlorine-abstraction and chlorine-migration reactions of silyl radicals, *J. Organomet. Chem.,* 39, C47, 1972.

166. **Davidson, I. M. T. and Matthews, J. I.,** Chlorine abstraction from silicon-chlorine bonds by trimethylsilyl radicals, *J. Chem. Soc., Faraday Trans. 1,* 77, 2277, 1981.

167. **Gammie, L., Sandorfy, C., and Strausz, O. P.,** Photochemistry of silicon compounds. VI. The 147-nm photolysis of tetramethylsilane, *J. Phys. Chem.,* 83, 3075, 1979.

168. **Morris, E. R. and Thynne, C. J.,** Hydrogen atom abstraction from silane, trimethylsilane, and tetramethylsilane by methyl radicals, *J. Phys. Chem.,* 73, 3294, 1969.

169. **Dainton, F. S. and McElcheran, D. E.,** The reaction $CH_3 + CD_4 \rightarrow CH_3D + CD_3$, *Trans. Faraday Soc.,* 51, 657, 1955.

170. **Berkley, R. E., Safarik, I., Gunning, H. E., and Strausz, O. P.,** Arrhenius parameters for the reactions of methyl radicals with silane and methylsilanes, *J. Phys. Chem.,* 77, 1734, 1973.

171. **Berkley, R. E., Safarik, I., Strausz, O. P., and Gunning, H. E.,** Arrhenius parameters for the reactions of higher alkyl radicals with silanes, *J. Phys. Chem.,* 77, 1741, 1973.

172. **Ingold, K. U.,** Rate constants for free radical reactions in solution, in *Free Radicals,* Vol. 1, Kochi, J. K., Ed., Wiley-Interscience, New York, 1973, chap. 2.

173. **Davies, A. G. and Roberts, B. P.,** Bimolecular homolytic substitution at metal centers, in *Free Radicals,* Vol. 1, Kochi, J. K., Ed., Wiley-Interscience, New York, 1973, chap. 10.

174. **Ingold, K. U. and Roberts, B. P.,** Free-Radical Substitution Reactions, *Wiley-Interscience,* New York, 1971.

175. **Band, S. J. and Davidson, I. M. T.**, Gas-phase reaction between hexamethyldisilane and iodine, *Trans. Faraday Soc.*, 66, 406, 1970.

176. **Hosomi, A. and Sakurai, H.**, Cleavage of silicon-silicon and germanium-germanium bonds with 1,2-dibromoethane by a free-radical mechanism. Evidence for bimolecular homolytic substitution at silicon and germanium, *J. Am. Chem. Soc.*, 94, 1384, 1972.

177. **Hosomi, A. and Sakurai, H.**, Polar effects on the cleavage reaction of silicon-silicon bonds of 1,2-diaryltetramethyl-disilane with 1,2-dibromoethane. Structure and reactivity in S_H2 reaction, *Chem. Lett.*, 193, 1972.

178. **Safarik, I., Pollock, T. L., and Strausz, O. P.**, Model calculation of the hydrogen/deuterium kinetic isotope effect in the H + Si_2H_6 reaction, *J. Phys. Chem.*, 78, 398, 1974.

179. **Nay, M. A., Woodall, G. N. C., Strausz, O. P., and Gunning, H. E.**, The mercury 6(^{3}P) photosensitization of the methylsilanes and silane, *J. Am. Chem. Soc.*, 87, 179, 1965.

180. **Kamaratos, E. and Lampe, F. W.**, A mass spectrometric study of the mercury-photosensitized reactions of silane and methylsilane with nitric oxide, *J. Phys. Chem.*, 74, 2267, 1970.

181. **Avar, G. and Neumann, W. P.**, Capture reactions of short-lived radicals. XVI. Reactions of trimethylstannyl and trimethylsilyl radicals with tetraphenyldiphosphine, phosphorus esters and peresters, *J. Organomet. Chem.*, 131, 207, 1977.

182. **Gara, W. B. and Roberts, B. P.**, An ESR study of the reactions of trialkylsiloxyl and trialkylsilyl radicals with dialkylsulfides, *J. Organomet. Chem.*, 135, C20, 1977.

183. **Jackson, R. A., Malek, F., and Ozaslan, N.**, The reaction of trichlorosilyl radicals with cyclohexyl octyl ether. An S_H2 reaction at an ether oxygen atom, *J. Chem. Soc., Chem. Commun.*, 956, 1981.

184. **Adeleke, B. B., Wong, S.-K., and Wan, J. K. S.**, Electron spin resonance study of the arylsilyl adducts of phenyl *tert*-butyl nitrone and their decomposition kinetics, *Can. J. Chem.*, 52, 2901, 1974.

185. **Riviere, P., Richelme, S., Riviere-Baudet, M., Satgé, J., Riley, R. I., Lappert, M. F., Dunogues, J., Calas, R.**, Addition 1-3 d'hydrosilanes et stannanes sur diverses nitrones: radicaux nitroxydes α-metalles

$$R_3M\text{–}C\text{–}N.\text{–}O \text{ et } O\text{– et } C\text{– metalla-hydroxylamines } R_3M\text{–}O\text{–}N\text{–}C\text{–}H \text{ et } R_3M\text{–}C\text{– } N\text{ –}, \quad OH$$

J. Chem. Res. (M), 1663, 1981.

186. **Chandra, H., Davidson, I. M. T., and Symons, M. C. R.**, Unstable intermediates. CCII. Use of spin traps to study trialkylsilyl and related species, *J. Chem. Soc., Perkin Trans. 2*, 1353, 1982.

187. **Chandra, H., Davidson, I. M. T., and Symons, M. C. R.**, Use of spin traps in the study of silyl radicals in the gas phase, *J. Chem. Soc. Faraday Trans. 1*, 79, 2705, 1983.

188. **Aurich, H. G., and Czepluch, H.**, Aminyl oxide (nitroxide). XXXI. Investigation of spin density distribution in various types of aminyl oxides using ^{17}O-labelled radicals, *Tetrahedron*, 36, 3543, 1980.

189. **Planinic, J.**, Spin trapping of silyl radicals, *Radiochem. Radioanal. Lett.*, 56, 205, 1983.

190. **Planinic, J.**, Rate constant in spin trapping: triethylsilyl radical and *tert*-nitrosobutane, *Radiochem. Radioanal. Lett.*, 58, 147, 1983.

191. **Cammagi, C. M., Leardini, R., and Placucci, G.**, Substituent effects in heterocyclic rings. An electron spin resonance study of some 5-substituted 2-thienyl nitroxides and nitro-anions, *J. Chem. Soc. Perkin Trans. 2*, 1195, 1974.

192. **Lunazzi, L., Placucci, G., and Roncin, N.**, Alkoxy nitroxide radicals from photolysis of nitropyridines: a kinetic investigation by electron spin resonance spectroscopy, *J. Chem. Soc. Perkin Trans. 2*, 1132, 1977.

193. **Reuter, K. and Neumann, W. P.**, New nitrogen containing radicals from nitro and organometallic compounds: R'–NO˙–OMR$_3$ (R' = aryl, alkyl), Ar–N˙–OMR$_3$, and Ar–N(OMe$_3$)–N˙Ar (M = Si, Ge, Sn), *Tetrahedron Lett.*, 5235, 1978.

194. **Baban, J. A., Cook, M. D., and Roberts, B. P.**, An electron spin resonance study of trialkylsilyl radical addition to alkyl isocyanates, *J. Chem. Soc. Perkin Trans. 2*, 1247, 1982.

195. **Camaggi, C.-M., Caser, M., and Placucci, G.**, Addition of organometallic radicals to organic substrates. A useful spin-trap reagent for group 4 radicals, *J. Chem. Soc. Perkin Trans. 2*, 1675, 1979.

196. **Gaspar, P. P., Ho, C. T., and Choo, K. Y.**, Organosilicon iminamino radicals from the addition of silyl radicals to diphenyldiazomethane, *J. Am. Chem. Soc.*, 96, 7818, 1974.

197. **Scaiano, J. C. and Ingold, K. U.**, Kinetic applications of electron paramagnetic resonance spectroscopy. XXV. Radicals formed by spin trapping with di-*tert*-butyl thioketone, *J. Am. Chem. Soc.*, 98, 4727, 1976.

198. **Gasanov, R. G., Ivanova, L. V., and Freidlina, R. Kh.**, Rate constants of the addition of triethylsilyl radicals to spin traps, *Izv. Akad. Nauk SSSR Ser. Khim.*, 938, 1984.

199. **Gooden, R.**, Photooxidation of trichlorosilane in silicon tetrachloride, *Inorg. Chem.*, 22, 2272, 1983.

200. **Niki, H., Maker, P. D., Savage, C. M., and Breitenbach, L. P.**, FTIR spectroscopic observation of peroxyalkyl nitrates formed via ROO + NO_2 → $ROONO_2$, *Chem. Phys. Lett.*, 55, 289, 1978.

INDEX